# History, Philosophy and Theory of the Life Sciences

Volume 10

More information about this series at http://www.springer.com/series/8916

Steeves Demazeux • Patrick Singy
Editors

# The DSM-5 in Perspective

## Philosophical Reflections on the Psychiatric Babel

Springer

*Editors*
Steeves Demazeux
Université Bordeaux Montaigne
Pessac, France

Patrick Singy
Union College
Schenectady, NY, USA

ISSN 2211-1948 ISSN 2211-1956 (electronic)
History, Philosophy and Theory of the Life Sciences
ISBN 978-94-024-0610-8 ISBN 978-94-017-9765-8 (eBook)
DOI 10.1007/978-94-017-9765-8

Springer Dordrecht Heidelberg New York London

Softcover reprint of the hardcover 1st edition 2015


Printed on acid-free paper

Springer Science+Business Media B.V. Dordrecht is part of Springer Science+Business Media (www.springer.com)

# Contents

# Editors and Contributors

**David Cohen** is the Marjorie Crump Professor of Social Welfare at the Luskin School of Public Affairs, University of California, Los Angeles, and taught at Université de Montréal and Florida International University. He held the Distinguished Tocqueville-Fulbright Chair to France and received awards for his scholarship, research, teaching, and advocacy. In over 100 articles and chapters, he examines prescribed psychoactive drugs as social and cultural phenomena, theories of psychoactive drug action, conventional notions and tests of drug safety and efficacy, and the medicalization of life. He has also been a practicing clinical social worker and written about uses and misuses of drugs in helping relationships. He edited or co-authored *Challenging the Therapeutic State* (1990), *Tardive Dyskinesia and Cognitive Dysfunction* (1993), *Médicalisation et contrôle social* (1994), *Guide critique des médicaments de l'âme* (1995), *Your Drug May Be Your Problem* (1999/2007), *Critical New Perspectives on ADHD* (2006), and *Mad Science: Psychiatric Coercion, Diagnosis, and Drugs* (2013).

University of California, Los Angeles, CA, USA

**Rachel Cooper** studied for her Ph.D. in History and Philosophy of Science at Cambridge University, and is currently a Senior Lecturer in Philosophy at Lancaster University, UK. She works mainly on conceptual problems around the classification of mental disorders. Her publications include Diagnosing *the Diagnostic and Statistical Manual of Mental Disorders* (Karnac 2014), *Classifying Madness* (Springer 2005) and *Psychiatry and the Philosophy of Science* (Acumen 2007).

Lancaster University, Lancaster, UK

**Steeves Demazeux** is an Associate Professor of Philosophy at the Université Bordeaux Montaigne (France). He received his doctoral degree at the Paris-Sorbonne University (IHPST laboratory), and has been during two years a post-doctoral fellow at the CeRMES Institute (Paris Descartes University). His research interests include history and philosophy of psychiatry, philosophy of medicine, and philosophy of science, in general. He is the author of *Qu'est-ce que le DSM? Genèse* *et transformations de la bible américaine de la psychiatrie* (Ithaque 2013), and the co-author, with Françoise Parot and Lionel Fouré, of *Psychothérapie, fondements et pratiques* (Belin 2011).

Associate Professor, Université Bordeaux Montaigne, Pessac, France

**Luc Faucher** is Full Professor in Philosophy at the Université Du Québec à Montréal (Canada). After a Ph.D. in Montréal, he was postdoctoral fellow at Rugers University under the direction of Stephen Stich. His research interests cover philosophy of cognitive sciences, philosophy of race, philosophy of emotions and philosophy of psychiatry. He is the co-editor (with Christine Tappolet) of *The Modularity of Emotions* (2008) and the editor of *Philosophie et psychopathologies* (2006). He published essays in *Philosophy of Science*, *Philosophy of Social Sciences*, *Emotion Review*, *Journal of Social Philosophy*, *Synthese* and the *Monist.*

Université du Québec à Montréal, Montréal, Canada

**Tomi Gomory** is a Fulbright Scholar and Associate Professor at the College of Social Work, Florida State University. His research interests are homelessness, mental health, philosophy of science, and social work education. Before beginning his academic career, he spent ten years working as a social worker, as a clinician, director of the first adult homeless shelter in Brooklyn and a stint as the San Francisco project director of the Robert Wood Johnson and HUD coordinated federal Homeless Families Model Project. He has published extensively on issues related to mental health. His most recent completed publication is the book, *Mad Science: Psychiatric Coercion, Diagnosis, and Drugs in America,* published by Transaction Publishers in 2013.

Florida State University, Tallahassee, FL, USA

**Simon Goyer** is currently research assistant at the Research Centre of the Centre hospitalier de l'Université de Montréal (CHUM) where he is involved in a research on clinical reasoning in psychiatry. He is a doctoral student at the Université du Québec à Montréal (UQAM) where he specializes in the fields of philosophy of psychiatry, philosophy of science and ethics.

Université de Montréal (CHUM), Montréal, Canada

**Andrew Hinderliter** is a Ph.D. candidate in the Department of Linguistics at the University of Illinois at Urbana-Champaign, completing his dissertation on the evolution on online asexual discourse. He also has an M.A. in mathematics from the University of Missouri-Columbia. His research interests include computational and corpus-based sociolinguistics, using computational methods for analyzing the evolution of terminology and concepts in specific discourse domains, asexuality, and the philosophy of psychiatry.

University of Illinois, Chicago, IL, USA

**Stuart A. Kirk** is a Distinguished Professor Emeritus in Social Welfare at the Luskin School of Public Affairs, University of California, Los Angeles. His research critically examines the conventional wisdom of the helping professions, focusing on the interplay of science, social values and professional politics to illuminate the evolution of professional beliefs and practices. For example, he has written extensively about the effectiveness of attempts by the profession of social work to make practice more scientifically based, summarized in the co-authored book with William J. Reid, *Science and Social Work: A Critical Appraisal*, 2002. Professor Kirk has been particularly interested in mental health policy, politics and services. In scores of articles and three co-authored books (*The Selling of DSM: The Rhetoric of Science in Psychiatry*, 1992; *Making Us Crazy: DSM–The Psychiatric Bible and the Creation of Mental Disorders*, 1997; and *Mad Science: Psychiatric Coercion, Diagnosis and Drugs*, 2013) he traces the evolution and challenges the underlying assumptions and scientific claims of the guiding document of the psychiatric enterprise, the *Diagnostic and Statistical Manual of Mental Disorders* (DSM).

University of California, Los Angeles, CA, USA

**Baptiste Moutaud** is a postdoctoral researcher at the CERMES3 (Research center on Medicine, Sciences, Health, Mental Health and Society, Paris Descartes University, Ecole des Hautes Etudes en Sciences Sociales, CNRS, INSERM). He received his Ph.D. in Social Anthropology at the Paris Descartes University in 2009 (CESAMES research center). He has also been postdoctoral fellow at the Montréal University (MéOS research team). His research interests stand at the crossroads of medical anthropology and science studies. Since 2004, he is exploring the rise of neuroscience and its consequences on the construction of contemporary individualism. He is the co-editor with Brigitte Chamak of *Neurosciences et société. Enjeux des savoirs et pratiques sur le cerveau* (Armand Colin 2014).

Centre de recherche, médecine, sciences, santé, santé mentale, société (CERMES3), INSERM U988, CNRS UMR 8211, École des hautes études en sciences sociales, Université Paris Descartes, Paris, France

**Dominic Murphy** (Ph.D. Rutgers 1999) is Associate Professor and Director of the Unit for History and Philosophy of Science at the University of Sydney. He has previously taught at Washington University, St. Louis, and the California Institute of Technology, and held visiting fellowships at Pittsburgh (Center for HPS) and Tours. His research has centred on the philosophy of psychiatry, chiefly in his 2006 book, *Psychiatry in the Scientific Image*, which discussed the conceptual issues raised by treating psychiatry as a form of cognitive neuroscience. He has also published papers on the medical model, explanation and reduction in psychiatry, as well as further papers in philosophy of medicine and philosophy of mind. Murphy currently holds an Australian Research Council Grant to work on the challenges raised by contemporary cognitive science to traditional philosophical pictures of the self. He is also working on concepts of disorder and homeostasis and cultural variation in mental illness.

Unit for History and Philosophy of Science, University of Sydney, Sydney, Australia

**Jeffrey Poland** is a Senior Lecturer in History, Philosophy, and Social Science at RISD and a Visiting Professor in History at Brown University. He has degrees in philosophy of science (Ph.D., MIT 1983) and clinical psychology (MA, Southern Connecticut State University 1982). His primary research interests concern problems with contemporary approaches to psychiatric classification, research, and clinical practice. Other research interests focus on philosophical materialism and the unity of science. He is author of *Physicalism: The Philosophical Foundations* (Oxford University Press 1994), co-author (with William Spaulding and Mary Sullivan) of *Treatment and Rehabilitation of Severe Mental Illness* (Guilford 2003), co-editor (with William Spaulding and William Shuart) of *Modeling Complex Systems: Motivation, Cognition and Social Processes* (University of Nebraska Press 2007), and co-editor (with George Graham) of *Addiction and Responsibility* (The MIT Press 2011). Articles include: "Problems with the DSM Approach to Classification of Psychopathology" (1994), "How to Move Beyond the Concept of Schizophrenia" (2007), and "Mapping the Domain of Mental Illness" (with Barbara Von Eckardt 2013).

Rhode Island School of Design and Brown University, Providence, RI, USA

**John Z. Sadler** M.D. is currently a Professor of Psychiatry and Clinical Sciences and The Daniel W. Foster, M.D. Professor of Medical Ethics at the University of Texas Southwestern Medical Center. Dr. Sadler directs the Division of Ethics in the Department of Psychiatry and is the institution-wide Director of the Program in Ethics in Science and Medicine at UT Southwestern. Dr. Sadler's main research area has been in the philosophy and ethics realm – from clinical ethics, to research ethics, to the philosophy of psychiatry. He is a Co-Editor of an international journal, *Philosophy, Psychiatry, and Psychology*, the *Current Opinion in Psychiatry* section on History and Philosophy, and the Oxford University Press book series, "International Perspectives on Philosophy and Psychiatry." He recently finished working on a large textbook for Oxford University Press, *the Oxford Handbook of Philosophy and Psychiatry* (2013), and is lead editor of the forthcoming *Oxford Handbook of Psychiatric Ethics,* aimed for a July 2015 release. Dr. Sadler has edited or co-edited four books, several special issues of professional journals, and authored a comprehensive monograph, *Values and Psychiatric Diagnosis*, published in 2005 and in collaboration with philosopher Jennifer Radden, *The Virtuous Psychiatrist: Character Ethics in Psychiatric Practice* in 2009. His follow-up book to *Values and Psychiatric Diagnosis, "Vice and Psychiatric Diagnosis,"* is in preparation.

University of Texas Southwestern Medical Center, Dallas, TX, USA

**Patrick Singy** is currently an Adjunct Professor at Union College in Schenectady, NY. He received his Ph.D. in History and Philosophy of Science from the University of Chicago in 2004, and has been a postdoctoral fellow at Northwestern University and Columbia University, as well as a scholar-in-residence in the Bioethics Center at Union Graduate College. His research interests include the history of medicine and sexuality, the historiography of science, and the history and philosophy of psychology and psychiatry. He has published essays in several American and international journals, including *Representations*, *Journal of Modern History*, *Journal of the History of Sexuality*, and *Journal of the History of Medicine and Allied Sciences*. He is the author of *L'Usage du sexe. Lettres au Dr Tissot, auteur de* L'Onanisme *(1760)* (BHMS 2014).

Union College, Schenectady, NY, USA

**Şerife Tekin** is currently an Assistant Professor of Philosophy at Daemen College in Buffalo, NY. She received her Ph.D. in Philosophy from York University in Canada, and has been a postdoctoral research fellow at Dalhousie University, Canada, and the University of Pittsburgh, USA. Her research interests include the history and philosophy of cognitive science and psychiatry, narrative approaches to the self, and feminist philosophy of science. She has published articles in several American and international journals, including *Philosophy, Psychiatry, and Psychology; Philosophical Psychology;* and *American Journal of Bioethics*. She is currently co-editing *Extraordinary Science: Overcoming the Crisis in Psychiatric Research* with Jeffrey Poland, under contract with the MIT Press.

Department of Philosophy and Religious Studies, Daemen College, Amherst, NY, USA

**Jonathan Y. Tsou** is an Associate Professor of Philosophy in the Department of Philosophy and Religious Studies at Iowa State University. He completed his Ph.D. in History and Philosophy of Science at the University of Chicago in 2008 and was a Postdoctoral Fellow in Philosophy at the University of British Columbia in 2008–2009. He has a wide range of research interests in philosophy of science, including philosophy of psychiatry, history and philosophy of psychology, and history of twentieth century philosophy of science. His research has examined issues such as natural kinds in psychiatry, the role that pharmacological drugs play as experimental instruments for identifying the neurobiological causes of mental disorders, and the looping effects of human science classifications. His articles have appeared in journals such as *Philosophy of Science*, *Synthese*, *Studies in History and Philosophy of Biological and Biomedical Sciences*, *International Journal of Law and Psychiatry*, *Theory & Psychology*, and *International Studies in the Philosophy of Science*.

Department of Philosophy and Religious Studies, Iowa State University, Ames, IA, USA

**Jerome Wakefield** is University Professor, Professor of Social Work, and Professor of the Conceptual Foundations of Psychiatry at New York University. He is also an Affiliate Faculty in Bioethics and an Honorary Faculty Member of the Institute for Psychoanalytic Education. He holds a Ph.D. in Philosophy, a DSW in Clinical Social Work, and an MA in Mathematics (Logic and Methodology of Science), all from University of California, Berkeley. After post-doctoral fellowships in Women's Studies (Brown University), Cognitive Science (University of California, Berkeley), and Mental Health Services Research (Rutgers University), he held faculty positions at University of Chicago Columbia university, and Rutgers University, before coming to NYU in 2003. He has published over 200 scholarly articles on the conceptual foundations of diagnosis and clinical theory, with recent publications focusing on controversies over proposed changes to DSM-5, especially issues concerning the relationship between depression and grief and the boundary between normal distress and mental disorder. He is the co-author (with Allan Horwitz) of "The Loss of Sadness: How Psychiatry Transformed Normal Sorrow into Depressive Disorder" (Oxford 2007), named the best psychology book of 2007 by the Association of Professional and Scholarly Publishers, and "All We Have to Fear: How Psychiatry Transforms Natural Fear into Mental Disorder" (Oxford 2012). He is currently completing a book on Freud's case history of Little Hans and its significance in the history of psychoanalysis, to be published by Routledge.

School of Social Work, Department of Psychiatry, and Center for Bioethics, New York University, New York, NY, USA

# Introduction

For over three decades, the *Diagnostic and Statistical Manual of Mental Disorders* (DSM) has dominated psychiatry all over the world. American psychiatry was the fertile ground on which grew this new kind of classification. The latter, thanks to its hegemonic status, returned the favor by making American psychiatry the leading voice in the international field of psychiatric research. More than the mere symbol of the late but very fast development of American psychiatry since the second half of the twentieth century, the DSM played the role of a catalyst for American psychiatry: it accelerated its prominence and motivated its main choices and assumptions.

Let us first recall a few historical facts. Clinical psychiatry emerged in nineteenth-century France, Germany, and Switzerland, and until World War II almost all the most influential psychiatrists, laboratories, and scientific journals were European. American psychiatry did not yet have an identity of its own. It was, at best, a borrowed psychiatry. Benjamin Rush, who in the USA is often considered the founding father of American psychiatry, was very heavily influenced by European psychiatrists, especially by Philippe Pinel and his "moral treatment." Rush's own influence outside the USA has always been minimal. It is then no surprise that when at the beginning of the twentieth century American statisticians and psychiatrists planned to create a national classification, they looked first at what France and Germany had done before them.

The first American classifications of diseases, such as the *Statistical Manual for the Use of Institutions for the Insane* (1918) and the *Standard Classified Nomenclature of Diseases* (1933), lacked both rigor and originality. These classifications could have become useful tools for statistical inventories and the homogenization of diagnostic practices, but most clinicians dismissed them. Take for instance Alfred Meyer (1866–1950), a Swiss psychiatrist who immigrated to the USA in 1892 and who greatly contributed to a type of psychiatry that is both scientific and humane. Meyer, arguably the first great professor of American psychiatry, never hid his scorn for all nosological endeavors. He never showed any interest in the formal aspect of diagnosis, finding it too rigid and too easily subject to reification (Grob 1991). As a counter-example, one might mention the more accommodating position toward classifications of the brothers Menninger, Karl (1893–1990) and William (1899–1966). William, in particular, was behind the *War Department Technical Bulletin, Medical 203*, also called the *Nomenclature of Psychiatric Disorders and Reactions*. This military classification of mental disorders was published in 1946 and directly influenced the text of the first edition of the DSM (1952). But while William did not live to see the publication of the DSM-II (1968), his brother Karl came to deplore the direction taken by this second edition, accusing it of sacrificing simplicity at the altar of international standardization (Spitzer and Wilson 1975).

The rise of American clinical psychiatry only began in the 1960s. Around the world, a profound positivist transformation of psychiatry took place and was signaled by the invention of new statistical tools and experimental procedures, and by the development of research in psychopharmacology, genetics, epidemiology, and eventually neuroscience (Castel et al. 1979). In this new game, American psychiatrists quickly learned to play their cards right. In collaboration with British psychiatrists, they began, during the revisions of the sixth edition of the WHO's *International Classification of Diseases* (ICD-6), to influence the debates about nosological issues. They launched epidemiological studies of an unprecedented size, such as the *Midtown Manhattan Study* at the beginning of the 1960s or the *Epidemiological Catchment Area* at the end of the 1970s, the results of which were analyzed and discussed all over the world. In American scientific journals, quantitative clinical research replaced long narrative reports of individual cases, while the spectacular progress of genetics and neurophysiology pushed even further the reduction of patients to their biological components.

The first very visible success of the *National Institute of Mental Health* (NIMH), and more generally of American neuropsychiatry, was the Nobel Prize won by Julius Axelrod in 1970 (the prize was shared with Ulf von Euler and Bernard Katz). Later, the 1990s were called the "Decade of the Brain" by President George Bush. In 2000, the Nobel Prize won by the neuropsychiatrist Eric Kandel (shared with Paul Greengard and Arvid Carlsson) confirmed one of the great implicit assumptions made by American psychiatry: psychiatry can become more *scientific* only by becoming more *neurobiological*. Despite resistances and long-lasting internal conflicts (about which Kandel has offered very lucid reflections (Kandel 1998)), the identity of American psychiatry has been constructed by reaffirming a very old conviction that appears throughout the history of psychiatry: mental disorders are first and foremost brain diseases. This is indeed how American psychiatry is most usually perceived in the rest of the world.

Given this general evolution of American psychiatry, the history of the DSM is paradoxical. The third edition (DSM-III), published in 1980, constitutes according to historians a true nosological revolution. But this revolution did not consist in having fully embraced neurological or genetic factors, as we might have expected given the general evolution of psychiatry, and especially of American psychiatry. Rather, it consisted in remaining, or trying to remain, atheoretical. The DSM-III offered a classification that voluntarily ignored the etiological models of mental disorders, in order to focus instead on the task of providing unambiguous descriptions of these disorders by means of precise and exhaustive diagnostic criteria. It is often said that the DSM-III broke with psychoanalysis, which was dominant in large American cities. It is less often said that it also broke with the essentially biological direction of its predecessors (while the influence of psychoanalysis on the DSM-I and DSM-II is regularly stressed by historians, we should not forget that the first two editions of the DSM gave an important role to biology, as illustrated by the key category of "Organic Brain syndrome"). With its more than 300 different diagnostic categories, the DSM-III had the ambition of covering all the mental disorders accepted by the psychiatric community, regardless of their explanatory models and hypotheses. In the history of psychiatry, never had a classification been created that was so detailed and exhaustive; never had a classification worried so much about clinical standardization. By its breadth, the DSM-III was more than a useful classification. It was almost a clinical treatise, and it is this ambiguous status that has often remained the main target of the DSM's critics.

There is another paradox in the history of the DSM: its extraordinary success. While the DSM-III did not offer any theoretical innovation, but only a stupendous methodological audacity, its impact on contemporary psychiatric discourse and practice has been considerable. The first printing of the DSM-III, in 1980, was quickly sold out. It was soon translated into many languages: Chinese, Danish, Dutch, Finnish, French, German, Greek, Italian, Japanese, Norwegian, Portuguese, Spanish, Swedish, etc. (APA 1987, p. xvii). The successive editions, the DSM-III-R (1987), the DSM-IV (1994), and the DSM-IV-TR (2000), only reinforced the world domination of the DSM. The DSM quickly supplanted most national classifications, wiped them from memory, and contributed to an unprecedented homogenization and universalization in the practice of psychiatry.[1] It even influenced the WHO's ICD, which, from 1992 on, came to adopt a structure and methodological principles that are very similar to the DSM's (with the crucial difference that the ICD distinguishes between different levels of use by offering distinct systems for everyday practice and for research). Since countries that are members of the WHO are bound by treaty to use the ICD (or a system close to the ICD), they are forced to gradually abandon their national classifications. This is what happened for instance with the Chinese classification, the *Chinese Classification and Diagnostic Criteria of Mental Disorders* (CCMD) published by the *Chinese Society of Psychiatry*. While important structural and clinical differences continue to distinguish all classificatory systems, and while there still remain some significant national idiosyncrasies in classifications and clinical practice, clinical standardization has been massive in the past decades and might well constitute one of the most striking events in the recent

[1]One of the great difficulties resulting from the DSM's claim to universality comes from the fact that it requires leaving aside the different national traditions in health care systems. Yet the *theoretical* issue of the validity of the diagnostic categories is inextricably intertwined with the *practical* issue of the usefulness of the classification. The DSM Work Groups are almost exclusively constituted of American psychiatrists, and as a result it is naturally the American health care system that is invoked when practical constraints are taken into consideration to modify diagnostic criteria. A different health care system could determine different theoretical choices. For instance, in the context of French health insurance, many of the difficulties that American psychiatrists have to face become incomprehensible. In particular, the debate about the rigidity and authority of the DSM criteria is difficult to understand in France, where there is no requirement to receive an official diagnosis in order for a patient to be treated and reimbursed.

history of psychiatry. Without a doubt, today no psychiatrist in the world is unaware of the meaning of the acronym "DSM," even when many clinicians continue to resist the imperative to use this manual. Most certainly, all scientifically ambitious researchers in psychiatry must refer to the DSM, however unsatisfied they might be with the manual. Today, with the publication of the DSM-5 in 2013, and despite all the criticisms and complaints directed against this new edition even before it was published (and often formulated by American psychiatrists themselves), the hegemony of the American system remains intact.

One of the most important historical consequences of the DSM has been that the manual replaced or displaced the clinicians' clinical judgment. With its gnawing authority, the DSM has profoundly transformed everyday clinical practice, both by desacralizing the diagnostic assessment and by sanctioning the use of symptom checklists and diagnostic questionnaires. Perhaps irreversibly, the DSM has replaced an opaque, traditional, and almost regal clinical art, with a transparent, public, and debatable technique. Perfectly standardized criteria have replaced clinical experience and insight.[2]

The DSM's influence reached much beyond the psychiatric world itself. Its criteria are regularly mentioned in health magazines, newspapers, and advertisements (in the USA and New Zealand, "Direct-to-Consumer Advertising" is legal), and of course they are easily accessible online. With the Internet, it is in fact very easy for anyone to acquire a legal or illegal copy of the DSM-5, only a few months after its publication. It is not uncommon to see psychiatric patients recite the exact list of DSM criteria that are used to identify their disorder. In *The Marriage Plot*, novelist Jeffrey Eugenides captured perfectly this profound influence:

> Leonard paid sympathetic attention as people spoke about their lives. He tried to take comfort in what they said. But his main thought was of how much worse off they were than he. This belief made him feel better about himself, and so he clung to it. But then it was Leonard's turn to tell *his* story, and he opened his mouth and out came the most nicely modulated, well-articulated bullshit imaginable. He talked about the events that led up to his breakdown. He recited swaths of the *DSM III* that he'd apparently committed to memory without trying. He showed off how smart he was because that was what he was used to doing. He couldn't stop himself. (Eugenides 2011, p. 254)

Based on its editorial success alone, the DSM certainly deserves to be called a "Bible," as it is often done. The APA spent nearly 25 million dollars on the DSM-5[3] (almost five times more than it did on its previous edition, the DSM-IV[4]). One can only imagine the return on investment that is expected through the selling of the manual, of its translations, and through merchandising.

---

[2]DSM developers have repeatedly insisted on the importance of clinical experience for the correct use of their manual. Yet it is clear that this requirement is purely formal, since all clinicians are meant to identify and understand the criteria in the same manner.

[3]http://www.dsm5.org/about/pages/faq.aspx#11

[4]http://www.psychologytoday.com/blog/dsm5-in-distress/201206/follow-the-money

However, one could also argue that the proper metaphor for the DSM is not the Bible, but one of its stories: the Tower of Babel. Seen in this light, the DSM looks like an imperfect and unachievable monument, which was originally built to celebrate the new unity of clinical psychiatric discourse, and which ended up creating, as a result of its hubris, ever more profound practical divisions and theoretical difficulties. For while the DSM, in its fifth incarnation, has never been so well ingrained institutionally, it is fair to say that its cracks are also becoming increasingly visible. Only a few weeks before the DSM-5 was published in May 2013, an alternative project that had started in 2009, the *Research Domain Criteria* financed by the NIMH, was publicized in the media. The director of the NIMH, Thomas Insel, publicly denounced the lack of validity of the DSM-5 (Insel 2013). The ripple effect of this declaration was immediate: the adversaries of the DSM picked up on Insel's critique and repeated it in popular and scientific journals worldwide.

For many, the publication of the DSM-5 was supposed to be a turning point in the history of the DSM. The subtle change in the numeral (from Roman to Arabic) was the promise of a profound historical change. But this promise remains unfulfilled. The DSM-5 undoubtedly continues the nosological tradition started with the DSM-III. Short of the paradigm shift that was prophesized, the Arabic numeral signals only a deliberate coming together of the DSM with the ICD (whose eleventh edition, the ICD-11, is due in 2017), and the desire to encourage progressive transformations (DSM-5.1, DSM-5.2, etc.). This is fixing a crumbling edifice with paper mâché.

*

* *

The combination of unprecedented methodological ambition, undisputed scientific and cultural success, and conspicuous theoretical weaknesses could not fail to attract the interest of many academic critics, in particular of philosophers, sociologists and anthropologists. Some of the most influential works published on the DSM include Blashfield (1984), Kirk and Kutchins (1992), Caplan (1995), Hacking (1995), Young (1995), Kutchins and Kirk (1997), Cooper (2005), Sadler (2005), Horwitz and Wakefield (2007), Lane (2008), Horwitz and Wakefield (2012) and Cooper (2014). Several edited volumes have also greatly contributed to the scholarly literature on the DSM: Sadler et al. (1994), Sadler (2002), Beutler and Malik (2002), Fulford et al. (2006, 2013), Radden (2007), Kendler and Parnas (2008, 2012), Cooper and Sartorius (2013) and Paris and Philips (2013). In this substantial body of literature, we can roughly identify two types of critiques, which are closely related and often combined with each other, although they remain conceptually distinct: an *epistemological* critique that focuses on the theoretical principles and assumptions of the classification, on its impasses and alternative strategies, and on its implied conception of the normal and the pathological, and a *sociological* critique of the (harmful) effects of the DSM on psychiatric practice and on the patients' understanding and experience of their suffering.

The history of the making of the DSM also begins to be better known. As it happens, three authors in this volume (Rachel Cooper, Steeves Demazeux and Jonathan Tsou) have dealt extensively with the history of the DSM in their philosophy of science dissertations. While they have not written books specifically on the history of the DSM, three other contributors (Jerome Wakefield, John Sadler, and Dominic Murphy) have given a central role to this history in their respective works. One must also mention three recent books: Hannah Decker's rich historical study of the DSM-III (2013), the meticulous account of the making of the DSM-5 offered by the author and psychotherapist Gary Greenberg (2013), and the relentless and well-publicized attack on the DSM-5 written by Allen Frances, the chair of the DSM-IV Task Force (2013).

If the owl of Minerva, according to Hegel, spreads its wings only with the falling of the dusk, then today is the time to evaluate philosophically and methodologically the strengths and weaknesses of the DSM. The originality of the present book is to have mobilized psychiatrists, philosophers, historians and anthropologists, who were all close observers of the revisions that led to the DSM-5.

Part I of this book includes six chapters that deal with general issues associated with the DSM-5. All with a different perspective, the first four chapters focus on the failure of the scientific effort made by the writers of the DSM to provide a purely descriptive classification of mental disorders. In chapter "The Ideal of Scientific Progress and the DSM," Steeves Demazeux offers a historical account of the rational and scientific principles that have shaped the DSM since its third edition. With its innovative content and, most importantly for this chapter, its innovative revision methodology, the DSM-III (1980) represents a milestone in the history of modern psychiatry. In order to characterize the general modalities of the DSM revision process through time, Demazeux contrasts two principles: the "permanent innovation principle" (PIP) and the "prudential conservatism principle" (PCP). He demonstrates that the first principle characterizes the DSM-III/DSM-III-R while the second principle characterizes the DSM-IV/DSM-IV-TR. He highlights the fact that the DSM-5, contrary to its original intention, did not reconnect with the PIP, but instead ended up perpetuating the PCP. This switch from a methodologically innovative attitude to a more cautious and consensual revision process calls for a philosophical examination of the idea of "scientific progress" supported by the DSM revision in the past four decades.

In chapter "DSM-5 and Research Concerning Mental Illness," Jeffrey Poland examines the role of the DSM in research. Starting from the observation that DSM-based research has overall not yielded any solid result, he proceeds to explain this disappointing performance by the fact that the DSM provides researchers with de-contextualized diagnoses that are symptom-based, atheoretical, polythetic, and not associated with well-confirmed tests and models. Simply put, DSM categories provide artificial groupings of individuals experiencing mental illness. Poland further argues that the DSM-5 only continues the failures of the previous DSMs, since the flawed epistemological structure of the classification remains unchanged and many problematic categories have been grandfathered through, for fear that clinical and research practice be disrupted. Poland concludes his chapter with suggestions for improving research on mental illness. After acknowledging that two recent research developments, including the much talked-about RDoC initiative (see chapter "RDoC: Thinking Outside the DSM Box Without Falling into a Reductionist Trap" below), are steps in the right direction, he defends an approach that would divest itself entirely of the categories and commitments of the DSM, and be more directly responsive to the various features of the domain of mental illness.

In chapter "DSM-5 and Psychiatry's Second Revolution: Descriptive vs. Theoretical Approaches to Psychiatric Classification," Jonathan Tsou offers a meticulous appraisal of the relative merits of purely descriptive and theoretical approaches to psychiatric classification. He argues that the purely descriptive approach, favored by the DSM since the DSM-III, is superior for improving communication between the users of the manual. But it is inferior for the two other main goals of the DSM, i.e., treatment and, as Poland argued in chapter "DSM-5 and Research Concerning Mental Illness," research. Consequently, Tsou suggests a hybrid model that combines the benefits of the two approaches and avoids their pitfalls. Eager to avoid the intellectual morass of the current purely descriptive approach, but worried about the danger of theoretical dogmatism, Tsou defends the values of theoretical pluralism, where a multiplicity of scientific theories can coexist, including conflicting ones.

In chapter "DSM-5: The Delayed Demise of Descriptive Diagnosis," Stuart Kirk, David Cohen and Tomi Gomory note that the descriptive approach embraced by the DSM since the DSM-III was meant to increase reliability, which in turn was supposed to permit research into the biological causes of madness. For the authors of this chapter, this scientific effort clearly failed, since reliable biomarkers for mental disorders have never been found, and so-called "mental disorders" continue to be defined ambiguously, redundantly, or tautologically, among other problems. At the same time, the political, cultural and economic success of the DSM guarantees that it will remain immune to profound criticisms. In a tongue-in-cheek conclusion to their chapter, Kirk, Cohen and Gomory offer to rewrite the "Introduction" to the DSM-5. In this utopian "Introduction," the writers of the DSM-5 warn their readers of the nature and possible harms of the DSM-5, and present it as "an American political document."

In chapter "Must Disorders Cause Harm? The Changing Stance of the DSM," Rachel Cooper traces the evolution of the official definition of "mental disorder" in the past editions of the DSM and focuses on the importance of the "harm" criterion in the "Introduction" of each edition. She raises a central yet neglected philosophical question: are all mental disorders in the DSM considered to be harmful for the individual by virtue of a logical necessity, or is it only considered so based on repeated clinical observations? In DSM-III and DSM-IV, the link between "disorder" and "harm" was quite strong. Cooper deplores that this link has been downgraded in the DSM-5: mental disorders are now said to be only "generally associated" with harm. This shift over time, the author concludes, has considerable philosophical as well as ethical implications.

One novelty in the fourth edition of the DSM was the introduction of the notion of "culture-bound syndromes." In chapter ""Deviant Deviance": Cultural Diversity in DSM-5," Dominic Murphy examines the reasons why the DSM-5 has adopted a new model based on three cultural concepts: syndromes, idioms, and local explanations.

He argues that this shift represents an improvement in the sense that the DSM-5 can better take into account the important role of culture in the shaping of certain forms of mental disorder. However, he deplores the persistent tendency of the DSM to think in terms of universal categories, especially when it deals with conditions that are found to be quite common in North America. As Murphy ironically notes, "American mental illness is universal, and other cultures have specific conditions." This statement raises an important epistemological problem: how can we distinguish true universal syndromes from culture-bound ones? The risk on ethnocentrism is patent, since most studies in psychology and psychopathology rely on implicit Western norms. Yet, psychiatrists are not condemned to adopt cultural relativism: as the author remarks in his conclusion, it is possible to take into account cultural variations without undermining the legitimacy of the "universalizing project" of psychiatric classification.

Part II includes seven chapters dedicated to the examination of specific issues related to the DSM-5.

In the history of the DSM, one event marked the starting point of most philosophical reflections on mental disorder: the decision to remove homosexuality from the DSM-II in 1973. Although today sexuality is no longer the focus of most criticisms against the DSM, it remains a hotly debated issue, as chapters "Danger and Difference: The Stakes of Hebephilia" and "Sexual Dysfunctions and Asexuality in DSM-5" demonstrate. In chapter "Danger and Difference: The Stakes of Hebephilia," Patrick Singy offers an analysis of the recent debate that took place around the category of "hebephilia," which is defined as the erotic preference for pubescent children. This category was suggested for inclusion in the DSM-5, but was ultimately rejected after a contentious back-and-forth between the proponents of the category and psychiatrists, psychologists and philosophers who deemed the category absurd. Singy takes a step back. Instead of debating the validity of the category of "hebephilia," he questions the legal assumptions that have made this debate so important in the first place. He argues that the problem with hebephilia is not that it might not be a valid mental disorder, but that it is used in a legal game that is clearly unjust.

While Singy focuses on the most violent and therefore forensically relevant types of sexuality, in chapter "Sexual Dysfunctions and Asexuality in DSM-5" Andrew Hinderliter turns to the other extreme in the continuum of sexual expressions: asexuality. One central difficulty for asexuals is to distinguish asexuality (considered to be an identity category) from disorders involving a lack of sexual desire. Influenced in part by the asexual community, the DSM-5 itself uses the category of asexuality as an exclusionary clause for the diagnoses of Female Sexual Interest/Arousal Disorder and Male Hypoactive Sexual Desire Disorder: if an individual exhibits all the symptoms for one of these two diseases but self-identifies as an asexual, then the diagnosis should not be made. The self-identification of asexuals is therefore crucial, and Hinderliter describes the importance of the Internet for the creation of asexual communities, which in the past 10–15 years have grown significantly. The general public's interest in asexuality remains however rather low. Asexuality is above all a category created from the ground up, and its relation with psychiatry in general, and with the DSM in particular, is both complicated and tenuous.

Paraphilic disorders such as hebephilia are one instance of disorders defined by single symptoms or single symptom clusters, of which kleptomania, pyromania and other categories are further examples. It is to this particularly problematic type of disorders that John Z. Sadler turns to in chapter "The Crippling Legacy of Monomanias in DSM-5." Tracing back the history of these disorders to the early nineteenth-century concept of "monomania," Sadler looks at the reasons behind the rejection of this concept in the second half of the nineteenth century, applies this historical critique to the monomania-like disorders of the DSM-5, and reinforces it with a conceptual critique of his own. Sadler shows that an impoverished, monosymptomatic description of a disorder cannot possibly suffice to tease out superficially similar conditions. Instead of focusing exclusively on the "loudest" symptom – the symptom that no one can miss – he suggests that we also pay attention to more clinically subtle symptoms or signs.

In chapter "The Loss of Grief: Science and Pseudoscience in the Debate over DSM-5's Elimination of the Bereavement Exclusion," Jerome C. Wakefield offers some new decisive arguments for a public debate in which he has been personally involved since the publication of *The Loss of Sadness* (2007, co-authored with Allan Horwitz). In this influential book the two authors attacked the DSM criteria for major depression by showing that these criteria were partly responsible for the over-medicalization of normal sadness. They advocated for a better acknowledgment of the clinical context in the criteria of depression. Their conclusions had an impact on the experts involved in the revision of the DSM-5, but against all expectations these experts decided to move in the complete opposite direction than the one encouraged by Horwitz and Wakefield: they removed the only criterion that was sensitive to the clinical context in the diagnostic of depression, i.e., the "bereavement exclusion" criterion. In this chapter, Wakefield explores the debate about the category of depression that surrounded the revision of the DSM-5, and provides some intriguing insights on the reasons why experts can be misguided.

While Wakefield denounced the removal of the bereavement exclusion from the DSM, in chapter "Against Hyponarrating Grief: Incompatible Research and Treatment Interests in the DSM-5" Şerife Tekin looks at another side of the depression issue. She argues that even if it were true that there are no significant symptomatic differences between people who are grieving and people who suffer from depression, still these two categories of people should not be treated in the same manner. While Tekin concedes that the symptom-based approach of the DSM might have some advantages for research, she is concerned with the clinical implications of this approach. By prioritizing the outward manifestation of symptoms, the DSM-5 distracts the clinicians from the subjective details of grief. Yet these details are an integral part of grief and should not be pushed aside, lest the treatment remain ineffective by limiting the grieving person's ability to develop coping strategies. Tekin concludes her chapter by challenging a *credo* of the DSM, namely, that a single representation of mental disorders can be useful in a wide variety of contexts. Future DSMs would benefit from identifying their primary purpose and from developing models of mental disorders that fit this specific purpose, whether the latter is therapy, research, or something else still.

When in 2009 the National Institute for Mental Health (NIMH) elaborated a new experimental classification project entitled "Research Domain Criteria" (RDoC), many commentators in the psychiatric field were surprised. The DSM-5 revision was still far to be finalized, and the emergence of this new and generously funded project raised much speculation: was it aimed at constituting a rival classification to the DSM-5? Was the existence of such a project tantamount to an official acknowledgment that the DSM's model was in a deadlock? Was it possible to conceive that the RDoC project could in the near future supplant the DSM-5, for both research and clinical purposes? In chapter "RDoC: Thinking Outside the DSM Box Without Falling into a Reductionist Trap," Luc Faucher and Simon Goyer expose the strengths and weaknesses of the RDoC project, and compare them with the DSM. Is the RDoC project more reductionist than the DSM? If so, in what sense? And what kind of new fears (real or false) does this new project bring along? By exanimating in detail the rationale and implicit assumptions of the two classificatory projects, the authors shed light on the future challenges posed by neurosciences for psychiatric classification.

In chapter "DSM-5 and the Reconceptualization of Obsessive-Compulsive Disorder," Baptiste Moutaud offers an intriguing examination of the profound yet rather unnoticed theoretical and clinical transformation of a mental disorder, obsessive-compulsive disorder (OCD). A superficial comparison between the DSM-IV and the DSM-5 would make us believe that this disorder has remained stable, and indeed the revision of the category did not raise much controversy among experts. Yet, as Moutaud demonstrates, the creation in the DSM-5 of a new broader category, "Obsessive-Compulsive and Related Disorders" (OCRD), represents an important conceptual and clinical shift in the model of OCD: from affect to behavior, and from emotion to control motor. In other terms, what was traditionally considered genuinely "obsessive" in OCD is no longer the central feature: anxiety has been replaced by impulse control. In order to offer some general reflections on the meaning of symptoms in the light of neurophysiology, Moutaud also relies on the ethnographic investigation that he has conducted during several years in a French neuroscience laboratory. He highlights how much the emergence of new conceptual models and new therapeutic strategies (in the case of OCD: deep brain stimulation) can have a profound impact on the conceptualization of the disorder as well as on its clinical description.

Instead of using the DSM-5 as a convenient springboard to tackle the broad philosophical issues that are the lot of any classificatory system, the 13 chapters in this volume have in common to provide analyses of the difficulties that are specific to the methodology and assumptions of the DSM. We hope that as a whole the volume will do justice to the complexity and richness of the philosophical, political, and ethical stakes of the DSM-5.

**Acknowledgment** The two editors of this volume are particularly grateful to Philippe Huneman, the series co-editor, who suggested that we work together on this project.

Pessac, France Steeves Demazeux
Schenectady, NY, USA Patrick Singy

## References

American Psychiatric Association. (1952). *Diagnostic and statistical manual of mental disorders* (1st ed.). Washington, DC: American Psychiatric Association.
American Psychiatric Association. (1968). *Diagnostic and statistical manual of mental disorders* (2nd ed.). Washington, DC: American Psychiatric Association.
American Psychiatric Association. (1980). *Diagnostic and statistical manual of mental disorders* (3rd ed.). Washington, DC: American Psychiatric Association.
American Psychiatric Association. (1987). *Diagnostic and statistical manual of mental disorders* (3rd ed.). Washington, DC: American Psychiatric Association.
American Psychiatric Association. (1994). *Diagnostic and statistical manual of mental disorders* (4th ed.). Washington, DC: American Psychiatric Association.
American Psychiatric Association. (2000). *Diagnostic and statistical manual of mental disorders* (4th ed.). Washington, DC: American Psychiatric Association.
American Psychiatric Association. (2013). *Diagnostic and statistical manual of mental disorders* (5th ed.). Arlington, VA: American Psychiatric Association.
Beutler, L. E., & Malik, M. L. (2002). *Rethinking the DSM: a psychological perspective.* Washington, DC: American Psychological Association.
Blashfield, R. K. (1984). *The classification of psychopathology: Neo-Kraepelinian and quantitative approaches.* New York: Plenum Publishing Corporation.
Caplan, P. (1995). *They say you're crazy: How the world's most powerful psychiatrists decide who's normal.* Reading: Addison-Wesley.
Castel, F., Castel, R., & Lovell, A. M. (1982). *The psychiatric society.* New York: Columbia University Press.
Cooper, R. (2005). *Classifying madness; a philosophical examination of the diagnostic and statistical manual of mental disorders.* Dordrecht: Springer.
Cooper, R. (2014). *Diagnosing the diagnostic and statistical manual of mental disorders.* London: Karnac.
Cooper, J. E., & Sartorius, N. (2013). *A companion to the classification of mental disorders.* Oxford: Oxford University Press.
Decker, H. (2013). *The making of DSM-III: A diagnostic manual's conquest of American psychiatry.* New York: Oxford University Press.
Eugenides, J. (2011). *The marriage plot.* New York: Farrar, Straus and Giroux.
Frances, A. (2013). *Saving normal: An insider's revolt against out-of-control psychiatric diagnosis, DSM-5, big pharma, and the medicalization of ordinary life.* New York: William Morrow.
Fulford, K. W. M., Thornton, T., & Graham, G. (2006). *Oxford textbook of philosophy and psychiatry.* Oxford/New York: Oxford University Press.
Fulford, K. W. M., Davies, M., Gipps, R. G., Graham, G., Sadler, J. Z., Stanghellini, G., & Thornton, T. (2013). *The Oxford handbook of philosophy and psychiatry.* Oxford: Oxford University Press.
Greenberg, G. (2013). *The book of woe: the DSM and the unmaking of psychiatry.* New York: Penguin Books.

Grob, G. N. (1991). Origins of DSM-I: a study in appearance and reality. *American Journal of Psychiatry, 148*(4), 421–431.
Hacking, I. (1995). *Rewriting the soul. Multiple personality and the sciences of memory*. Princeton: Princeton University press.
Horwitz, A. V., & Wakefield, J. C. (2007). *The loss of sadness: How psychiatry transformed normal sorrow into depressive disorder*. New York: Oxford University Press.
Horwitz, A. V., & Wakefield, J. C. (2012). *All we have to fear: Psychiatry' s transformation of natural anxieties into mental disorders*. New York: Oxford University Press.
Insel, T. (2013). Director's blog: Transforming diagnosis. Retrieved on April 29, 2013, from http://www.nimh.nih.gov/about/director/2013/transforming-diagnosis.shtml
Kandel, E. R. (1998). A new intellectual framework for psychiatry. *American Journal of Psychiatry, 155*, 457–469.
Kendler, K. S., & Parnas, J. (Eds.). (2008). *Philosophical issues in psychiatry I: Explanation, phenomenology, and nosology*. Baltimore: John Hopkins University Press.
Kendler, K. S., & Parnas, J. (Eds.). (2012). *Philosophical issues in psychiatry II: Nosology*. Oxford: Oxford University Press.
Kirk, S., & Kutchins, H. (1992). *The selling of DSM. The rhetoric of science in psychiatry*. New York: De Gruyter.
Kupfer, D. J., First, M. B., & Regier, D. A. (Eds.). (2002). *A research agenda for DSM-V*. Washington, DC: American Psychiatric Association.
Kutchins, H., & Kirk, S. (1997). *Making us crazy. DSM: The psychiatric bible and the creation of mental disorders*. London: Constable.
Lane, C. (2008). *Shyness: How normal behavior became a sickness*. New Haven: Yale University Press.
Murphy, D. (2006). *Psychiatry in the scientific image*. Cambridge, MA/London: MIT Press.
Paris, J., & Phillips, J. (2013). *Making the DSM-5: Concepts and controversies*. Dordrecht: Springer.
Radden, J. E. (2004). *The philosophy of psychiatry: A companion*. New York: Oxford University Press.
Sadler, J. Z. (Ed.). (2002). *Descriptions and prescriptions: Values, mental disorders, and the DSMs*. Baltimore: The Johns Hopkins University Press.
Sadler, J. Z. (2005). *Values and psychiatric diagnosis*. Oxford: Oxford University Press.
Sadler, J. Z., Wiggins, O. P., & Schwartz, M. A. (Eds.). (1994). *Philosophical perspectives on psychiatric diagnostic classification*. Baltimore: The Johns Hopkins University Press.
Spitzer, R. L., & Wilson, P. T. (1975). Nosology and the official psychiatric nomenclature. In A. M. Freedman, B. J. Sadock, & H. I. Kaplan (Eds.), *Comprehensive textbook of psychiatry* (pp. 826–845). Baltimore: Williams & Wilkins.
Young, A. (1995). *The harmony of illusions: Inventing post-traumatic stress disorder*. Princeton: Princeton University Press.

# Part I
# General Issues

# The Ideal of Scientific Progress and the DSM

**Steeves Demazeux**

**Abstract** In this chapter I explore the ways in which the *Diagnostic and Statistical Manual of Mental Disorders* has been informed by the ideal of scientific progress since the 1980s. My aim is not to evaluate the arguments for or against the scientific nature of the DSM as they have been brought forward and discussed in the public debate for purely promotional or polemical purposes. It is rather to highlight the kind of scientific optimism that has been a driving force in the recent history of the DSM. I argue that while the nature of the scientific ideal of the DSM through its three last editions has remained the same, its effects and its rationale have dramatically changed. Whereas scientific progress required that the DSM-III firmly espouse what I call here a "permanent innovation principle," it would later paradoxically motivate, in the DSM-IV and especially in the DSM-5, an opposed principle that I call the principle of "prudential conservatism." I conclude that, while in the past decades the DSM has made an important effort to improve the scientific quality of its revision process, development from the DSM-III to the DSM-5 can hardly be said to have been scientifically progressive.

## The Ideal of Scientific Progress and the DSM

The third edition of the *Diagnostic and Statistical Manual of Mental Disorders* (DSM-III) was soon proclaimed by its proponents to have achieved a "scientific revolution," while its opponents accused it of being "unscientific." Even if the terms "science" or "scientific" are mostly absent from the DSM (and even from much of the official proceedings surrounding the development of the several editions), it is clear that scientific progress, along with clinical utility, has been a key issue for assuring the legitimacy and the credibility of the American classification system. In particular, the original project of the DSM-III was to develop an atheoretical classification of mental disorders that would be progressively enriched and consolidated by the best empirical data available in the scientific literature.

S. Demazeux (✉)
Associate Professor, Université Bordeaux Montaigne, Pessac, France
e-mail: sdemazeux@gmail.com



S. Demazeux, P. Singy (eds.), *The DSM-5 in Perspective*, History, Philosophy and Theory of the Life Sciences 10, DOI 10.1007/978-94-017-9765-8_1

In this chapter I show in what ways the DSM, dating back to the 1980s, has been informed by the ideal of scientific progress. My aim is not to evaluate the arguments for or against the scientific nature of the DSM as they have been brought forward and discussed in the public debate for purely promotional or polemical purposes. It is rather to highlight the kind of scientific optimism that has been a driving force in the recent history of the DSM. I argue that the nature of the scientific ideal of the DSM has remained the same over its three last editions, but that its effects and rationale have dramatically changed. Whereas scientific progress required that the DSM-III firmly espouse what I call here a "permanent innovation principle," it would later paradoxically motivate, in the DSM-IV and especially in the DSM-5, an opposed principle that I call the principle of "prudential conservatism."

## Scientific Progress and the DSM

The role of scientific progress in the DSM can be approached from at least three different perspectives: by focusing on *fundamental, methodological*, and *particular* issues. First, one may focus on the overall project of applying the scientific method to psychiatric nosology. Since the turning point in 1980, when the third edition was published, the DSM has been hailed as "a significant reaffirmation on the part of American psychiatry of its medical identity and its commitment to scientific medicine" (Klerman et al. 1984, 539). More specifically, according to Melvin Sabshin, Medical Director of the American Psychiatric Association (APA) from 1974 to 1997, the aim of the DSM-III was to help "to transform psychiatry into a reputable field based on sound scientific principles" (Sabshin 2008, 37). Its main strategy was to adopt an atheoretical approach (explained below) coupled with the provision of specific clinical criteria for each mental disorder listed in the manual. Instead of just mirroring the "ideologies" that were dividing the psychiatric arena into conflicting schools of thought (Sabshin 2008, 37), the DSM attempted to achieve a more objective and rational approach by describing mental disorders in a manner that would be beneficial and reinforce psychiatry as a genuine branch of medicine.

A second way to consider scientific progress is to focus more specifically on the *methodology* for revising each edition of the manual over time. The novelty of the DSM-III was to encourage an ongoing process of revision by integrating better scientific information as soon as it became available in the literature. The DSM-III was primarily based on expert consensus, but the developers of the following editions have sought to adopt a more data-driven revision process. As we will see, the modalities of the revision process have evolved considerably with the different editions of the DSM, from the DSM-III-R to the DSM-5.

Finally, a third way to evaluate the scientific soundness of the DSM is to pay attention to all the particular decisions that have been made by the architects of the manual. When the time comes for revision, many proposals are discussed amongst the experts from the several subgroups of the Task Force committee. Fortunately or unfortunately, the final decisions are never made on solely scientific grounds. On the one hand, members of the Task Force are quick to recognize that they have to take into account several important non-scientific factors, such as the clinical utility of the modification, its accordance with the clinical tradition, its compatibility with *International Classification of Diseases* (ICD) codes, its clarity and simplicity for administrative purposes, and its potential harmful impact on public health as well as its potential impact on the health economy (American Psychiatric Association 1987, xx–xxi). On the other hand, it must be acknowledged that misconceptions and distortions of different kinds (corruption, intellectual shortsightedness, cultural biases, etc.) are likely to affect the final decisions, as they regularly do affect the results of any scientific process. The sector of psychiatry, which since the 1970s has become a major market in the health economy with strong ties to the pharmaceutical industry, is undoubtedly more prone to corruption today than in past decades.

The accusation that the DSM is "unscientific" has been made repeatedly in terms of each of these perspectives. Early on, several authors questioned the most basic assumptions of the DSM, arguing that the American classification system was grounded on shaky philosophical foundations (Zubin 1978; Schacht 1985; Galatzer-Levy and Galatzer-Levy 2007), or that the overall project was based on bad science and merely used the "rhetoric of science" (Kirk and Kutchins 1992). There have also been many criticisms focusing on the rationale and general methodology for revising the DSM, and of course many criticisms regarding particular decisions, both tiny and major, made by the experts recruited by the Task Force.

In the rest of this chapter, I focus on fundamental and methodological criticisms. Of primary importance is an examination of the causes of specific pitfalls or pseudo-scientific judgments that have led to erroneous local decisions (a good example of such a critique is provided by Wakefield in this volume). But it should be noted that these failures take place within the normal functioning of a scientific revision process. As long as this process is not systematically biased, the fact that some local decisions are contentious does not compromise the legitimacy of the scientific enterprise as a whole.

## The DSM-III "Revolution"

The publication of the DSM-III in 1980 is often presented as an "event of capital importance" in the history of modern psychiatry in the sense that it gave a "redirection of the discipline toward a scientific discourse" (Shorter 1997, 302). A quote, reported by many commentators, summarizes quite well the scientific enthusiasm that surrounded the publication of the DSM-III. Its author is Gerald Maxmen, professor of clinical psychiatry at Columbia University, in a 1985 publication:

> On July 1, 1980, the ascendance of scientific psychiatry became official. For on this day, the APA published a radically different system for psychiatric diagnosis called ... DSM-III. By adopting the scientifically based DSM-III as its official system for diagnosis, American

> psychiatrists broke with a fifty-year tradition of using psychoanalytically based diagnoses. Perhaps more than any other single event, the publication of DSM-III demonstrated that American Psychiatry had indeed undergone a revolution. (Maxmen 1985, 35)

The widespread belief that the DSM-III brought about "a small revolution in American psychiatry" was defended early on (see Andreasen 2007; Decker 2013, 108). In a 1984 debate on the DSM-III, Gerald Klerman claimed "the decision of the American Psychiatric Association first to develop DSM-III and then to promulgate its use represents a significant reaffirmation on the part of American psychiatry of its medical identity and its commitment to scientific medicine" (Klerman et al. 1984, 539). Later, he went on to characterize the DSM-III as the culmination of a "paradigm shift in general psychiatry" (Klerman 1990, 29).

It is clear that the DSM-III, developed from 1974 to 1980, was a project of an unprecedented scale in the history of psychiatry, mobilizing dozens of experts and hundreds of external collaborators (Kendell 1980). The project appeared primarily as a response to a profound crisis of legitimacy that American psychiatry was facing in the 1970s (Demazeux 2013; Decker 2013). The field was deeply divided between supporters of a biological approach, proponents of a psychodynamic approach, and proponents of a more psychosocial and community-based approach (Strauss et al. 1964). Violent criticism expressed by antipsychiatrists (Szasz 1960) and sociologists (Scheff 1966) aggravated an already deteriorated public image of psychiatry. Large psychiatric hospitals and asylums were generally perceived as repressive and as contributing to the stigmatization of mental patients (Goffman 1961). At the beginning of the 1970s, many debates – such as the quarrel over homosexuality, which would lead in 1973 to its depathologization (Bayer 1981), the Rosenhan study (Rosenhan 1973), or numerous empirical studies that highlighted the profound inconsistency of psychiatric diagnoses (see Kendell 1975; Blashfield 1984) – brought into question the scientific character of clinical psychopathology.

Two important original changes characterized the DSM-III project from an epistemological point of view. The most important decision was to adopt an atheoretical perspective with regard to etiology (American Psychiatric Association 1980, 7), i.e. a "purely descriptive" perspective in organizing the classification system, in contrast to traditional systems, which are based on theoretical (and often ideological) hypotheses concerning etiology. The idea was to focus on stable clinical features instead of venturing into etiological hypotheses, since no consensus was possible on this level given the current state of knowledge. The second important decision was to rely on operational criteria in order to clinically characterize each mental disorder listed in the manual. Taken together, these two related methodological decisions were expected to impel psychiatric nosology into a more scientific direction.[1]

---

[1]There were other methodological novelties in the DSM-III, like the introduction of the multiaxial system of diagnosis. Although the multiaxial system was often considered to be an important feature of the new classification (Millon 1983, 810) and its use seen as "an important step in achieving greater scientific precision" (Sabshin 2008, 277), it has recently been removed from the DSM-5.

## *The DSM's Theory-Neutral Strategy*

There has been much criticism of the so-called "atheoretical" perspective adopted by the DSM-III. In a 1996 paper, William Follette and Arthur Houts claimed that there is a deep inconsistency inside the DSM: while the whole project was built on the explicit adoption of an atheoretical perspective, it implicitly relied on a specific theoretical model, i.e. the medical model – a model that considers all mental disorders without exception to be *diseases* in the general medical sense of the term. According to the authors, this inconsistency undermines "scientific credibility for the DSM," since "DSM adherents seem oblivious to the need for theory to organize classification and, thus, appear to go blithely along believing that the DSM is atheoretical" (Follette and Houts 1996, 1122).

It is clear that the DSM is theory-laden and that it is not atheoretical in the strict sense of the term. But it has never, in fact, pretended to be such. As Wakefield (1999) rightly pointed out in response to Follette and Houts's paper, the term "atheoretical," if strictly intended, would imply "that the DSM diagnosis makes no reference whatever, direct or indirect, to unobservable processes or etiological variables." However, this is explicitly denied in the introduction to DSM-III, which allows that when etiology is scientifically established, then it can be used in the definition of the disorder:

> The approach taken in DSM-III is atheoretical with regard to etiology or pathophysiological process except for those disorders for which this is well established and therefore included in the definition of the disorder. (American Psychiatric Association 1980, 7)

But in most cases, when etiology is unknown or controversial, the definition of a disorder must only be at the "lowest order of inference necessary to describe the characteristic features of the disorder" (American Psychiatric Association 1980, 7), which may involve some inference to unobservables.[2] Instead of speaking about "atheoretical strategy," Wakefield concludes, the authors of the DSM would have been more rigorous if they had spoken of "theory neutrality," in the sense that they aimed at providing "*theory neutral* categories or criteria that are neutral with respect to plausible theories of etiology" (Wakefield 1999, 966).

The atheoretical approach aimed at fulfilling three different goals: first, it aimed at moving beyond ideological divisions and toward science[3]; second, it aimed at temporarily setting aside the etiological issue in order to focus on acute descriptions

---

[2]The DSM-III itself offers the example of the criterion for borderline personality disorder of "identity disturbance" (American Psychiatric Association 1980, 7).

[3]Melvin Sabshin, the medical director of the APA during this period, and co-author of an influential book in 1964 called *Psychiatric Ideologies and Institutions* (Strauss et al. 1964), considered that the success of the DSM-III was a victory of science over ideology (Sabshin 2008). It is nevertheless interesting to note that the opposition between science and ideology has different meanings in Sabshin's work: in 1964, Sabshin and his colleagues defined "ideology" as including any "theoretical models of mental disorders" (see Strauss et al. 1964, 8). This conciliatory definition tends to consider that science and ideology are always inextricably intertwined. But in his 2006 book, Sabshin defines "ideology" as all "scientifically unsupported dogmas" (Sabshin 2008, 36).

of mental disorders[4]; and third, it aimed at reforming the usual clinical vocabulary by avoiding, as much as possible, any claims about mental events.[5] But, by adopting the atheoretical approach, the developers of the DSM-III neglected the overall form and structure of the classification as a scientific system (Demazeux 2013, 154). This explains why the DSM-III resembles a useful nomenclature or a "trouble-shooting guide" (Wakefield 1999, 969) more than a true nosological system.

## *From the "Professor Model" to the "Expert Model"*

The hallmark advance of DSM-III was the provision of specific diagnostic criteria for each mental disorder listed in the manual. DSM developers made a conscious effort to provide clinical definitions based on testable clusters of symptoms. This strategy, which provides long lists of specific diagnostic criteria, is what gives the manual its famous and particular aspect of a "Chinese menu."

In nineteenth century psychiatry, nosological systems were created and promoted by great professors of psychiatry. Whereas the committees that drafted the DSM-I and (to a lesser extent) DSM-II were mainly composed of distinguished professors of psychiatry – most of them had served during WWII – the DSM-III Task Force was made up of a new generation of psychiatrists, the so-called "neo-Kraepelinian" generation of psychiatrists, most of whom were scholars with academic credentials in epidemiology or psycho-biometry. The ambition to establish a classification system informed by the best available clinical data instead of one based on the opinions of great clinicians marks the move from a "professor model" to an "expert model" (Frances and Egger 1999, 162). It is crucial to understand this change in perspective: it moved the weight of epistemic authority from the dominant academic opinion to the analysis by an expert community of all the scientific literature. As prestigious as a clinician might be, his sole opinion no longer counted; most important was the evidence that he could mobilize in order to defend his position.

Two main scientific indicators were and are still used: reliability and validity. In order to measure reliability, members of the committee used a statistical measure designed by Jacob Cohen in 1960 and called "kappa" (Cohen 1960). As for evaluating validity, they relied on a widely cited medical model developed by Robins

---

Here, ideology is clearly set outside the range of science (See Demazeux 2013, 152). With distance and hindsight, it is clear that, according to Sabshin, the atheoretical perspective of the DSM-III is scientific in opposition to ideological in the second sense.

[4]In a famous 1959 conference in front of an audience of psychiatrists, the philosopher Carl Hempel first put forward the idea of taking a step back in order to propose a purely descriptive classification, which could, in a second phase, foster the construction of an etiological classification (Hempel 1965). (For a discussion concerning the influence of Hempel's ideas over the process of the DSM-III, see Kendler et al. 2010, 140 and Fulford et al. 2013, 6sq).

[5]Yet, Spitzer insisted on the fact that the Task Force "never believed that it was possible or desirable to limit diagnostic criteria to 'attributes that could be readily observed'" (Spitzer 1991, 295).

and Guze in a 1970 paper. This five-step model of clinical validation served as a reference point during the elaboration of the Feighner criteria (Feighner et al. 1972) as well as during the development of the Research Diagnostic Criteria (Spitzer et al. 1978). The idea was to take into account the most recent data from clinical research, laboratory data, follow-up studies, and familial aggregation, in order to consolidate the validity of the clinical criteria adopted in the manual. As Spitzer put it: "The DSM-III committee shared the view that progress in psychiatric nosology (improving the validity of psychiatric diagnosis) will come primarily from data collected in empirical research studies" (Spitzer 2001, 354). Robert Spitzer himself, considered the leader of the "neo-Kraepelinian circle," which gathered around a common credo (Klerman 1978; Blashfield 1982), would only admit one as his own: that of being "data oriented" (Spitzer 1982, 592).

In practice, though, the move from the professor to the expert model was not very efficiently handled during the construction of the DSM-III. The main difficulty was in providing good scientific support for each diagnostic category. At a time when empirical data were insufficient and scarce, most of the decisions were in fact makeshift solutions, and many were taken on the grounds of the "best clinical judgment" that resulted in consensus between the members of the Task Force or by majority votes. This fragile decision-making process was explicitly recognized, as is clear from the following passage from the DSM-III introduction:

> In attempting to resolve various diagnostic issues, the Task Force relied, as much as possible, on research evidence relevant to various kinds of diagnostic validity. For example, when discussing a problematic diagnostic category, the Task Force considered how the disorder, if defined as proposed, provided information relevant to treatment planning, course, and familial pattern. It should come as no surprise to the reader that even when data were available from relevant research studies, Task Force members often differed in their interpretations of the findings. (American Psychiatric Association 1980, 3)

Harold Pincus comically summarized this general method as the BOGSAT method: "A bunch of guys sitting around a table" (Kendler and Parnas 2012, 141). When sufficient support was not available, clinical descriptions provided by the DSM-III using operational criteria sought to provide at least "face validity," that is, a consensual validity between experts. Spitzer recognized that "limiting DSM-III to only those categories that had been fully validated by empirical studies would be at the least a serious obstacle to the widespread use of the manual by mental health professionals" (Spitzer 1991, 294). In other words, the practical purposes of the classification outweighed its scientific achievement. But Robert Spitzer himself deplored that this kind of validity usually varies according to the "face" of an expert.

## DSM-III-R as an Illustration of the "Permanent Innovation Principle"

Soon after the DSM-III was released in 1980, a new revision process was planned. The need for a revision of the newly-minted classification system was already expressed in the DSM-III's own introduction: "In the several years that it has taken to develop DSM-III, there have been several instances when major changes in initial drafts were necessary because of new findings. Thus, this final version of DSM-III is only one still frame in the ongoing process of attempting to better understand mental disorders" (American Psychiatric Association 1980, 12). The conclusion of a 1982 paper by Williams and Spitzer perfectly illustrated this new attitude of constant revision. The authors concluded with the formulation of good and bad news: "The good news is that with the advent of specified diagnostic criteria in the 1970s, research investigators for the first time can have some assurance that different researchers using these criteria mean the same thing" (Williams and Spitzer 1982, 1289). But, the authors continue:

> The bad news is that these new DSM-III criteria are, for the most part, based on limited knowledge. Inevitably, new sets of diagnostic criteria will evolve based on more experience and validity studies. Already, in fact, the newly developed DSM-III criteria are being studied with a view towards their eventual modification. But such is the price of progress! (Williams and Spitzer 1982, 1289)

Constant innovation is the price to be paid for scientific progress. In fact, there was a need to rectify a large number of errors that had quickly been identified in the DSM-III. More than just a need for refinement or clarification, there were even some conceptual inconsistencies in some parts of the manual that could be detrimental in clinical practice. Nevertheless, with this constant effort to keep the DSM up-to-date with the best available data, the DSM-III-R (R for "revision"), published in 1987, illustrates the pursuit of the ideal of a classification that sticks to scientific research, both by being informed by the most recent data and by *stimulating* new data. Only seven years after the DSM-III, many changes were adopted, some of which were very significant. Robert Spitzer (Chair of the work group for the DSM-III) and his wife Janet Williams ("Text Editor") supervised and intervened in each of the 26 advisory groups that composed the Task Force.

The DSM-III revision process that led to the DSM-III-R, an optimistic progressive and amelioration of the classification, perfectly illustrates what we can call a 'permanent innovation principle' (PIP). We can summarize this principle as a maxim of action: "We should modify the clinical criteria of the current nosology whenever it is justified by the best available scientific data, even though these changes may disturb the practices of clinicians." Instead of according a great epistemic authority to the weight of clinical tradition – as was common in most nosological systems – the DSM relied explicitly on scientific literature, and tried furthermore to stimulate clinical studies in new directions. Long before the emergence of evidence-based medicine in the 1990s (Sackett et al. 1996), the developers of the DSM decided to give the preference to data over clinical expertise as a rational justification for their decisions.

The PIP quickly showed its limitations once the manual was in use. First of all, a gap appeared between the two main functions that the classification was supposed to fulfill: to be scientifically sound, and to be useful for clinicians (and for administrators). It was quite inconvenient for a reference classification that was standard across the field to deal with constant and rapid changes: psychiatrists' usual modes of practice were perturbed, and they complained that the classification was not only complex but also unstable; comparisons between data became more difficult to perform for administrative purposes; and even in research, the recording of diagnosis was uncomfortable and the comparisons between different studies were made less easy.

But the PIP also showed its internal limitations. An upside of the DSM-III was that it stimulated the production of many new experiments and clinical analyses. But the downside was that by the end of the 1980s, no clear etiological pathway had been discovered for any mental disorder, and the data available did not promise substantial progress: many studies showed conflicting results, notably in clinical research and in epidemiology. While the revision process was increasingly guided by scientific standards, the risk was that the experts could pick out from this new abundant literature whatever studies and conclusions best supported their personal opinions.[6] Additionally, many commentators criticized the fact that the DSM-III-R had been introduced so soon after the introduction of DSM-III.

## DSM-IV and IV-TR: From Innovation to Conservatism

Only 1 year after the publication of the DSM-III-R, the APA officials appointed the next round of experts to take charge of the next revision process. But contrary to the reason given for the move from the DSM-III to the DSM-III-R – that of urgently updating the classification – the official reason alleged by the APA officials for such a premature move forward was the need for the next version to be in line with the World Health Organization's ICD-10, the publication of which was planned for 1992. To re-appoint Robert Spitzer (who by then had become a very famous and influential psychiatrist) and his wife Janet Williams to the head of the revision process would have been seen by public opinion as an oligarchical move. The APA preferred to nominate a triumvirate of less colorful characters: Allen J. Frances (chairperson), Harold A. Pincus (vice-chairperson) and Michael B. First ("Editor, Text and Criteria").

Soon after Frances was appointed in May 1988 as the new chairperson of the DSM-IV Task Force, he declared that "it is in fact possible that the major innovation in DSM-IV will not be in its having surprising new content but rather it will reside in the systematic and explicit method by which DSM-IV will be constructed and

[6]There were some harsh controversies over some new labels proposed for inclusion in the DSM-III-R, like the Premenstrual Dysphoric Disorder (lately renamed Late Luteal Phase Dysphoric Disorder), the Sadistic Personality Disorder, the Self-defeating (or Masochistic) Personality Disorder, the Paraphilic Coercive Disorder, etc. Given the important coalition that formed to protest these new labels (particularly coming from feminist groups inside and outside the APA), an Ad Hoc Committee to Review DSM-III-R was constituted in 1985. Its goal was to discuss and decide on controversial diagnoses. Some of these controversial categories were finally included in the DSM-III-R Appendix A and were presented as "proposed diagnostic categories needing further studies" (American Psychiatric Association 1987, 367sq).

documented" (Frances et al. 1989, 375). So it was clear from the start of the revision process that the expected improvement would be more *methodological* than *substantial*. The main reason was easy to understand and Frances was willing to take on the task: the content of the DSM had to be stabilized for at least a decade or two.[7]

Frances argued that the DSM-IV revision process should be driven by a new methodology. As David Shaffer, an expert on the DSM-IV Task Force, reported: "Frances instructed us to keep differences between DSM-III-R and DSM-IV to a minimum. We should make only those changes that would correct some generally recognized mistake or could be justified by sound empirical evidence" (Shaffer 1996). Frances thought that an important step to take in the revision process was to rely more on data, and less on the personal opinions of experts. According to him, the solution was to move toward more science and less ideology (Goleman 1994). The outcome should no longer depend on "who's in the room and who argues the loudest," he declared, implicitly suggesting that such had been the rule during the DSM-III revision process (Goleman 1994).

An easy way to constrain the normal tendency among experts to innovate was to increase the threshold of scientific evidence required in order to make a change, and to require experts to make explicit statements of the rationale for each proposal. This new strategy, clearly conservative, did not satisfy everyone. The major weakness of this method is that strong evidence is required only to make changes or add new categories, and not to keep current criteria and categories. If there was no strong evidence for a disorder to be adopted in the DSM-III, it would remain the same – unless some strong evidence showed that the category should be removed, which has only very occasionally occurred.

Spitzer and Williams strongly disagreed with the new methodology adopted by the DSM-IV developers from the start. According to them, the soundest scientific strategy for improving the current classification was not to rely on the available literature but to stimulate alternative proposals for diagnostic criteria and to test them. So the idea was to develop a research "multiplex study design" (Spitzer and Williams 1988, 45), i.e., a multicenter field trial of alternative sets of diagnostic criteria.[8] There was no room for complacency, and the classification had not merely to consolidate its current criteria but also to look for new directions: "no one, including us, is entirely happy with our current nosologic system" (Spitzer and Williams 1989). Spitzer and Williams were harsh about the new tone of the revision process introduced by Frances, especially the idea that progress in the classification

[7]An additional constraint was the necessity for improvement in compatibility with ICD-10. The United States, along with all other members of the WHO, is supposed to officially adopt the ICD as its standard classification. In psychiatry, this should become official in October 2015 with the adoption of the ICD-10-CM (CM for clinical modification – given that American clinicians are allowed to adapt some ICD categories and codes in order to meet the clinical and administrative particularities of the American health system; see Reed 2010).

[8]This interesting idea is close to the "polydiagnostic strategy" developed in Germany by psychiatrists Peter Berner and Heinz Katschnig at the beginning of the 1980s (Berner and Katschnig 1984).

system could be implemented automatically by relying on literature reviews: "It is silly to suggest that if the process of developing DSM-III and DSM-III-R were truly based on 'data' there would have been no controversies or need to vote" (Spitzer and Williams 1989, 960). "Our view," they wrote, "is that real progress can only come from new data collected to test alternative promising hypotheses about how to define and classify mental disorders. That is our dream" (Spitzer and Williams 1989, 960).

Yet the strategy put forward by Frances prevailed and was chosen for the next edition of the DSM. Beforehand, however, he decided that in order to conduct the revision process with more transparency the appointment of the work group members would be different. Experts would not be recruited only for their scientific achievement or dominant influence in the field but also for their capacity to gather, discuss, interpret and objectively evaluate the available scientific data from different perspectives:

> We took a number of precautions to ensure that the Work Group recommendations would reflect the breadth of available evidence and opinion and not just the views of the specific members. After extensive consultations with experts and clinicians in each field, we selected Work Group members who represented a wide range of perspectives and experiences. Work Group members were instructed that they were to participate as consensus scholars and not as advocates of previously held views. Furthermore, we established a formal evidence based process for the Work Groups to follow. (American Psychiatric Association 1994, xv)

Once 13 work groups were constituted, Frances and his colleagues decided to organize a three-stage review of the current state of psychiatric knowledge, "using documentation derived from comprehensive literature reviews, re-analysis of existing data sets, and rigorous and extensive field trials" (Frances and Egger 1999, 162).

So, during the first part of the elaboration of the DSM-IV, all relevant data were carefully and exhaustively analyzed and discussed within various committees of experts of the working group. The results of this huge literature review analysis were published in three important sourcebooks.[9] A fourth book provided data re-analyses and a last book contained "reports of the field trials and a final executive summary of the rationale for the decisions made by each Work Group" (American Psychiatric Association 1994, xx).

The publication of the DSM-IV in 1994 marked a turning point in the history of the DSM. Never had a psychiatric classification been so attentive to the objective data provided by the scientific literature as the DSM-IV. Yet, paradoxically, the innovative strategy adopted by the former edition of the DSM was replaced by a completely opposite strategy, explicitly said to be conservative. This improved revision model process, in which "the empirical information is systematically and

[9]Jeffrey Poland (see his contribution in this volume) has offered a general review of the DSM sourcebooks. Concerning the first volume, he remarks: "The papers collected in this volume exhibit a number of shortcomings that lead one to conclude that the development process did not live up to its own standards and has not achieved some of its most important stated objectives" (Poland 2001).

objectively reviewed and used to inform the nosologic process," can be called an "advisory model" (Kendler 1990, 973). The role of the expert is to *advise* the committee, not to influence it. The professor model had died out, but Frances was convinced that it was already time to *move beyond the expert model*: "The innovation of DSM-IV was to reach beyond the expert model to attempt to develop an evidence-based psychiatric nosology based on a systemic review of the available scientific data about psychiatric disorders. It was not meant to be paradigm shift but rather a good workman-like improvement informed by a comprehensive review of available scientific knowledge" (Frances and Egger 1999, 162).

Robert Spitzer remained unconvinced that this new emphasis on objective reviews of the literature by putting constraints on the work of the experts could be fruitful. As he declared in a 1991 paper, "I am troubled by the tendency (intended or not) to play down the major role that expert consensus will have in the final decision-making process for DSM-IV" (Spitzer 1991, 294). He pointed out that most concerns in the DSM were conceptual rather than empirical, and that at the end of the process, decisions would "still be based primarily on expert consensus, rather than on data, as was the case with DSM-III and DSM-III-R" (Spitzer 1991, 294).[10] The DSM-IV was deliberately non-innovative in the proposed content. But Spitzer doubted that the DSM-IV was even *methodologically* innovative.

Apart from this evidence-based decision-making ideal, another novelty in the DSM-IV process was the better attention paid to clinical utility (the DSM-IV sought to become "more user-friendly") and to all the "unintended consequences" (Andreasen 2007) of the DSM on the psychiatric field, like the risk of over-medicalizing mental suffering. In a nutshell, DSM-IV sought to be more conservative and at the same time more consensual than its previous manifestation. It aimed to find equilibrium between historical tradition, clinical utility, and scientific improvement.[11] The final result, the DSM-IV, was declared to be the work of the whole group. These are the very first words of the DSM-IV, and one should concede that they are not just the usual and expected words of introduction: "DSM-IV is a team effort" (American Psychiatric Association 1994, xiii).

In 1997, the mid-term review of the DSM-IV that would give birth to the DSM-IV-TR began. The process became ever more anonymous, without any singular leader: Allen Frances was no longer part of the revision process, replaced by Michael B. First (co-chairperson and Editor) and Harold A. Pincus (co-chairperson). The result, the DSM-IV-TR, published in 2000, is a perfect illustration of what one might call, in opposition to the PIP, an illustration of the prudential conservatism principle (PCP). Behind the subtle difference in the acronym that passes almost

[10]Frances replied to Spitzer that this was a "false dichotomy," in the sense that "DSM-IV decisions were the result of expert consensus on how to best interpret the data" (Frances et al. 1995, 34).

[11]One important issue for the DSM-IV revision was to limit, as much as possible, the tendency to add new categories in the manual. On this issue, Harold Pincus remembers "more than 150 different proposals were received for adding categories for DSM-IV" (Kendler and Parnas 2012, 151).

unnoticed between DSM-III-R and DSM-IV-TR (the initials "TR," "Text Revision," mean that only the text that accompanies the clinical descriptions is changed, and not the clinical criteria), a profound change in the whole procedure for revising the DSM was made. The race for nosological innovation gave way to the need to stabilize the reference classification (that had come to be used throughout the mental health profession as a textbook) by keeping the clinical data (prevalence, history, prognosis, etc.) up to date with the explosion of research. The criteria changes were few and concerned only a few diagnostic labels. The DSM-IV developers wanted to allow some time before giving nosology its next big change.

## The DSM-5 Process

### *A Chaotic Revision Process*

At the end of the 1990s, the plan for the development of the next version of the DSM was still under discussion. What should be the correct orientation of the future revision process that would give birth to the DSM-V?[12] Experts from the APA were divided between two fundamental options: should we continue conservatism (PCP) or should we return to innovation (PIP)? From September 1999 to October 2000, three DSM-V Research Planning Conferences were organized under the joint sponsorship of the APA and the National Institute of Mental Health (NIMH). In the *Research Agenda for DSM-V,* published in 2002 as a result of this series of conferences, David Kupfer, Michael First, and Darrel Regier passed severe judgment on the accomplishments of the DSM-III and -IV:

> In the more than 30 years since the introduction of the Feighner criteria by Robins and Guze, which eventually led to DSM-III, the goal of validating these syndromes and discovering common etiologies has remained elusive. Despite many proposed candidates, not one laboratory marker has been found to be specific in identifying any of the DSM-defined syndromes. Epidemiologic and clinical studies have shown extremely high rates of comorbidities among the disorders, undermining the hypothesis that the syndromes represent distinct etiologies. Furthermore, epidemiologic studies have shown a high degree of short-term diagnostic instability for many disorders. With regard to treatment, lack of treatment specificity is the rule rather than the exception. (Kupfer et al. 2002, xviii)

The authors conclude that the time has come to consider the possibility of "a paradigm shift" for the manual (Kupfer et al. 2002, xviii).

But the main defect of the classification, according to the authors, lay in the atheoretical methodology adopted by the DSM-III, which had not proved to be scientifically fruitful: "the DSM's descriptive approach may have outlived its usefulness and is in fact potentially misleading" (Kupfer et al. 2002, 31). For the

[12]The decision to switch from a Roman (DSM-V) to an Arabic numeration (DSM-5) was taken in 2008.

first time since the DSM-III, official experts in the APA shared a conviction common to many commentators. Indeed many analysts had been convinced for a long time that psychiatry would progress much better scientifically if clinicians could rely on explicit theory-based models of diagnosis and test them (Follette and Houts 1996; Cooper 2005; Murphy 2006; Tsou 2008).

Unlike the process of revising the DSM-IV, which was undertaken mostly from a retrospective perspective through a review of all the existing literature, the authors of the DSM-5 bet on the ability to stimulate clinical research in order to explore alternative proposals long before the revision process began. Unlike DSM-IV, the idea was that we should seize the opportunity to look forward, not backward, in order to create the space for important potential changes. Following this line of thinking, the developers of the DSM-5 claimed "no a priori constraints should be placed on the degree of change between DSM-IV and DSM-5" (American Psychiatric Association 2013, 7).

In 2006, David J. Kupfer and Darrel A. Regier were respectively nominated as Chair and Vice-chair of the revision process. They brought together a task force composed of 28 members and recruited more than 130 work group experts (American Psychiatric Association 2013, 7). It was clear from the beginning that, contrary to the previous editions, the DSM-5 revision process would be pursued in the public spotlight. The revision process would be scrutinized, just as the background and careers of all recruited experts were, including most particularly their potential links with the pharmaceutical industry. A well-documented analysis of the various links with the pharmaceutical industry of many members of the DSM-IV Task Force (Cosgrove et al. 2006), led the leaders of the APA to require each expert involved in the revision process to disclose all their sources of income, including their income from the pharmaceutical industry. This was the first time in the history of the DSM – and perhaps in the whole history of psychiatric nosology – that a conflict-of-interest policy, in addition to an income cap were imposed on experts as a condition for participation. Kupfer and Regier were proud to declare that the revision process would be the "most inclusive and transparent" in the 60-year history of DSM (Kupfer and Regier 2009).[13]

Yet, also for the first time in the history of psychiatry, experts were required to comply with a confidentiality clause. Robert Spitzer, the chair of the DSM-III, wrote a letter to the blog site Psychiatric News in July 2008 to complain about what he considered to be a historic reversal in the methods of revising the DSM that violated the basic principle of scientific publicity (Spitzer 2008, 2009). Purportedly in order to protect the revision process from any smear campaigns and from premature misinterpretations, the several drafts of the new classification were presented only after a few months to the general public on the official DSM-5 website. Three drafts were available for consultation in 2010, 2011 and 2012. Analysts, clinicians and even lay-people were allowed to send their comments or criticisms concerning

[13]Yet Cosgrove and Krismky (2012) have demonstrated that a lot of financial conflicts of interest of DSM-5 experts remained despite the adoption of this disclosure policy.

the new classification by email.[14] During the revision process, thousands of emails were sent to the Task Force and many complaints were taken into account through this new channel of communication. 2009 was a bad year for the DSM-5. It was this year that Allen Frances began to publicly criticize what he considered as the weak methodology of the review process (Frances 2009a, b). In July 2009, the two chairmen of previous editions of the DSM, Spitzer (DSM-III and III-R) and Frances (DSM-IV), co-signed a letter to the Board of Trustees where they asked that the American Psychiatric Association take responsibility for what could lead, according to them, to some "disastrous unintended consequences" if the experts continued with their planned methodology (Frances and Spitzer 2009). In December 2009, DSM developers announced that the official publication of the DSM-5 would be postponed to 2013.

Divisions also arose on the inside. Some experts resigned from the Task Force. Jane Costello publicly complained about the eagerness to change the paradigm without sufficient empirical basis and without being able to measure all the potential consequences. But the most dismal failure came from the new system proposed for personality disorders. Many controversies erupted when a first proposal, composed of an extremely complex system that was supposed to reconcile the categorical and dimensional approaches, was presented in 2011. John Livesley and Roel Verheul walked out on the Task Force; they would not sacrifice scientific rigor at the altar of a consensual hybrid system that had never been tested.

Ultimately, the approval of the future text of DSM-5 by the Board of Trustees of the APA in December 2012 was a rebuttal of the most innovative changes. The 10 personality disorder categories from the DSM-IV-TR were eventually kept and the most controversial new labels were not included in the official part of the classification. David Kupfer attempted to save face by saying: "We tried to take a very conservative approach in our revision of the DSM" (Moran 2013). This final attitude of renunciation is reflected in the introduction of DSM-5:

> Although the need for reform seemed apparent, it was important to respect the state of the science as well as the challenge that overly rapid change would pose for the clinical and research communities. In that spirit, revision of the organization was approached as a conservative, evolutionary diagnostic reform that would be guided by emerging scientific evidence on the relationships between disorder groups. (American Psychiatric Association 2013, 10)

The release of the DSM-5 in May 2013 at the annual APA conference was not a triumphant or celebratory occasion. The new manual received a fairly cold reception from mental health professionals and from the media, and there was public discussion about the many weaknesses of the new classification. Nassir Ghaemi complained that the DSM-5 revision process, like the revision process of

[14]Critical feedback provided by letters sent by clinicians and lay users of the DSM was already experimented in the DSM-III and DSM-IV. At the time of the DSM-IV-TR, hundreds of letters were sent to the Task Force and then discussed by experts (American Psychiatric Association 2000, introduction).

its predecessors, had been "unscientific" in the sense that the vast majority of the mental disorders listed in the DSM are still poorly supported by empirical evidence (Ghaemi 2013). More emblematic, Thomas Insel, the director of NIMH, wrote in April 2013 that the DSM-5 was lacking in scientific validity and that it was still looking "at best" like "a dictionary" (Insel 2013). In January 2014, the public image of the new manual deteriorated even further when public accusations of conflict of interest were made against David Kupfer, the chairman of the DSM-5 revision process, concerning a dimensional instrument that he helped to design and that could become an important source of income if it becomes widely used (Gibbons et al. 2014).

## *The Establishment of the Scientific Review Committee (SRC)*

The first three DSM revisions had been pursued on the assumption that it was possible to make prudential ongoing progress in the nosological system. The revision process from DSM-III to DSM-5, however, was progressively paralyzed by the fear of endorsing important changes without sufficient empirical support, allowing for all the harmful consequences that can occur with over-hasty changes.

The DSM-5 endorsed the idea put forward by the DSM-IV that the solution was to move beyond the expert model. A major novelty was introduced during the DSM-5 revision process: an independent expert panel that would evaluate the scientific support for proposed changes in diagnostic criteria in DSM-5 outside the official Task Force team. In 2010, the president of the APA, Carol Bernstein, in response to the heated criticism of the DSM-5 process, asked Kenneth S. Kendler, a distinguished psychiatrist, to chair a Scientific Review Committee (SRC). The role of the SRC was just advisory but the idea was it would provide an independent scientific control of the proposed changes. Each expert involved in the SRC had to score the quality of the scientific support for each proposed change sent in by the different work groups of the DSM-5 Task Force from 1 (strong support) to 6 (insufficient data). In the introduction of the DSM-5, the goals of the SRC are explained:

> The Scientific Review Committee (SRC) was established to provide a scientific peer review process that was external to that of the work groups [....] Each proposal for diagnostic revision required a memorandum of evidence for change prepared by the work group and accompanied by a summary of supportive data organized around validators for the proposed diagnostic criteria (i.e., antecedent validators such as familial aggregation, concurrent validators such as biological markers, and prospective validators such as response to treatment or course of illness). The submissions were reviewed by the SRC and scored according to the strength of the supportive scientific data. (American Psychiatric Association 2013, 9)

According to Kendler, the goal of the SRC was to help the DSM move from a "scientifically assisted expert consensus" to a "scientifically driven expert consensus model" (Kendler 2013). In other words, Kendler agrees with Spitzer that it is not possible to get rid of the consensus model completely, but he also agrees with Frances that experts have to be *driven* by science, and not only *assisted* by it. The revision process must become more focused on robust scientific evidence and less vulnerable to transient opinions and the human tendency of the experts to willingly "put their mark" on the classification (Kendler 2013). The SRC certainly helped exclude personal opinion in the decision process by exerting an external control on the experts' decisions. But this strong emphasis on a pure evidence-based process of revision rested on the following questionable assumptions: (1) weak empirical support is a sufficient ground for the exclusion of a suggested change, even if this change appears conceptually and clinically sound and may eventually have positive effects on clinical practice; (2) there is no risk of a circular and systematic bias between the available data in the literature and the current diagnostic criteria that serve as reference in research studies; (3) the revision process is stable over time and there are no fundamental trends in the current literature toward a dominant false conception of clinical entities. Finally, one can conclude that the SRC tends *inherently* to endorse the Prudential Conservatism Principle (PCP) rather than the Permanent Innovation Principle (PIP). In that regard, it is interesting to note that Kendler acknowledges, over time, that "the SRC was among the most conservative of voices, typically (although not always) arguing against the inclusion of changes advocated by the WGs [Work Groups]" (Kendler 2013).[15]

To elaborate his defense of the SRC from a theoretical point of view, Kendler draws on the concept of epistemic iteration proposed by philosopher Hasok Chang. The progressive acquisition of knowledge is possible, based on a sequential (or iterative) revision through an ongoing process of clarification and increase in accuracy in the clinical descriptions. It opposes what Kendler characterizes as the "random walk" of great historical professors of psychiatry, which he illustrates with the very different conceptualizations of schizophrenia that Kraepelin, Bleuler, Langfeldt, and Schneider proposed, respectively, during the first half of the twentieth century (Kendler and Parnas 2012, 310). Such a view of scientific progress in psychiatric nosology is appealing but it raises many difficulties. In particular, it raises theoretical questions, such as the question of realism vs. coherentism (see the very interesting philosophical critique by Schaffner in his commentary of Kendler's chapter, in Kendler and Parnas 2012, 324sq). It also raises more immediate but important scientific considerations: are we certain that there is a "real" target toward which iteration proceeds in DSM revisions? What if, as many authors fear, the DSM is completely wrong in its atheoretical perspective? Finally, where is the evidence that the epistemic iteration model embodied by the DSM has been effective so far?

[15]It was still possible for the work group experts, at the end of the process, to override the SRC's judgment (and in fact, it happened with important proposals).

## Conclusion

The DSM-5 developers were forced to lower their scientific ambitions. How the next revision process (DSM 5.1, DSM 5.2 … DSM-6) might be strategically conducted appears to be in deadlock as of now. This is not because psychiatrists are short of ideas or proposals for innovation, but rather because to innovate would mean to abandon an unsatisfactory yet well-accepted nosological system and replace it with an untested and non-validated new system. On the one hand, everybody agrees that to require strong scientific evidence for supporting proposed modifications will necessarily continue to result in accepting only few minor changes in future editions. This will tend to perpetuate the PCP indefinitely. On the other hand, the only way to reconcile the DSM with the PIP would be to take responsibility for dramatic innovations in the classification without knowing what the consequences of such a dramatic move would be on the mental health system.

The DSM has lastingly shaped the field of psychiatric nosology. But we may wonder if it has produced scientific progress. Simply to claim that the DSM is scientific or unscientific is meaningless and can only serve promotional arguments or ad-hoc attacks. We do not follow Houts when he claims that "after 25 years of following changes in the various editions of the DSMs, I have concluded that there is far more pseudoscience than real science in the modern DSMs" (Houts 2002, 17), for he merely begs the question of what a clear definition of "pseudoscience" means in the case of the DSM. And it is important to recall that the developers of the DSM have never pretended that their decision-making process was *merely* scientific in nature: they have always assumed that there were also important political, financial, forensic, and mostly clinical factors to take into consideration. In any case, it is more interesting, I think, to investigate the DSM's scientific merit not as an achieved state, but as a process.

It would be further unfair to deny that the DSM, through its past editions, has tried to endorse a more rigorous scientific process of revision even if (and this is partly due to its success) the risk of biases and corruption have become more preponderant during the last decades. This reason may be sufficient to explain why the DSM, while endorsing an ideal of scientific progress, has tended to push toward a more conservative and prudential process. Finally, we must ask: has the sequence leading from the DSM-III, DSM-III-R, DSM-IV, DSM-IV-TR to the DSM-5 been wholly progressive?

In *The Advancement of Science* (1993), the philosopher Philip Kitcher addresses the difficult notion of "scientific progress." He distinguishes two varieties of progress potentially of interest for the project of evaluating psychiatric classification. There is *practical* progress, and there is *cognitive* progress (Kitcher 1993, 92). Here, I did not address the former, i.e., I did not ask whether the DSM has been useful for clinicians or has permitted practical progress in the field (such as facilitating communication between clinicians, reaffirming the legitimacy of psychiatry as a genuine branch of medicine, etc.). I have restricted my interest to *cognitive progress*, i.e., to the way in which the classification has offered the tools for a better scientific conceptualization and understanding of mental disorders. Concerning *cognitive progress* specifically, Kitcher divides it into two main components: conceptual progress and explanatory progress. "Conceptual progress," Kitcher claims, "is made when we adjust the boundaries of our categories to conform to kinds and when we are able to provide more adequate specifications of our referents" (Kitcher 1993, 95–96). The problem in the light of the DSM is that we are still not sure what the "referents" of the diagnostic categories really are. It would be very difficult to defend the idea that there has been a general upward movement on this issue during the past decades, given that the most fundamental questions are still strongly debated – such as, are mental disorders natural kinds? Or should they rather be thought of as practical kinds (Zachar 2000; Haslam 2002; Kendler et al. 2011)? Should nosology espouse a categorical or a dimensional model of mental disorders (Widiger and Trull 2007)? And so on. It would be no less difficult to defend the idea that the DSM has offered psychiatry *explanatory progress*, i.e., a better understanding of the causal structure of the pathological conditions that we call mental disorders. There has been only very modest advancement (and certainly no decisive advancement) in the discovery of the etiological factors behind any of the clinical categories listed in the DSM. Causes and pathophysiological mechanisms remain largely unknown in psychiatry.

So whether the science has progressed from the DSM-III to the DSM-5 is, at the very least, doubtful. The implications of this conclusion seem to me twofold. First, it should be clearly acknowledged that the atheoretical approach employed in the DSM (or, to use Wakefield's expression, the "theory neutrality" perspective) has not yet paid off. There is no more "random walk" in nosology, but the *epistemic iteration model* has not proved that it could permit real scientific progress. The second implication is more radical. Perhaps the fact that the DSM has stayed unsuccessful thus far is not the result of some methodological shortcoming but instead may depend on the general model the DSM has endorsed from the 1980s on, i.e., the medical model of mental illness. Many categories listed in the DSM may not be true diseases after all. This skeptical claim does not necessarily involve any form of anti-psychiatrist stance. It just reminds us that the question "what is a mental disorder?" has never reached any consensual solution, and continues to divide philosophers and clinicians.

**Acknowledgments** I would like to thank Patrick Singy, Kathryn Tabb and Jerome Wakefield for their helpful comments and suggestions on this chapter.

## References

American Psychiatric Association. (1980). *Diagnostic and statistical manual of mental disorders* (3rd ed.). Washington, DC: American Psychiatric Association.

American Psychiatric Association. (1987). *Diagnostic and statistical manual of mental disorders* (3rd rev. ed.). Washington, DC: American Psychiatric Association.

American Psychiatric Association. (1994). *Diagnostic and statistical manual of mental disorders* (4th ed.). Washington, DC: American Psychiatric Association.
American Psychiatric Association. (2000). *Diagnostic and statistical manual of mental disorders* (4th ed. (Text Revision)). Washington, DC: American Psychiatric Association.
American Psychiatric Association. (2013). *Diagnostic and statistical manual of mental disorders* (5th ed.). Arlington, VA: American Psychiatric Association.
Andreasen, N. (2007). DSM and the death of Phenomenology in America: An example of unintended consequences. *Schizophrenia Bulletin, 33*(1), 108–117.
Bayer, R. (1981). *Homosexuality and American psychiatry: The politics of diagnosis*. New York: Basic Books.
Berner, I., & Katschnig, H. (1984). Polydiagnostic approach in psychiatric research. *Annales Médico-Psychologiques, 142*(6), 825–831.
Blashfield, R. K. (1982). Feighner et al. invisible colleges and the Matthew effect. *Schizophrenia Bulletin, 8*(1), 1–6.
Blashfield, R. K. (1984). *The classification of psychopathology: Neo-Kraepelinian and quantitative approaches*. New York: Plenum Publishing Corporation.
Cohen, J. A. (1960). Coefficient of agreement for nominal scales. *Educational and Psychological Measurement, 20*(1), 37–46.
Cooper, R. (2005). *Classifying madness; A philosophical examination of the diagnostic and statistical manual of mental disorders*. Dordrecht: Springer.
Cosgrove, L., & Krimsky, S. (2012). A comparison of DSM-IV and DSM-5 panel members' financial associations with industry: A pernicious problem persists. *PLoS Medicine, 9*(3), e1001190.
Cosgrove, L., Krimsky, S., Vijayaraghavan, M., & Schneider, L. (2006). Financial ties between DSM-IV panel members and the pharmaceutical industry. *Psychotherapy and Psychosomatics, 75*, 154–160.
Decker, H. (2013). *The making of DSM-III: A diagnostic manual's conquest of american psychiatry*. New York: Oxford University Press.
Demazeux, S. (2013). *Qu'est-ce que le DSM? Genèse et transformations de la bible américaine de la psychiatrie*. Paris: Ithaque.
Feighner, J., Robins, E., Guze, S. B., Woodruff, R. A., Winokur, G., & Muñoz, R. (1972). Diagnostic criteria for use in psychiatric research. *Archives of General Psychiatry, 26*, 57–63.
Follette, W. C., & Houts, A. C. (1996). Models of scientific progress and the role of theory in taxonomy development: A case study of the DSM. *Journal of Consulting and Clinical Psychology, 64*, 1120–1132.
Frances, A. J. (2009a). A warning sign on the road to DSM-V: Beware of its unintended consequences. *Psychiatric Times, 26*(8), 1–9.
Frances, A. J. (2009b). Alert to the research community – be prepared to weigh in on DSM-V. *Psychiatric Times, 27*(2), 1–5.
Frances, A. J., & Egger, H. L. (1999). Whither psychiatric diagnosis. *Australian and New Zealand Journal of Psychiatry, 33*, 161–165.
Frances, A. J., Spitzer, R. L. (2009). Letter to the APA board of trustees. Retrieved from http://fr.scribd.com/doc/17172432/Letter-to-APA-Board-of-Trustees-July-7-2009-From-Allen-Frances-and-Robert-Spitzer. 6 July 2009.
Frances, A. J., Widiger, T. A., & Pincus, H. A. (1989). The development of DSM-IV. *Archives of General Psychiatry, 46*, 373–375.
Frances, A. J., First, M. B., & Pincus, H. A. (1995). *DSM-IV guidebook*. Washington, DC: American Psychiatric Association.
Fulford, K. W. M., Davies, M., Gipps, R. G., Graham, G., Sadler, J. Z., Stanghellini, G., & Thornton, T. (2013). *The Oxford handbook of philosophy and psychiatry*. Oxford: Oxford University Press.
Galatzer-Levy, I. R., & Galatzer-Levy, R. M. (2007). The revolution in psychiatric diagnosis: Problems at the foundations. *Perspectives in Biology and Medicine, 50*(2), 161–180.

Ghaemi, S. N. (2013). Why DSM-III, IV, and 5 are unscientific. *Psychiatric Times.* Retrieved from http://www.psychiatrictimes.com/blogs/couch-crisis/why-dsm-iii-iv-and-5-are-unscientific. 14 Oct 2013.

Gibbons, R. D., Weiss, D. J., Pilkonis, P. A., Frank, E., & Kupfer, D. J. (2014). Failure to report financial disclosure information. *JAMA Psychiatry, 71*(1), 95.

Goffman, E. (1961). *Asylums: Essays on the social situation of mental patients and other inmates.* New York: Doubleday Anchor.

Goleman, D. (1994). Scientist at work: Allen J. Frances; revamping psychiatrist's Bible. *New York Times*, 19 Apr.

Haslam, N. (2002). Kinds of kinds: a conceptual taxonomy of psychiatric categories. *Philosophy, Psychiatry and Psychology, 9*(3), 203–217.

Hempel, C. (1965). Fundamentals of taxonomy. In *Aspects of scientific explanation and other essays in the philosophy of science*. New York: The Free Press.

Houts, A. C. (2002). Discovery, invention, and the expansion of the modern diagnostic and statistical manuals of mental disorders. In M. L. Malik & L. E. Beutler (Eds.), *Rethinking the DSM: A psychological perspective*. Washington, DC: American Psychological Association.

Insel, T. (2013). Director's blog: Transforming diagnosis. Retrieved from http://www.nimh.nih.gov/about/director/2013/transforming-diagnosis.shtml. 29 Apr 2013.

Kendell, R. E. (1975). *The role of diagnosis in psychiatry.* Oxford: Blackwell Scientific Publications.

Kendell, R. E. (1980). DSM-III: A British perspective. *American Journal of Psychiatry, 137*(12), 1630–1631.

Kendler, K. S. (1990). Toward a scientific psychiatric nosology: Strengths and limitations. *Archives of General Psychiatry, 47*, 969–973.

Kendler, K. S. (2013). A history of the DSM-5 scientific review committee. *Psychological Medicine, 43*(09), 1793–1800.

Kendler, K. S., & Parnas, J. (Eds.). (2012). *Philosophical issues in psychiatry II: Nosology*. Oxford: Oxford University Press.

Kendler, K. S., Muñoz, R. A., & Murphy, G. (2010). The development of the Feighner criteria: A historical perspective. *American Journal of Psychiatry, 167*(2), 134–142.

Kendler, K. S., Zachar, P., & Craver, C. (2011). What kinds of things are psychiatric disorders? *Psychological Medicine, 41*(6), 1143–1150.

Kirk, S., & Kutchins, H. (1992). *The selling of DSM. The rhetoric of science in psychiatry*. New York: De Gruyter.

Kitcher, P. (1993). *The advancement of science*. New York: Oxford University Press.

Klerman, G. (1978). The evolution of a scientific nosology. In J. Shershow (Ed.), *Schizophrenia: Science and practice*. Cambridge, MA: Harvard University Press.

Klerman, G. (1990). Paradigm shifts in USA psychiatric epidemiology since World War II. *Social Psychiatry and Psychiatric Epidemiology, 25*(1), 27–32.

Klerman, G. L., Vaillant, G. E., Spitzer, R. L., & Michels, R. (1984). A debate on DSM-III. *The American Journal of Psychiatry, 141*, 539–553.

Kupfer, D. J., & Regier, D. A. (2009). Toward credible conflict of interest policies in psychiatry: Counterpoint. *Psychiatric Times, 26*, 40–41.

Kupfer, D. J., First, M. B., & Regier, D. A. (Eds.). (2002). *A research agenda for DSM-V*. Washington, DC: American Psychiatric Association.

Maxmen, J. (1985). *The new psychiatrists*. New York: New American Library.

Millon, T. (1983). DSM-III: An insider's perspective. *American Psychologist, 38*, 804–814.

Moran, M. (2013). APA board of trustees approves DSM-5; Release scheduled for May 2013. *Psychiatric News Alert, 47*(24), 1.

Murphy, D. (2006). *Psychiatry in the scientific image*. Cambridge/London: MIT Press.

Poland, J. (2001). Review of "DSM Sourcebook, volume 1." Metapsychology. Retrieved from http://metapsychology.mentalhelp.net/poc/view_doc.php?type=book&id=557. Spring 2001.

Reed, G. M. (2010). Toward ICD-11: Improving the clinical utility of WHO's international classification of mental disorders. *Professional Psychology: Research and Practice, 41*(6), 457–464.

Robins, E., & Guze, S. B. (1970). Establishment of diagnostic validity in psychiatric illness: Its application to schizophrenia. *American Journal of Psychiatry, 126*(7), 983–987.

Rosenhan, D. L. (1973). On being sane in insane places. *Science, 179*(4070), 250–258.

Sabshin, M. (2008). *Changing American psychiatry: A personal perspective*. Washington, DC: American Psychiatric Publishing.

Sackett, D., Rosenberg, W. M., Gray, M., Haynes, R. B., & Richardson, W. S. (1996). Evidence based medicine: What it is and what it isn't. *BMJ, 312*(7023), 71–72.

Schacht, T. E. (1985). DSM-III and the politics of truth. *American Psychologist, 40*(5), 513–521.

Scheff, T. J. (1966). *Being mentally Ill: A sociological theory*. Chicago: Aldine.

Shaffer, D. (1996). A participant's observations: Preparing DSM-IV. *Canadian Journal of Psychiatry, 41*(6), 325–329.

Shorter, E. (1997). *A history of psychiatry. From the era of the asylum to the age of Prozac*. New York: Wiley.

Spitzer, R. L. (1982). To the editor. *Schizophrenia Bulletin, 8*(4), 592.

Spitzer, R. L. (1991). An outsider-insider's views about revising the DSMs. *Journal of Abnormal Psychology, 100*(3), 294–296.

Spitzer, R. L. (2001). Values and assumptions in the development of DSM-III and DSM-III-R: An insider's perspective and a belated response to Sadler. Hulgus and Agich's 'on values in recent American psychiatric classification'. *The Journal of Nervous and Mental Disease, 189*(6), 351–359.

Spitzer, R. L. (2008). DSM-V: Open and transparent? *Psychiatric News, 43*(14), 26.

Spitzer, R. L. (2009). DSM-V transparency: Fact or rhetoric? *Psychiatric Times, 26*(3), 8.

Spitzer, R. L., & Williams, J. B. (1988). Having a dream: A research strategy for DSM-IV. *Archives of General Psychiatry, 45*(9), 871–874.

Spitzer, R. L., & Williams, J. B. (1989). The vote on DSM-IV-Reply. *Archives of General Psychiatry, 46*(10), 959–960.

Spitzer, R. L., Endicott, J., & Robins, E. (1978). *Research diagnostic criteria (RDC) for a selected group of functional disorders*. Bethesda, MD: National Institute of Mental Health.

Strauss, A., Schatzman, L., Bucher, R., Ehrlich, D., & Sabshin, M. (1964). *Psychiatric ideologies and institutions*. New York: Free Press of Glencoe.

Szasz, T. (1960). The myth of mental illness. *American Psychologist, 15*, 113–118.

Tsou, J. Y. (2008). *The reality and classification of mental disorders: Issues in the philosophy of psychiatry*. Ph.D. dissertation, Chicago, University of Chicago.

Wakefield, J. C. (1999). Philosophy of science and the progressiveness of the DSM's theory-neutral nosology: Response to Follette and Houts, part 1. *Behaviour Research and Therapy, 37*, 963–999.

Widiger, T. J., & Trull, T. J. (2007). Plate tectonics in the classification of personality disorder: Shifting to a dimensional model. *American Psychologist, 62*(2), 71–83.

Williams, J. B., & Spitzer, R. L. (1982). Research diagnostic criteria and DSM-III: An annotated comparison. *Archives of General Psychiatry, 39*(11), 1283–1289.

Zachar, P. (2000). Psychiatric disorders are not natural kinds. *Philosophy, Psychiatry and Psychology, 7*(3), 167–182.

Zubin, J. (1978). But is it good for science? *Clinical Psychologist, 31*(2), 1–7.

# DSM-5 and Research Concerning Mental Illness

**Jeffrey Poland**

**Abstract** It is widely agreed that the DSM-IV categorical framework (and its predecessors) have a number of problems (e.g., questionable reliability in the field, questionable validity, heterogeneity, unexplained comorbidity, an unsound concept of mental disorder) that have compromised its utility in research concerning mental illness. At the root of these problems is a substantial "lack of fit" between the DSM framework and the domain of mental illness. With the publication of DSM-5, it is appropriate to ask whether the process of revision leading from DSM-IV to DSM-5 has been sufficiently responsive to the problems with DSM-IV to justify continued use of DSM categories in either basic research concerning psychopathology or more applied clinical research. In this paper, I argue that the revision process has not been responsive to these problems and that, hence, DSM-5 categories ought not to be used in research concerning mental illness. Rather, alternative approaches should be developed, and I conclude with a discussion of three such alternatives.

## Introduction

It is widely agreed that the DSM-IV categorical diagnostic framework (and its predecessors) has problems (e.g., questionable reliability in the field, questionable construct and predictive validity, poor phenotypic definitions, heterogeneity, comorbidity, an unsound concept of mental disorder) that have compromised its utility in research concerning mental illness. Critics of the DSM (Cromwell 1982; Blashfield 1984; Eysenck 1986; Carson 1991; Kirk and Kutchins 1992; Poland et al. 1994; Murphy 2006; Poland and Von Eckardt 2013) have argued to this effect over the past three decades and the view that the DSM has serious shortcomings is now emerging as a consensus view, even in the research community which has relied upon the DSM since 1980 (see Kendall and Jablensky 2003; Andreassen 2007; Kendler et al. 2009; Hyman 2010; Insel et al. 2010).

J. Poland (✉)
Rhode Island School of Design and Brown University, Providence, RI, USA
e-mail: jpoland@risd.edu



S. Demazeux, P. Singy (eds.), *The DSM-5 in Perspective*, History, Philosophy and Theory of the Life Sciences 10, DOI 10.1007/978-94-017-9765-8_2

In the light of this emerging consensus, it is appropriate and timely to ask whether the revision process culminating in the publication of DSM-5 has been sufficiently responsive to the problems afflicting the DSM-IV that compromise its research utility. In the following, I will present a hypothesis for explaining why the DSM-IV has exhibited its many problems and argue that the DSM-5 represents no significant improvement over DSM-IV with respect to research utility. I will conclude with a discussion of alternative approaches to research concerning mental illness.

## Why Does the DSM-IV Lack Research Utility?

Our starting point is the consensus view that the DSM-based research program has not yielded the sorts of results that were expected (viz., validation of the diagnostic categories). In addition, the categories exhibit substantial heterogeneity, confusing comorbidities, and poorly defined phenotypes, each of which is problematic for research purposes. As a consequence, research has tended to produce findings that are negative, non-replicable, inconsistent, weak, non-specific, or uninterpretable. In this section I shall argue that the reason DSM-based research has been non-productive in these ways is that there is a *lack of fit* between the conceptual resources (broadly speaking) available in the DSM classification system and the domain of mental illness. To make good on this claim, four steps are required: (1) an overview of the known features of the domain of mental illness; (2) a review of the representational resources and assumptions provided by the DSM classification system; (3) an argument that DSM-based representational resources and associated assumptions are insufficient for representing and managing the various features of the domain (i.e., there is a "lack of fit"); and, (4) an argument that, as a result, they are unlikely to aid in the pursuit of research questions concerning mental illness.

### *Features of the Domain of Mental Illness*

The domain of mental illness includes phenomena involving individual lifeproblems, distress, disability, deviance, failures to perform social functions, and maladaptation. It is a domain in which both mental and behavioral capacities are centrally involved, and it is obviously a domain of considerable human interest that can be a target of scientific research. For present purposes, the most important point is that this domain, like all domains of human functioning, exhibits considerable complexity of process and structure at many levels of analysis, as well as normative and perspectival dimensions. The following list summarizes many of the relevant features (see Poland 2014 for further discussion of these features):

- *Causal Ambiguity*: features in the domain of mental illness can be derived from many different causal processes.

- *Hierarchical organization*: human biological systems consist of many levels of organization ranging from low level genetic, biochemical, and neuroanatomical to high level cognitive, behavioral, and socio-cultural. These levels are also present in mental illness.
- *Multi-dimensionality*: the state or condition of a person with mental illness at a time consists of features and processes of many different sorts, within and across levels of organization.
- *Interactivity and context sensitivity*: the features and processes of mental illness are typically interactive with each other, and each is sensitive to the context in which it is embedded.
- *Dynamics*: the features and processes of mental illness evolve over time at various time scales along varying trajectories, and they can exhibit phase dependence and a variety of distinctive causal patterns.
- *Perspective and agency*: individuals suffering mental illness are persons who are agents and who have a first person perspective on themselves, the world, their past, and their future.
- *Normativity*: the identification of conditions as problematic, deviant, maladaptive, dysfunctional, diseased, distressing, or disabling presupposes background norms, values, or interests that may be theoretical, personal, social or of some other sort ("normative pluralism").
- *Normal and abnormal conditions and processes*: although there may be conditions or processes in mental illness that violate some specified norms, there are also conditions and processes that are normal by the same or different standards ("normative diversity").
- *Relational and non-relational problems*: the kinds of problems that people with mental illness can suffer can be both non-relational (i.e., conditions of the individual) and relational (i.e., conditions involving relationships between an individual and other people or between an individual and some aspect of the non- personal environment).
- *Individual variability*: individuals suffering a mental illness vary widely and tend to exhibit relatively unique combinations of problems, functional profiles, embedding contexts, and causal processes.

These features characterize the general outlines of the domain of mental illness,[1] whereas scientific inquiry is required to flesh out how each is concretely manifested by the various phenomena in the domain. At issue in the present context is whether it is productive to do this research within the framework of the DSM category system.

[1]These features are, of course, not specific to the domain of mental illness, but might be found in many areas of normal human functioning as well as in chronic and complex physical diseases.

## *Resources Provided by the DSM-IV*

The DSM-IV-TR (American Psychiatric Association 2000) is a categorical scheme comprised of over 300 categories of mental disorder, conceived of as harmful dysfunctions or prototypical patterns of symptoms,[2] that can be identified on the basis of atheoretical,[3] polythetic[4] diagnostic criteria framed in terms of clinically salient features (signs and symptoms) or other characteristics (e.g., of history or context) that can be readily determined by the clinician. The disorders in the DSM are specified as individualistic (i.e., non-relational); each is viewed as a condition of a person consisting of a pattern (or set of patterns) of symptoms along with a putative (but unspecified) biological, psychological, or behavioral dysfunction of the individual, a dysfunction that manifests itself in terms of the signs and symptoms specified by the diagnostic criteria. In mainstream psychiatry, such conditions are typically assumed to be brain diseases, an assumption tied to the medicalization of mental health research and clinical practice (see Poland 2014; Poland and Von Eckardt 2013).

It should be noted that in DSM-IV there are also resources for representing conditions that can be the objects of clinical concern but are not mental disorders (e.g., relationship problems, employment problems); these are coded in what are called "V-Codes" and they represent an important body of information.[5] In addition there is an Axis 4 coding of psychosocial and environmental problems that may be the occasion for a mental disorder or a consequence of a mental disorder. Finally there is an Axis 5 coding of a "Global Assessment of Functioning," which is a subjective rating by the clinician on a scale of 0–100 of occupational, social, and psychological functioning (not due to physical or environmental limitations).[6]

The aggregation of the above types of information (categorical diagnosis based on a clinical interview, plus any relevant V, Axis 4 and Axis 5 codes) provides the clinician with a basis for: identifying a person's condition and theproblems

---

[2]Although there are competing conceptions of the DSM categories (viz., harmful dysfunctions, clinical prototypes), the arguments presented in the text apply equally to both. In what follows, I will formulate the issues in terms of the harmful dysfunction view (see Wakefield 1992).

[3]"Atheoretical" criteria do not refer to either pathology or etiology.

[4]"Polythetic" criteria are disjunctive and their use is supposed to reflect the idea that mental disorders can manifest themselves in various ways across individuals with the same disorder.

[5]At a minimum, V codes draw attention to significant aspects of the context in which a putative mental disorder arises; but perhaps more important is that such problems are essentially implicated in a person's current mental health condition, and are a critical component in understanding what is wrong (if anything) and what is likely to help. Arguably, from the point of view of research, information picked out by V codes is required for a realistic scientific analysis of the problems and processes involved in mental illness.

[6]Note that, in DSM-5, this multi-axial approach has been dropped, although V-codes have been retained.

they face; developing a case formulation drawing on the clinician's background knowledge, understanding and experience; and, combined with standards of care for specific conditions, proposing a treatment plan.[7]

In contrast, in pursuing DSM-based research, researchers typically employ only the DSM diagnosis, specifically to identify the subjects for a study; the other information is typically left behind. In other words, for research purposes, DSM diagnoses (which are symptom-based, atheoretical, polythetic, and taken out of context) are assumed to represent meaningful targets of research and to provide a useful identification of subjects with the (more or less) same condition. In studies in which DSM categories are used to create subject groupings, researchers also typically leave behind the raw clinical data (e.g., the specific symptoms exhibited by each individual) upon which the subjects' diagnoses were based. Finally, given that DSM categories are not associated with specific diagnostic tests and that DSM-based research has not provided well-confirmed specific models and findings concerning the categories, researchers cannot supplement a diagnosis with a consensually validated model of pathology or etiology (although many hypotheses abound.) So, to sum up, the representational resources provided by the DSM to researchers working within the conventional psychiatric research tradition, are limited to de-contextualized diagnoses that are symptom-based, atheoretical, polythetic, and not associated with well-confirmed tests and models. Of course, researchers can and do introduce other representational resources in their research; at issue here is what resources the DSM provides, what roles these resources play, and whether they contribute anything of value.

## *Lack of Fit Between the DSM-IV Framework and the Domain of Mental Illness*

The phrase 'lack of fit' will be used to refer to the idea that DSM categories provide artificial groupings of individuals experiencing mental illness, that DSM representational resources do not map well onto the features of the domain of mental illness, and that the DSM approach makes problematic assumptions about the domain. Alternatively put, for there to be a "good fit" with the domain of mental illness the DSM framework should satisfy the following conditions: (1) it should

[7]The reasons for being suspicious of the use of DSM diagnoses in research are also very good reasons for being suspicious of their use in the clinic, although I will not pursue that line of argument here (see Poland et al. 1994; Poland 2003; Spaulding et al. 2003). However, it should not be supposed that supplementing the diagnostic categories with V-codes and Axis 4 and 5 codes is sufficient for meeting either clinical or research challenges and hence for retaining the DSM categories for use in those contexts. The argument below suggests why this is the case with respect to research.

exhibit construct as well as predictive validity[8] of the categories (i.e., they should not be artificial); (2) it should have the resources for representing and managing important features, conditions, problems, processes, and groups in the domain; and (3) it should not make any problematic assumptions about the domain. The DSM framework fails to satisfy these conditions.

First, as noted above, it is widely acknowledged that DSM categories lack established validity. Such a lack is an inevitable consequence of the features of the domain of mental illness and the DSM categories being de-contextualized, atheoretically conceived, and defined in terms of polythetic diagnostic criteria focusing on superficial aspects of clinical phenomenology: when such criteria are applied to the domain of mental illness exhibiting the characteristics identified above, any patterns of behavior and other clinically identified features will mask a wide range of distinct causal processes and a wide range of distinct features at all levels of analysis. Such heterogeneity combined with the dynamic interactivity and context sensitivity of such processes and features make it highly likely that de-contextualized, atheoretical and polythetic criteria focused on clinical phenomenology will lead to DSM categories that are lacking in both construct and predictive validity (i.e., they are artificial) (see Poland et al. 1994).[9]

Second, the use of decontextualized, atheoretical, polythetic, and symptom-focused DSM diagnostic categories does not provide the descriptive resources required for representing important features, conditions, problems, processes, and groupings in the domain of mental illness. A DSM diagnosis, for example, is essentially blind to both the hierarchical, multi-dimensional complexity of mental illness and the dynamic interactivity and context sensitivity of the causal processes involved in a hierarchically organized biological system. Instead, such clinical diagnostic categories are artificial impositions on a domain with these characteristics and mask the identity and variability of key features and processes of research significance. For instance, clinical symptoms such as delusional speech, impulsivity,

[8]Many conceptions of validity have been employed in the evaluation of research in psychology and psychiatry. Roughly speaking for present purposes, validity concerns (1) the empirical or theoretical integrity of a construct establishing that it picks out what it is supposed to be picking out (i.e., construct validity) and (2) empirical or theoretical relations between a construct and other variables of interest (i.e., predictive validity and related concepts.) This approach is a general approach widely employed in psychological research. In psychiatry, it is standard to employ a notion of validity introduced by Robins and Guze (1970) that concerns the establishment of empirical relations between a diagnostic category (or syndrome) and various "validators" established in five phases of validation research (viz., clinical description, laboratory study, exclusion of other disorders, follow-up study, and family study.) Subsequently, additional validators (e.g., response to treatment) have been included (see Kendler 1980; Andreassen 1995). This second approach is supposed to imply the construct and predictive validity of diagnostic categories. In any event, the problems with the validity of DSM categories discussed in the text can be framed in terms of either approach to understanding what validity consists in. See also Kendell and Jablensky 2003.

[9]This line of argument strongly suggests that, at the time of the development and publication of DSM-III (American Psychiatric Association 1980), it was quite predictable that the strategy adopted by the developers of DSM-III would fail.

or confusion can be the result of a variety of quite different causal processes and reflect very different conditions.[10] Symptom-based DSM diagnostic categories do nothing to resolve such causal ambiguity, and thus in a research context, the use of such categories creates artificial subject groupings that mask such variation. Finally, the widely acknowledged heterogeneity of DSM diagnostic groupings is largely unmanaged by the use of DSM categories in research. This heterogeneity is typically recognized in terms of symptomatology, but in reality the heterogeneity of DSM categories concerns a wide range of features at all levels of analysis and is especially significant with respect to process heterogeneity (viz., the variability of causal structures and processes, both normal and abnormal, across individuals with the same DSM diagnosis).[11] In sum, a decontextualized, atheoretical, polythetic, and symptom-focused DSM diagnosis lacks the resources for representing any person's relatively unique mix of problems, capacities, deficits, perspective, and psychosocial and biological context and, hence, is insufficient for conceptually managing the variability, complexity, ambiguity, dynamic interactivity, context sensitivity, and resulting uncertainty that constitute the domain of mental illness.

Third, the conventional psychiatric understanding of the DSM makes a number of problematic assumptions concerning the domain of mental illness. One of these is the assumption of individualism, explicitly made with respect to DSM categories (American Psychiatric Association 1994, xxi–xxii): all mental disorders are constituted by bodily states and processes (viz., behavioral, psychological, or biological dysfunctions and the symptomatic patterns they produce) within the individual who has the disorder. This assumption leaves out a host of relational problems (e.g., interpersonal relationship problems, social problems) that are partially constitutive of the domain of mental illness. But more importantly, the assumption of individualism fails to take into account the complex dynamically interactive relations between an individual and aspects of the physical or social world; both positive and negative feedback loops can be present and the relevant processes can only be understood in terms of the relations involved, as in escalating arousal in two individuals interacting with each other. Another example concerns the evaluation of cognitive states as delusional, something that typically involves an essential reference to the socio-cultural context in which the delusional individual is embedded.[12]

Further, the various interpretations of this individualism, viz., mental illnesses consist of mental disorders (i.e., harmful dysfunctions) or of brain diseases, have not found much research support in the sense of discovering well confirmed

[10]See Wiecki et al., in press for discussion of strategies and techniques for resolving causal ambiguity (e.g., of the clinical symptom of impulsivity) using the resources of computational cognitive neuroscience.

[11]See Fair et al. 2012 for research concerning ADHD and heterogeneity of cognitive profiles in both normal and clinical populations.

[12]As observed by an anonymous reviewer, there may be states that are manifestly delusional in any cultural context; but nonetheless the delusional character of any such state is constituted by presupposed epistemic norms characteristic of the local culture and variation in such local norms can lead to variation in the character and significance of the delusion.

associations of DSM categories with internal dysfunctions or brain diseases (e.g., a pathophysiology). Although mechanisms and processes associated with specific symptoms or other traits (e.g., cognitive deficits) are being studied with some success, this is not the same as identifying a dysfunction or disease associated with the diagnostic categories. In addition, the postulation of either disorders or diseases requires a framework of norms of functioning and a rigorous specification of the range of normal variation that has not been worked out. Thus the assumption that DSM categories are either harmful dysfunctions or brain diseases[13] presupposes a set of facts (e.g., concerning putative deviations from the range of normal functioning) and defensible norms (e.g., social, empirical, or theoretical norms that demarcate the normal from the abnormal) that are not currently available. As a consequence, such an assumption is an unsubstantiated ideological projection onto the domain, a projection that obscures many of the real problems and processes from which people suffer.

This last point is especially important for research: many of the problems and processes in the domain of mental illness are not discontinuous from normal function and persons suffering mental illness do not necessarily have diseased or disordered brains. For example, severe and persistent depressive symptoms consequent to a job loss is not necessarily a display of a brain disease in need of medical treatment so much as a personal employment problem calling for a new job.[14] In addition, a pathology focus on disorder or disease tends to obscure the operation of normal processes and the availability of a person's individual strengths and capacities. Consequently, with respect to research, framing questions in terms of the DSM categories may result in a neglect of such normal processes, their range of variation, and their context sensitivity.

In sum, since the DSM-IV has failed to satisfy the three conditions (viz., validity, representational adequacy, no problematic assumptions) for fitting the domain of mental illness, we must conclude that DSM-IV does not fit that domain.

## *The DSM-IV Lacked Research Utility*

As described above, it is widely agreed that the DSM-based research agenda has not delivered the long sought-for validation of the categories and has not provided well confirmed and well developed models of the etiology or pathology of the

[13]The idea that DSM diagnostic categories pick out brain diseases is not explicitly assumed within the DSM, whereas the idea that DSM categories pick out harmful dysfunctions is explicitly expressed. However, the idea that mental disorders (as conceived in the DSM) are medical diseases, and indeed brain diseases, is widely assumed among "biologically oriented" psychiatric clinicians and researchers.

[14]This point does not mean that research concerning brain processes associated with depressive symptoms is not important; rather, it is that a pathology focus on the individual can lead to disproportionate emphasis on causal processes in the brain relative to those in the environment.

putative mental disorders identified in the DSM-IV. Nor has it, for the most part, produced a body of findings to substantiate the predictive validity of the categories concerning response to treatment, course and outcome. These results are readily understood if one recognizes the poor fit between the categories and the domain: the DSM-IV categories and associated criteria were simply ineffective in representing important features, conditions, processes, problems, and groups, and, hence, in managing causal ambiguity, multi-dimensional complexity, individual variability and other features of the domain. As a consequence, they could not support a progressive research program concerning mental illness because they were ill suited for representing significant variables, for controlling systematic and unsystematic sources of error, for managing heterogeneity, and for grouping subjects with similar features and processes. Rather, they were superficial categories that created artificial and heterogeneous groupings, poorly defined phenotypes, and unexplained comorbidities[15] that compromised the research.

Such problems impacted research at every stage (e.g., sampling, subject grouping, measurement, design, analysis, and interpretation of results). As a consequence, and as noted earlier, the research agenda associated with DSM-IV categories has yielded a body of findings that are largely negative, unreplicated, inconsistent, weak, non-specific, or uninterpretable (see Heinrichs 2001 for a review of research findings concerning "schizopohrenia" which exhibits these patterns). The bottom line is that the research program based on the DSM-IV categorical system has not progressed, the categories have not been validated, and, thus, the DSM-IV categorical framework has exhibited very limited research utility.

## The DSM-5 Is No Improvement

In this section we turn to the question of whether the DSM process of revision has produced a DSM-5 that is responsive to the problems and limitations of DSM-IV with respect to research. I shall argue that it has not and that, therefore, DSM-5, like DSM-IV, has very limited utility for research on mental illness.

The DSM process of revision exhibited a number of features that limited the possibilities of serious reform of the diagnostic manual and hence the possibilities of responding to the problems of using the DSM in research concerning mental illness. First, the process was controlled by the American Psychiatric Association

[15]The prevalence of comorbidity of DSM diagnoses (e.g., ADHD and learning disorders) is problematic at least because the co-occurrence of multiple disorders is causally ambiguous: i.e., two conditions can co-occur because they are two independent conditions or because they share a pathogenic cascade and one is the downstream consequence of the other or because they are each consequences of some common cause or because they share overlapping diagnostic criteria, etc. Such causal ambiguity is not resolvable within the atheoretical DSM framework, compromises research and clinical practices, and may point to the necessity of radically re-conceiving the domain in terms of a framework based on causal structure and processes.

and substantially guided by considerations and constraints that promote the guild interests of psychiatrists. These included a strong commitment to the medicalization of mental health practice (e.g., mental illnesses are brain diseases that are diagnosed and treated by physicians in a psychosocial context of medical roles, identities, settings, etc.), and hence to a specific form of clinical practice. These commitments are reflected in the essentially conservative guidelines for change of the DSM-IV embodied in the following four principles (American Psychiatric Association 2013, p. 7):

1. DSM-5 is primarily intended to be a manual used by clinicians, and revisions must be feasible for routine clinical practice
2. recommendations for revisions should be guided by research evidence
3. where possible, continuity should be maintained with previous editions of DSM
4. no a priori constraints should be placed on the degree of change between DSM-IV and DSM-5.

Putting aside that (1) and (4) seem to be in serious tension with each other, the "routine clinical practice" mentioned in (1) refers to routine psychiatric practice as delimited by the training and experience of psychiatrists and the demands of various clinical psychiatric settings (e.g., time pressure, resources, institutional constraints, finances, etc.). The "research evidence" mentioned in (2) predominantly refers to DSM-based research that bears upon DSM diagnostic categories and associated criteria (e.g., research concerning their reliability and validity).

As a consequence of these features (viz., hierarchical control by the APA, influence of guild interests, focus on supporting routine clinical practice, continuity with DSM-IV,[16] employment of DSM-based research), the various aspects of the revision process (e.g., the literature reviews, the field trials, the review processes of the various works groups and the task force, the ultimate decision process) were all strongly anchored in a way that kept more fundamental questions regarding the domain of mental illness and the core assumptions of the DSM framework off the table. Hence, despite protestations that there were no a priori restrictions on the sorts of change that might be made (i.e., 4 above), the revision process was for the most part focused on revising the existing DSM-IV by reviewing DSM-based evidence that bears upon questions of whether to add or delete categories or to revise criteria for existing categories or to modify the accompanying text. The result is a DSM-5 that is, in essential respects, little changed from DSM-IV, that is heavily geared for preserving existing clinical practices and the framework of assumptions

[16]Continuity with DSM-IV is typically deemed important because radical changes would be too disruptive to both clinical and research practice; point 3 above is an acknowledgement of this that was added to the criteria for change late in the process. Early in the DSM-5 development process (see Kupfer et al. 2002) it was recognized to some extent that radical changes may well be required to be responsive to the problems of DSM-IV, as is partially acknowledged in point 4. What seems clear is that the tensions between these acknowledgements were never effectively resolved and that more conservative pressures were dominant.

that support them, and that retains categories exhibiting the same characteristics as those in previous DSMs (viz., de-contextualized, symptom-focused, atheoretical, polythetic).

This last point is especially significant. It is a characteristic of the past several revisions of the DSM that categories with no demonstrated validity when they were introduced in DSM-III in 1980 have been continuously grandfathered through on the grounds that there is no substantial evidence to justify their removal, and that their removal would be too disruptive of ongoing clinical and research practice (see Frances 2010a). As a consequence, essentially the same categories as were introduced in DSM-III are retained and augmented by more of the same, despite there being no solid evidence for the construct and predictive validity of the categories, the acknowledgement that they exhibit problems of heterogeneity, comorbidity, and poor phenotypic definition and hence that they have questionable value in research.[17] As suggested above, this grandfathering through of unvalidated clinical categories is the result of a process that is highly constrained by its task specification (to review and revise the existing DSM), the questions posed (which typically concern whether to add, retain, revise, or delete categories or criteria), the problematic knowledge base relative to which these questions are addressed (viz., existing DSM-based research), the stringent standards on any revision (e.g., that they be useful for routine clinical psychiatric practice, that they be supported by evidence), the aforementioned guild biases, and, in sum, the importance of keeping the DSM highly tuned to preserving clinical practice as it currently exists.

As a consequence, the process of revision was profoundly compromised in ways that make it highly unlikely that significant changes relevant to the issue of research utility have been made in DSM-5. In particular, and despite a few innovations to be discussed below, the DSM-5 revision process did not effectively engage the problem of the "lack of fit" of the DSM-IV with the domain of mental illness. It has retained (and augmented) the various artificial diagnostic categories in DSM-IV; it has not effectively addressed the failure of the DSM to provide sufficient representational resources for managing the features of the domain of mental illness (e.g., the diagnostic categories are still atheoretical, polythetic, and symptom focused); and it has retained various problematic assumptions concerning that domain (viz., individualism, medicalization and the assumption that mental illnesses are brain disorders.) Thus, the DSM-5 in all likelihood exhibits a lack of fit with the domain of mental illness and will not therefore contribute effectively to promoting a progressive research program.

It should be mentioned, however, that in addition to tinkering with the symptom-based categorical approach of the DSM-IV, the work groups of the DSM-5 process did consider other sorts of revision, some of which were included in

[17]Although not directly relevant to the present chapter concerning research utility, it should be noted that the idea that DSM categories are clinically useful is questionable. See Poland 2003 for a discussion of how diagnostic categories like "schizophrenia" function as harmful stereotypes in clinical settings.

the final document. For one thing, there was consideration of adding genetic and neuroscience based diagnostic criteria, thereby challenging the assumption of atheoreticity that has prevailed for the past 30 years (see Hyman 2007). However, it was widely agreed that this could not be done at present because the research didn't support it; none of the categories, after all, have been sufficiently validated in these ways. Short of introducing genetic or neuroscientific diagnostic criteria, information concerning genetic, neuroscientific, and other research findings related to various DSM categories has been introduced into the text for some categories (e.g., schizophrenia), although it is unclear how meaningful such discussions can be given the widely acknowledged problems with the DSM categories and the DSM-based research program.

There has also been a substantial addition of language and tools concerning dimensional assessment to the diagnostic process, thereby broadening diagnostic assessment beyond the assignment of individuals to a categorical diagnostic grouping. In addition to a dimensional severity coding that is part of the coding of many DSM diagnoses, DSM-5 has introduced in its Sect. 3 ("Emerging Measures and Models") a set of brief rating scales aimed at quantifying the severity of various symptoms, and, thus, the severity of diagnoses. Some of these scales focus on symptoms that are "cross-categorial" while others target specific diagnoses and symptom types. These scales, however, are highly constrained by the requirement that they be usable by clinicians in routine clinical psychiatric settings; hence they are tuned to the DSM-based training of clinicians and the pragmatics of conventional psychiatric practice more than they are tuned to the features of the domain of mental illness, the real epistemic demands of the clinic, and (most important for present purposes) the challenges of research, each of which requires a more nuanced and powerful dimensional approach.[18] The limits of the training of clinicians and the context of clinical use make more serious dimensional assessment approaches impracticable in routine clinical practice: consequently, the proposed dimensional tools are simple, superficial, and atheoretically conceived rating scales of clinical phenomenology that are likely to contribute little to the formidable assessment tasks posed by the features of the domain of mental illness for either clinical or research purposes. Among other limitations, they may lack important psychometric properties (see Frances 2010b), they focus on too narrow a range of features, and they are not theoretically related to underlying processes.

Another way in which a more "dimensional" approach has been introduced into DSM-5 is with respect to the higher order organizational structure of the categorical framework: chapter organization and ordering is now supposed to reflect significant dimensions along which categories of mental disorder fall. Shared similarities are supposed to include such factors as: neural substrates, family traits, genetic risk factors, specific environmental risk factors, biomarkers, temperamental antecedents, abnormalities of emotional or cognitive processing, symptom similarity, course of

[18] See Spaulding et al. 2003 for an example of how a more rigorous clinical assessment might proceed in the case of severe mental illness.

illness, high comorbidity, and treatment response (American Psychiatric Association 2013, p. 12). The idea is that grouping disorders with respect to similarities of these sorts, and locating them on some sort of dimension of similarity, will aid both clinicians and researchers in their respective tasks (Andrews et al. 2009; Bernstein 2011). Ultimately, given the putative clinical utility of developmental and lifespan considerations and empirical support for the value of the distinction between internalizing and externalizing factors, the final organizational structure of the DSM-5 involves the following ordering of categorical groupings: neurodevelopmental disorders, internalizing disorders, externalizing disorders, neurocognitive disorders, and other disorders. The stated purposes of this structure (APA 2013, pp. 12–13) are "to enable future researchers to enhance understanding of disease origins and pathophysiological commonalities between disorders," to thereby provide a base for assessing validity, to develop new diagnostic approaches, and to provide a basis for explaining heterogeneity and comorbidity of current categories.

Although the spirit of this change is of interest, and the identification of significant features and processes such as those listed above is on the right track with respect to research concerning the domain of mental illness, the idea of proceeding by creating super-groups and spectra of DSM categories based on putative similarities is problematic given the lack of validity and the heterogeneity of DSM categories; there is no reason to believe that super-groups or dimensional orderings of problematic categories will be any less problematic than the categories themselves. Hence, there is no reason to expect that the current DSM categories can be meaningfully said to fall along some theoretically or empirically significant dimensions. What is needed is an approach to research, free of DSM categories and commitments, that focuses more directly on significant features and processes of the domain.

A further idea being seriously considered is the idea of making the DSM-5 a "living document" (viz., one which will be updatable on much shorter timescales than previous DSM revisions). This, however, seems to be a move in the wrong direction, since, given the heavily biased and ineffective nature of the current revision process, what is needed is to slow the process down and reconstitute it so that it is unencumbered by the inertia of the DSM-5 framework and the associated commitments of the psychiatric tradition. Arguably, given the features of the domain of mental illness, a radical break from this framework and tradition is necessary if the challenges to research posed by those features are to be met. Making the DSM-5 a "living document" would likely deepen the entrenchment of the very framework and commitments that have led to the current crisis in research rather than provide a venue for the more radical changes required.

In any event, the proposed sorts of revision (viz., revision of categories/criteria, adding dimensional scales, adding genetic/neuroscientific information, identifying a metastructure of the categorical system, making DSM-5 a living document) do not effectively address the deeper problem with the DSM framework: its lack of fit with the domain of mental illness. The categories and criteria remain symptom-based, de-contextualized, atheoretical, polythetic, unvalidated, and hence artificial. The various attempts at making the DSM more dimensional in character fall well short of adequately augmenting the representational resources provided by the DSM because they are focused on (ambiguous) symptoms and symptom-based categories, and do not clearly provide a means of representing the hierarchical, multi-dimensional, dynamically interactive and context sensitive features of the domain. And, the DSM-5 continues to be associated with problematic commitments to individualism and medicalization. In sum, there is no reason to suppose that the DSM-5 will be any better than DSM-IV in fitting the domain of mental illness and helping to structure a research program focused on that domain. Hence, the DSM-5 ought not to play a role in such research programs.

## Significance and Outlook

In this paper I have argued that (1) DSM-IV-based research is compromised by the lack of fit between the DSM-IV and the domain of mental illness, (2) DSM-5, despite various revisions and innovations, is no better than DSM-IV with respect to research purposes, and (3) therefore DSM-5 ought not to play any serious role in research concerning mental illness. In light of these conclusions how should research on mental illness proceed?

To begin, despite the conclusions above, there may yet be two limited roles that the DSM might play in research concerning mental illness. The first is to review the existing DSM-based research record for ideas and findings that, despite being compromised by the employment of DSM categorical groupings, might be of heuristic value. For example, research aimed at explaining specific symptoms putatively associated with a given diagnostic category (e.g., anhedonia in schizophrenia) may identify important mechanisms and processes associated with the symptom (see Strauss et al. 2011 for an example), even if the research cannot be meaningfully interpreted as concerning something called "schizophrenia." Although the DSM-based research record has not been especially productive with respect to validating DSM categories, some (but not all) of such research may have this sort of heuristic value, although care must be taken in not taking the diagnostic categories too seriously.[19]

In addition, the inevitable problem of recruiting subjects into research may lead to a certain amount of reliance on DSM diagnoses although this should be quite minimal, and, as assessment and coding practices change, eliminated entirely. For example, research concerning "response inhibition" might proceed by recruiting subjects with such DSM diagnoses as ADHD, Tourette's syndrome,

[19]Note that the authors of DSM-III and DSM-IV issued various cautionary comments regarding the use of DSM categories; and see Hyman 2010 for a discussion of the mistake of reifying DSM categories. In research contexts this mistake takes the form of not recognizing that DSM diagnostic groupings are artificial in character. Given the various roles that DSM diagnostic categories have played in research, their toxic impact is not mitigated by either cautionary comments or discussions of the mistake of reification.

and OCD since problems of response inhibition appear to be associated with each of these categories. However, given the problems with DSM categories, the diagnostic associations cannot be presumed to have any deep significance. Once individuals have been recruited, more rigorous, individualized assessments should be employed to determine the specific features exhibited by subjects. In addition to the questionable value of the diagnostic association, a clinical symptom (e.g., impulsive behavior) is very likely to be causally ambiguous (viz., producible by many different causal processes) and the main focus of research should initially be to resolve such ambiguity, to employ more rigorous parameters of assessment, and to create more meaningful groupings. A DSM diagnosis might be tentatively used to get subjects in the door, but once inside a research context, other assessment tools should take over.

More generally, what is required is a loosening of the tight relation between DSM categories and research concerning mental illness; some examples of this highlight the fact that this is a matter of degree. First, research aimed at trying to identify sub-types of diagnostic groupings (e.g., sub-types of ADHD) is widely pursued, based on the recognition of the considerable heterogeneity of such groupings (see Durston et al. 2011 and Fair et al. 2012 for examples). Although such research may be heuristically useful in the way discussed above (e.g., by identifying possible mechanisms for specific symptoms; by identifying significant sources of variation in functioning), in the end this is a misguided approach due to taking the DSM categories too seriously (see Poland and Von Eckardt 2013 for discussion).

A second example of alternative research programs is embodied in the National Institute of Mental Health RDoC initiative (NIMH 2011; Insel et al. 2010), which is based on a matrix that identifies a range of domains of functioning (e.g., negative valence systems, positive valence systems, cognitive systems, systems for social processes) and encourages the pursuit of research concerning different units of analysis (e.g., genes, molecules, cells, circuits, physiology, behavior, self-reports). The goal is to fill in the matrix with research which will clarify, for the capacities and functions in each research domain, the range of normal functioning, various sorts of dysfunction, and associated mechanisms and causal processes. It is hoped that such research will lead to improved approaches to diagnostic classification and clinical intervention. In many respects this approach is on the right track, because it is directly responsive to various features of the domain of mental illness (e.g., multi-dimensional complexity, hierarchical organization) and it has dissociated itself from DSM diagnostic categories. However, the RDoC is also explicitly committed to a number of the commitments (e.g., medicalization) that have hampered the DSM approach (see Poland 2014). Consequently, it appears to be biased in the direction of lower levels of analysis, a focus on brain disease, and physical forms of intervention, each of which will tend to skew the research away from other aspects of the domain of mental illness (e.g., social and psychological processes; personal agency; non-physical forms of intervention).

Whereas these first two examples of research are potentially hampered by the over-reliance on DSM categories or other commitments that have compromised DSM-based research, a more progressive approach will divest itself of these categories and commitments and be more directly responsive to the various features of the domain of mental illness. One example of this sort of approach can be found in labs that employ the methods and representational resources of computational cognitive neuroscience to develop models of normal function, explore the range of individual differences with respect to a wide range of biological and psychological capacities and processes, and identify measurable deficits, dysfunctions and impairments (see Montague et al. 2011; Maia and Frank 2011). This sort of research is explicitly armed with resources for identifying and managing the various features of the domain. For example, computational models at multiple levels of analysis (e.g., neural, cognitive) are especially well suited for managing hierarchical, dynamic interactivity and the coordination of processes at different levels. In addition, computational modeling, parameter estimation, and clustering algorithms and techniques allow for the identification of measureable processes and features that promote the study of individual differences, the resolution of causal ambiguity (e.g., by identifying different parameter values that can lead to a common symptom), the management of context sensitivity (e.g., by creating multi-dimensional functional profiles built up from parameters estimated on the basis of a broad range of task performance data), and the identification of meaningful groupings of individuals. Of considerable importance is the promise of these techniques for reconceptualizing the phenomena in the domain of mental illness, thereby breaking the grip of conventional ways of conceptualizing such phenomena grounded in an unscientific clinical psychiatric tradition (e.g., traditional diagnostic categories, proto-scientific ways of conceiving symptoms).[20]

In conclusion, and drawing on Kuhn's classic work (Kuhn 1970, Chapter VIII) we are in a period of "extraordinary science" in which a crisis exists with respect to the dominant approach to research concerning mental illness; what is needed is a deep probing of the problems with the conventional DSM approach and associated research record, a loosening of the standards and models of research based on the DSM, and a proliferation of alternative approaches aimed at reconceptualizing the domain and managing its various features. A critical step in this direction involves changing patterns of thought, perception, and action concerning mental illness, and for this the first step is to stop thinking, perceiving, and acting in terms of the categories identified in DSM-5.

## References

American Psychiatric Association. (1980). *Diagnostic and statistical manual of mental disorders* (3rd ed.). Washington, DC: American Psychiatric Association.

American Psychiatric Association. (1994). *Diagnostic and statistical manual of mental disorders* (4th ed.). Washington, DC: American Psychiatric Association.

[20]See Wiecki et al., in press for a review of the various strategies and tools of "computational psychiatry" and examples suggesting the promise of the approach.

American Psychiatric Association. (2000). *Diagnostic and statistical manual of mental disorders* (4th ed. Text Revision). Washington, DC: American Psychiatric Association.

American Psychiatric Association. (2013). *Diagnostic and statistical manual of mental disorders* (5th ed.). Washington, DC: American Psychiatric Association.

Andreassen, N. C. (1995). The validation of psychiatric diagnosis: New models and approaches. *American Journal of Psychiatry, 152*, 161–162.

Andreassen, N. C. (2007). The DSM and the death of phenomenology in America. *Schizophrenia Bulletin, 33*(1), 108–112.

Andrews, G., Goldberg, D. P., Krueger, R. F., Carpenter, W. T., Hyman, S. E., Sachdev, P., & Pine, D. S. (2009). Exploring the feasibility of a metastructure for DSM-V and ICD-11: Could it improve utility and validity? *Psychological Medicine, 39*(12), 1993–2000.

Bernstein, C. (2011). Metastructure in DSM-5 process. *Psychiatric News, 46*(5), 7.

Blashfield, R. (1984). *The classification of psychopathology: Neo-Kraepelinian and quantitative approaches*. New York: Plenum Press.

Carson, R. (1991). Dilemmas in the pathway of the DSM-IV. *Journal of Abnormal Psychology, 100*, 302–307.

Cromwell, R. (1982). Eight decades focus on symptom classification is enough: A discussion. In M. Goldstein (Ed.), *Preventive intervention in schizophrenia: Are we ready?* Washington, DC: U. S. Government Printing Office.

Durston, S., van Belle, J., & de Zeeuw, P. (2011). Differentiating frontostriatal and fronto-cerebellar circuits in attention-deficit hyperactivity disorder. *Biological Psychiatry, 69*, 1178–1184.

Eysenck, H. (1986). A critique of contemporary classification and diagnosis. In T. Millon & G. Klerman (Eds.), *Contemporary directions in psychopathology* (pp. 73–98). New York: Guilford Press.

Fair, D., Bathula, D., Nikolas, M., & Nigg, J. (2012). Distinct neuropsychological subgroups in typically developing youth inform heterogeneity in children with ADHD. *PNAS, 109*(17), 6769–6774.

Frances, A. (2010a). DSM in Philosophyland: Curiouser and Curiouser. *Bulletin AAPP, 17*(2), 5–7.

Frances, A. (2010b). Rating scales: DSM5 bites off far more than it can chew. *Psychiatric Times*, May 7. http://www.psychiatrictimes.com/articles/rating-scales-dsm5-bites-far-more-it-can-chew. Accessed 3 Sept 2013.

Heinrichs, R. W. (2001). *In search of madness: Schizophrenia and neuroscience*. New York: Oxford University Press.

Hyman, S. (2007). Can neuroscience be integrated into the DSM-V? *Nature Reviews Neuroscience, 8*, 725–732.

Hyman, S. (2010). The diagnosis of mental disorders: The problem of reification. *Annual Review of Clinical Psychology, 6*, 155–179.

Insel, T., Cuthbert, B., Garvey, M., Heinssen, R., Pine, D. S., Quinn, K., Sanislow, C., & Wang, P. (2010). Research domain criteria (RDoC): Toward a new classification framework for research on mental disorders. *American Journal of Psychiatry, 167*(7), 748–751.

Kendell, R., & Jablensky, A. (2003). Distinguishing between the validity and utility of psychiatric diagnoses. *American Journal of Psychiatry, 160*(1), 4–12.

Kendler, K. S. (1980). The nosologica validity of paranoia (simple delusional disorder): A review. *Archives of General Psychiatry, 37*, 699–706.

Kendler, K., Kupfer, D., Narrow, W., Phillips, K., & Fawcett, J. (2009). Guidelines for making changes to DSM-V. [Online.] (Revised October 21, 2009). Available at: http://www.dsm5.org/ProgressReports/Documents/Guidelines-for-Making-Changes-to-DSM_1.pdf

Kirk, S., & Kutchins, H. (1992). *The selling of DSM: The rhetoric of science in psychiatry*. New York: Aldine De Gruyter.

Kuhn, T. (1970). *The structure of scientific revolutions*. Chicago: University of Chicago Press.

Kupfer, D. J., First, M. B., & Regier, D. A. (2002). *A research agenda for DSM-V*. Washington, DC: American Psychiatric Association.

Maia, T., & Frank, M. (2011). From reinforcement learning models to psychiatric and neurological disorders. *Nature Neuroscience, 14*(2), 152–162.

Montague, P. R., Dolan, R. J., Friston, K. J., & Dayan, P. (2011). Computational psychiatry. *Trends in Cognitive Sciences, 16*(1), 1–9.
Murphy, D. (2006). *Psychiatry in the scientific image*. Cambridge, MA: The MIT Press.
NIMH. (2011). NIMH research domain criteria (RDoC). NIMH website. http://www.nimh.nih.gov/research-funding/rdoc/nimh-research-domain-criteria-rdoc.shtml. Accessed 3 Sept 2013.
Poland, J. (2003). Bias and schizophrenia. In P. Caplan & L. Cosgrove (Eds.), *Bias in psychiatric diagnosis* (pp. 149–161). Lanham: Jason Aronson.
Poland, J. (2014). Deeply rooted sources of error and bias in psychiatric classification. In H. Kincaid & J. Sullivan (Eds.), *Classifying psychopathology: Mental kinds and natural kinds*. Cambridge, MA: The MIT Press.
Poland, J., & Von Eckardt, B. (2013). Mapping the domain of mental illness. In K. W. M. Fulford, M. Davies, R. Gipps, G. Graham, J. Sadler, G. Stanghellini, & T. Thornton (Eds.), *The Oxford handbook of philosophy and psychiatry*. Oxford: Oxford University Press.
Poland, J., Von Eckardt, B., & Spaulding, W. (1994). Problems with the DSM approach to classifying psychopathology. In G. Graham & G. L. Stephens (Eds.), *Philosophical psychopathology* (pp. 235–260). Cambridge, MA: The MIT Press.
Robins, E., & Guze, S. (1970). Establishment of diagnostic validity in psychiatric illness: Its application to schizophrenia. *American Journal of Psychiatry, 126*, 983–987.
Spaulding, W., Sullivan, M., & Poland, J. (2003). *Treatment and rehabilitation of severe mental illness*. New York: Guilford Press.
Strauss, G., Frank, M., Waltz, J., Kasanova, Z., Herbener, E., & Gold, J. (2011). Deficits in positive reinforcement learning and uncertainty driven exploration are associated with distinct aspects of negative symptoms in schizophrenia. *Biological Psychiatry, 69*(5), 424–431.
Wakefield, J. (1992). The concept of mental disorder: On the boundary between biological facts and social values. *American Psychologist, 47*(3), 373–388.
Wiecki, T., Poland, J., & Frank, M. in press. Model-based cognitive neuroscience approaches to computational psychiatry: clustering and classification. *Clinical Psychological Science*.

# DSM-5 and Psychiatry's Second Revolution: Descriptive vs. Theoretical Approaches to Psychiatric Classification

**Jonathan Y. Tsou**

**Abstract** A large part of the controversy surrounding the publication of DSM-5 stems from the possibility of replacing the purely descriptive approach to classification favored by the DSM since 1980. This paper examines the question of how mental disorders should be classified, focusing on the issue of whether the DSM should adopt a purely descriptive or theoretical approach. I argue that the DSM should replace its purely descriptive approach with a theoretical approach that integrates causal information into the DSM's descriptive diagnostic categories. The paper proceeds in three sections. In the first section, I examine the goals (viz., guiding treatment, facilitating research, and improving communication) associated with the DSM's purely descriptive approach. In the second section, I suggest that the DSM's purely descriptive approach is best suited for improving communication among mental health professionals; however, theoretical approaches would be superior for purposes of treatment and research. In the third section, I outline steps required to move the DSM towards a hybrid system of classification that can accommodate the benefits of descriptive and theoretical approaches, and I discuss how the DSM's descriptive categories could be revised to incorporate theoretical information regarding the causes of disorders. I argue that the DSM should reconceive of its goals more narrowly such that it functions primarily as an *epistemic hub* that mediates among various contexts of use in which definitions of mental disorders appear. My analysis emphasizes the importance of pluralism as a methodological means for avoiding theoretical dogmatism and ensuring that the DSM is a reflexive and self-correcting manual.

## Introduction

The question of whether mental disorders should be classified in a theoretical or purely descriptive manner is a philosophical issue embedded in the history of the *Diagnostic and Statistical Manual of Mental Disorders* (DSM). In the first two

J.Y. Tsou (✉)
Department of Philosophy and Religious Studies, Iowa State University, Ames, IA, USA
e-mail: jtsou@iastate.edu



S. Demazeux, P. Singy (eds.), *The DSM-5 in Perspective*, History, Philosophy and Theory of the Life Sciences 10, DOI 10.1007/978-94-017-9765-8_3

editions of the DSM (APA 1952, 1968), mental disorders were classified theoretically insofar as diagnostic categories reflected central distinctions of psychoanalysis (e.g., the distinction between neurosis and psychosis) and disorders were distinguished in terms of biological ('organic') and psychological ('functional') causes (see Tsou 2011). The publication of DSM-III (APA 1980) marked psychiatry's first revolution in psychiatric classification insofar as the DSM-III taskforce, chaired by Robert L. Spitzer, self-consciously replaced the psychoanalytic and etiological approach to classification adopted in DSM-I and DSM-II with a purely descriptive ("neo-Kraepelinian") approach that made no theoretical assumptions about the causes of mental disorders.[1] The purely descriptive approach to classification championed by the DSM since DSM-III has recently been brought into question with the publication of DSM-5.

Prior to the publication of DSM-5, there were some indications that DSM-5 would mark psychiatry's second revolution in classification by ushering a paradigm shift away from the DSM's purely descriptive approach. In particular, some reports from the DSM-5 taskforce, chaired by David J. Kupfer, indicated that DSM-5 would move the DSM away from the neo-Kraepelinian approach towards a theoretical and etiological approach to psychiatric classification informed by sciences such as genetics and neuroscience (e.g., see Kupfer et al. 2002, 2013, p. E2; Hyman 2007; Regier 2008; Regier et al. 2009; Kupfer and Regier 2011). In articulating a research agenda for DSM-5, Kupfer et al. (2002) contend that:

> [L]imitations in the current [neo-Kraepelinian] diagnostic paradigm suggest that research exclusively focused on refining the DSM-defined syndromes may never be successful in uncovering their underlying etiologies. For that to happen, an as yet unknown paradigm shift may need to occur. . . [An] important goal . . . is to transcend the limitations of the current DSM paradigm and *to encourage a research agenda that goes beyond our current ways of thinking to attempt to integrate information from a wide variety of sources and technologies.* (p. xix, emphasis added)

The publication of DSM-5 (APA 2013) would be disappointing to those who expected revisions of paradigm-shifting proportions. The main difference in DSM-5, compared to DSM-IV-TR (APA 2000), is the greater use of dimensional measures. Nonetheless, the recent publication of DSM-5 occasions a reconsideration of the prospects of theoretical versus purely descriptive approaches to psychiatric classification.

The aim of this paper is to critically evaluate the relative merits of purely descriptive and theoretical approaches to psychiatric classification for meeting the DSM's goals of providing a manual that can guide treatment, facilitate research, and improve communication. I argue that the DSM's purely descriptive approach is impoverished for meeting these aims and would benefit by shifting towards a theoretical system that integrated information about the causes of mental disorders

[1]For a more comprehensive discussion of the neo-Kraepelinian outlook of DSM-III, see Klerman (1978), Blashfield (1984), Wilson (1993), Compton and Guze (1995), Mayes and Horwitz (2005), Decker (2007, 2013), and Tsou (2011).

into its descriptive categories. The paper proceeds as follows. In the first section ("Aims of the DSM"), I discuss the aims associated with the purely descriptive approach to classification adopted since DSM-III. In the second section ("Descriptive vs. Theoretical Approaches to Psychiatric Classification"), I compare the relative benefits and costs of purely descriptive versus theoretical approaches for meeting these aims. I argue that descriptive approaches are best suited for coordinating communication among mental health professionals; however, theoretical approaches are better suited for meeting the DSM's goals of facilitating research and treatment. In the third section ("DSM-5 and Beyond: Steps towards a Theoretical Approach to Classification"), I argue that the DSM should adopt a hybrid model that integrates the benefits of descriptive and theoretical approaches. In articulating this argument, I suggest that the DSM ought to narrow its goals, such that it functions primarily as an *epistemic hub*, i.e., a reference point that can mediate among the various contexts in which definitions of mental disorders appear (Kutschenko 2011). To be a *useful* epistemic hub, I suggest that the DSM should classify mental disorders that are natural kinds (i.e., kinds associated with a distinctive biological causal structure) and its diagnostic categories should be informed by the best available theories on the biological causes of mental disorders. With respect to how the DSM could integrate causal theories into its descriptive categories, my argument emphasizes the importance of pluralism as a means to ensure that the DSM is informed by a multiplicity of, sometimes conflicting, scientific theories on psychopathology.

## Aims of the DSM

The aims of the DSM are ambiguous and multifaceted because they have been historically shaped by a variety of social forces, such as the US healthcare system, the pharmaceutical industry, and various lobbyist groups (see Cooper 2005; Tsou 2011; Sadler 2013). In the late 1970s, when DSM-III was being drafted, psychiatry (i.e., psychoanalytically oriented psychiatry) was in a state of crisis and the American Psychiatric Association (APA) was facing pressure to revise the DSM so it would provide clear and reliable diagnostic categories for purposes of medical insurance reimbursement (Tsou 2011). Moreover, the emergence of pharmacological drugs (e.g., antipsychotic drugs) in the 1950s and 1960s that could treat the symptoms of more severe mental disorders created a need for reliable diagnostic categories to ensure that patients were properly diagnosed before receiving pharmacological treatment. DSM-III addressed these practical problems by providing a manual with clear and reliable diagnostic classifications and introducing the now familiar diagnostic criteria that demand that a set of operationalized necessary and sufficient criteria (mainly behavioral criteria) be satisfied for a diagnosis to be made.

In the introduction to DSM-III, the purely descriptive approach to classification is justified in terms of its capacity to facilitate widespread use of the manual by clinicians of diverse theoretical orientations:

> The approach taken in DSM-III is atheoretical with regard to etiology . . . except for those disorders for which this is well established . . . *The major justification for the generally atheoretical approach . . . is that the inclusion of etiological theories would be an obstacle to use of the manual by clinicians of varying theoretical orientations* . . . Because DSM-III is generally atheoretical with regard to etiology, it attempts to describe comprehensively what the manifestations of the mental disorders are, and only rarely attempts to account for *how* the disturbances come about . . . This approach can be said to be "descriptive" in that the definitions of the disorders generally consist of descriptions of the clinical features of the disorders. . . . at the lowest order of inference necessary to describe the characteristic features of the disorder. (APA 1980, p. 7, emphasis added)

As alluded to in this passage, one of the chief motivations for adopting a descriptive and atheoretical approach was to remove speculative psychoanalytic theoretical assumptions that typified DSM-I and DSM-II. Consequently, the shift to a neo-Kraepelinian approach in DSM-III allowed mental health professionals with diverse theoretical orientations to utilize the DSM. In this regard, it is no accident that the DSM only became a highly influential, and increasingly entrenched, manual after 1980.

Besides permitting widespread usage of the manual, the DSM is presented as a manual that aims to guide treatment, facilitate research on mental disorders, and improve communication among mental health professionals. These goals are stated most explicitly in the introduction to DSM-IV-TR as follows:

> Our highest priority has been to *provide a helpful guide to clinical practice.* We hoped to make DSM-IV practical and useful for clinicians by striving for brevity of criteria sets, clarity of language, and explicit statements of the constructs embodied in the diagnostic criteria. An additional goal was to *facilitate research* and *improve communication* among clinicians and researchers. (APA 2000, p. xxiii, emphasis added)

> The purpose of DSM-IV is to provide clear descriptions of diagnostic categories in order *to enable clinicians and investigators to diagnose, communicate about, study, and treat people with various disorders.* (APA 2000, p. xxxvii, emphasis added)

As indicated here, the DSM aims to meet three primary goals: (1) to guide clinical treatment by providing operational definitions of mental disorders that allow clinicians to diagnose and make judgments about what treatment interventions are appropriate, (2) to facilitate research by providing standardized definitions that can be utilized in the study of mental disorders, and (3) to improve communication among mental health professionals presupposing disparate theoretical assumptions (cf. APA 2013, p. xli).

Despite its explicit articulation of its goals, the DSM is ambiguous in its intended aims, especially with respect to how—and to what extent—the manual is intended to facilitate treatment and research. Moreover, I contend that the DSM's assumption that the various goals and uses of the DSM "are compatible with one another" (APA 2000, p. xxviii) is misleading (cf. Frances and Widiger 2012, p. 110). My counter-suggestion is that there are tradeoffs between how well descriptive and theoretical manuals can facilitate these various goals.

## Descriptive vs. Theoretical Approaches to Psychiatric Classification

Given that the aims of the DSM are to facilitate treatment, research, and communication, what are the prospects of purely descriptive versus theoretical approaches for achieving these goals? While the DSM's purely descriptive approach does well at facilitating communication and promoting the widespread use of the manual, I argue that a theoretical approach would provide a superior method for guiding treatment and research. My argument assumes that the DSM ought to classify mental disorders that are natural kinds (as opposed to artificial kinds).

### *Descriptive Approaches*

The main virtue of purely descriptive approaches to classification is the facilitation of communication among mental health professionals. Purely descriptive approaches achieve this end by providing a system of standardized definitions of mental disorders, which can be utilized in a variety of contexts, including contexts of research and treatment. A related benefit of the DSM's neo-Kraepelinian approach is that its definitions can be widely used by mental health professionals working from a variety of theoretical orientations because the manual makes no metaphysical assumptions about the causes of mental disorders. Hence, the main benefits of purely descriptive approaches stem from the provision of a common language for communicating about mental disorders and the potential of its descriptive categories to be widely used.

While the DSM does well at facilitating communication and widespread usage, its purely descriptive approach is only minimally useful for research and treatment. With respect to research, the main disadvantage of the DSM's descriptive and atheoretical approach is the difficulty of testing its diagnostic categories. Since mental disorders are defined behaviorally, there is no principled way of determining whether a diagnostic category merits inclusion or exclusion in the manual. Moreover, although the authors of the DSM claim that the primary purpose of the DSM is to guide treatment, the DSM's fine-grained diagnostic categories are only minimally useful for guiding treatment decisions. For purposes of treatment, clinicians may find some use in broad DSM categories that distinguish depression from anxiety and psychosis; however, beyond these broad classificatory distinctions, DSM classifications do little to guide treatment interventions that go beyond clinicians' tacit knowledge (Brown 1987; Kirk and Kutchins 1988, 1992, Chap. 9; Whooley 2010). Moreover, as is well documented, one of the major difficulties with the DSM's descriptive approach is the high incidence of co-morbid diagnoses (Kessler et al. 2005). In the context of actual clinical practices, one of the most useful DSM tools is the "not otherwise specified" (NOS) diagnosis, which indicates that a cluster of symptoms do not neatly fit any of the DSM's diagnostic categories, but allows individuals seeking treatment to receive a DSM diagnosis for purposes of medical insurance (Hyman 2010, pp. 166–167). Hence, rather than functioning as a useful guide that facilitates treatment decisions, the DSM functions—in practice—as an administrative constraint that clinicians must satisfy to ensure that patients are reimbursed for treatment.

## *Theoretical Approaches*

Weaknesses of the DSM's descriptive approach are potential advantages of theoretical approaches to classification. The distinguishing feature of theoretical approaches is that its diagnostic categories would *incorporate information about the causes of disorders*.[2] My argument for a theoretical approach assumes that—as an ideal—the *DSM should classify natural kinds*.[3] On this view, some mental disorders (e.g., schizophrenia, depression, bipolar disorder) are natural kinds insofar as they are classes of abnormal behavior associated with a distinctive biological causal structure. More specifically, some mental disorders are natural kinds constituted by networks of identifiable biological mechanisms at multiple levels (e.g., molecular, developmental, neurobiological) that interact to produce the key features of the kind (see Kendler et al. 2011; Tsou 2012, 2013).[4] The significance of this view is that the common biological causal structure captured by a natural kind term allows for projectable inductive inferences (i.e., predictions) to be made about members of a kind.[5] This importantly includes inferences about the prognosis of a particular disorder and predictions regarding how an individual diagnosed with a disorder will respond to specific treatment interventions.

Compared to purely descriptive approaches, theoretical approaches would be superior for purposes of facilitating research. Currently the DSM guides research in a top-down manner by providing operational definitions of disorders, which

---

[2]It is important to note that theoretical approaches are compatible with descriptive approaches to psychiatric classification. Theoretical approaches are only incompatible with the purely descriptive (i.e., atheoretical) approach to classification associated with DSM-III (APA 1980).

[3]The question of whether the DSM intends to classify natural kinds is murky (cf. Cooper 2005; Tsou 2011). The DSM-III taskforce initially planned to include a statement in the introduction of DSM-III that stated that "mental disorders are a subset of medical disorders" (see Spitzer et al. 1977; Spitzer and Endicott 1978), which would suggest that the DSM does aim to classify natural kinds (cf. APA 1980, p. 6). However, this statement was ultimately not included in DSM-III due to protests from psychologists, social workers, and counselors who regarded it as a declaration that psychiatrists—with medical training—were solely responsible for the treatment of mental disorders (Mayes and Horowitz 2005).

[4]I assume that mechanisms are complex systems of entities and activities that are organized in a way to produce regular changes (see Bechtel and Richardson 1993; Glennan 1996, 2002; Machamer et al. 2000; Craver and Darden 2001; Machamer 2004; Tabery 2004).

[5]For discussion of the projectability of natural kind terms, see Quine (1969), Goodman (1983), Boyd (1985, 1999, 2010), and Khalidi (2013).

researchers employ to select homogenous populations of patients to study. Given the high inter-rater reliability of DSM categories, this strategy is useful for ensuring that researchers working in different locales are studying mental disorders in a uniform manner. A fundamental problem with this approach, however, is that while the DSM's diagnostic categories are reliable, its categories lack validity (Kendell 1989; Kendell and Jablensky 2003).[6] A theoretical approach to classification could individuate disorders by etiology, rather than behaviorally, which would provide a more effective method for identifying valid diagnostic categories. As Dominic Murphy (2006) has argued, the DSM's purely descriptive approach offers an incoherent methodology for psychiatric classification insofar as it "requires us to assume that a significant difference can exist between individuals at the level of surface symptoms that does not reflect an underlying causal difference" (p. 324). From this perspective, the DSM's purely descriptive approach is highly costly insofar as the manual functions—in practice—to *reify* and promote research on mental disorders (e.g., histrionic personality disorder, mathematics learning disorder) that may have no natural basis (Hyman 2010).

Theoretical approaches could also benefit research by providing a potentially testable and correctable system of psychiatric classification. The testability of theoretical approaches is an issue discussed by Carl Hempel in his classic analysis of psychiatric taxonomy (Hempel 1965).[7] Hempel (1965) argues that:

> [T]o be scientifically useful *a concept must lend itself to the formulation of general laws or theoretical principles which reflect uniformities in the subject matter under study, and which thus provides a basis for explanation, prediction, and generally scientific understanding*. This aspect of a set of scientific concepts will be called its *systematic import*, for it represents the contribution concepts make to *the systematization of knowledge in the given field by means of laws or theories*. (p. 146, emphasis added)

According to Hempel, scientific classification systems follow a regular progression. Initially, classification systems will aim to simply describe objects of

[6]While there is no agreed upon concept of validity in psychiatry, valid diagnostic categories are generally understood as classifications that pick out real natural phenomena, i.e., categories that 'carve nature at the joints.' For a more comprehensive discussion of various proposed definitions of validity (e.g., construct and content validity), see Robins and Guze (1970), Kendler (1990), Kendell and Jablenksy (2003), First et al. (2004), Murphy (2006, Chap. 6), Jablenksy (2012), and Shaffner (2012). While some theorists have argued that validity is best understood in terms of utility, I assume that these concepts are distinct, although it is important to recognize that valid diagnostic categories will be predictively useful (i.e., *projectable*), but not necessarily vice versa. Given the importance of making reliable predictions in psychiatry (and the difficulty in evaluating more general ideals of validity), predictive validity is arguably the most useful concept of validity to employ in evaluating diagnostic categories (Shaffner 2012).

[7]For a more comprehensive discussion of Hempel's analysis, see Schwartz and Wiggins (1986), Murphy (2006, Chap. 6), Bolton (2008), and Tsou (2011). Some authors have suggested that Hempel's paper played an influential historical role in DSM-III's adoption of an operationalized and purely-descriptive approach (e.g., see Bolton 2008, p. 3). In this paper, I argue that Hempel's emphasis on the *testability* of theoretical taxonomic systems offers compelling support for contemporary arguments in favor of theoretical approaches to psychiatric classification.

classification with the aid of operational definitions. However, as the system evolves, it ought to develop into a theoretical system where increased emphasis is placed on "the attainment of comprehensive theoretical accounts of the empirical subject matter under discussion" (Hempel 1965, p. 140). In arguing for a theoretical system of psychiatric classification, Hempel suggests that psychiatric classification could resemble other taxonomic systems in science, such as classification systems in chemistry and biology (Hempel 1965, pp. 147–149; cf. Hacking 2013). Importantly, a theoretical system of classification could make predictions and be testable in ways that purely descriptive systems are not.

The potential testability of theoretical approaches is an issue that the DSM-5 taskforce emphasize in their own arguments for a shift away from the DSM's neo-Kraepelinian approach. Regier et al. (2009) write:

> Mental disorder syndromes will eventually be redefined to reflect more useful diagnostic categories ('to carve nature at its joints') . . . [O]ur immediate task is to set up a framework for an evolution of our diagnostic system that can advance our clinical practice and *facilitate ongoing testing of the diagnostic criteria that are intended to be scientific hypotheses, rather than inerrant Biblical scripture*. (pp. 648–649, emphasis added)

Implicit in this statement is the idea that the DSM can advance to a theoretical and etiological system better suited for testing the validity of diagnostic categories. The DSM's atheoretical and purely descriptive orientation, by contrast, has functioned in practice to *protect* its diagnostic categories from revision given that—without appeals to etiology—there is no systematic way of determining when a disorder should be retained, revised, or removed from the manual. This aspect of the DSM has led to a trend of proliferation of diagnostic categories since the publication of DSM-III, which some authors have regarded as a sign of scientific degeneration (e.g., see Follette and Houts 1996; Houts and Follette 1998; Houts 2001; Horwitz 2002; cf. Wakefield 1998, 1999, 2001).

A theoretical approach to classification would also be beneficial for providing a manual that guides treatment decisions. As argued above, an advantage of theoretical approaches is that they could individuate mental disorders on the basis of shared biological regularities. This would ensure that the DSM's diagnostic categories are projectable in the sense that clinicians can make successful inductive inferences about treatment. Without a clear theoretical basis for the DSM's diagnostic categories, there is no assurance that the DSM's behavioral definitions are projectable. Paul Meehl (1995) has criticized the DSM on these grounds, arguing that:

> The advance-science medical model does not identify disease taxa with the operationally defined syndrome: the syndrome is taken as evidentiary, not as definitory. The explicit definition of a disease entity in non-psychiatric medicine is a conjunction of pathology and etiology and therefore applies to patients who are asymptomatic (which is why, e.g., one can have a silent brain tumor. . .). Perhaps we cannot blame psychologists ignorant of medicine for making this mistake, when some psychiatrists who are passionate defenders of the DSM don't understand how far it deviates from the optimal medical model. (p. 267)

As Meehl argues, psychiatric treatment would be better served by a theoretical and etiological approach to classification because it would incorporate information about the causes of disorders (also see Murphy 2006, pp. 324–326).

The main disadvantage of theoretical approaches is the potential compromise of the facilitation of communication and widespread usage of the DSM. Because the etiology of many mental disorders is unknown, theoretical approaches run the risk of introducing the kinds of speculative causal inferences that the authors of DSM-III sought to remove. However, while the shift to an atheoretical approach in DSM-III (APA 1980)—in the historical context of the 1970s—was justified because it helped to remove speculative psychoanalytic theories from the DSM, in our current situation, DSM's atheoretical stance has served to isolate the manual from the most promising scientific theories on psychopathology, including genetics, neurobiology, and the cognitive sciences (Murphy 2006, Chap. 9; Hyman 2007; Insel et al. 2010; Sanislow et al. 2010). This criticism is particularly salient given that one of the intended roles of the DSM is to reflect the best current scientific findings on psychopathology. Moreover, a fundamental pragmatic difficulty with theoretical approaches is that it is difficult to accommodate different sciences. As Rachel Cooper (2005, Chap. 3) has pointed out, a psychiatric classification system is only as good as the theory it is based on. From this perspective, a disadvantage of theoretical approaches to classification stems from the difficulty of choosing which theoretical sciences ought to be incorporated into a system of classification and the potential for excluding and marginalizing certain scientific theories. In the following section, I defend a hybrid approach to classification that recommends that the DSM's diagnostic categories are informed by the best available theories on the biological causes of disorders, while considering a plurality of competing theories. Compared to purely descriptive approaches, this (theoretical) hybrid approach would present an explicit basis for testing diagnostic categories and provide a more promising method for formulating valid diagnostic categories.

## DSM-5 and Beyond: Steps towards a Theoretical Approach to Classification

In what follows, I argue that the DSM ought to shift to a theoretical approach to classifying mental disorders and offer some recommendations for meeting this ideal. The theoretical approach that I advocate primarily seeks to integrate theoretical information about the causes of mental disorders into the DSM's descriptive diagnostic categories (see note 2). This hybrid approach to psychiatric classification could accommodate the benefits of descriptive approaches (i.e., facilitating communication) and theoretical approaches (i.e., facilitating treatment and research), while avoiding the costs of both.

## *The DSM as an Epistemic Hub*

One of the problems with the DSM is the ambitiousness of its aims. While the DSM aims to provide a manual that can simultaneously guide treatment, facilitate research, and improve communication, this stance fails to recognize that these aims will often conflict and that there are tradeoffs between how well descriptive and theoretical approaches can meet these goals. As a result, the DSM's purely descriptive approach has done well at improving communication among mental health professionals, but has only yielded minimal benefits for guiding treatment and facilitating research.

One way that the DSM could address this problem is to reconceive its aims in a more narrow manner. Rather than serving as a manual that is intended to guide treatment, facilitate research, and improve communication (with the facilitation of treatment regarded as its primary goal), the DSM should be reconceived as what Lara K. Keuck has called an *epistemic hub* (Kutschenko 2011). Keuck argues that broadly applied medical classification systems that are used in different settings by a variety of actors—such as the DSM—ought to serve primarily as hubs that can mediate between these various practices (cf. Pincus 2012, pp. 157–158). In this manner, epistemic hubs are *reference points* that allow for the exchange of information and the integration of explanations among researchers, practicing clinicians, and other mental health professionals. As Keuck points out, because epistemic hubs are intended for widespread use in different contexts, they ought to primarily strive for *connectivity with alternative descriptions of disorders,* rather than precision per se. While the DSM does not represent an ideal epistemic hub because its diagnostic categories are defined at a level of precision unnecessary for purposes of coordinating various practices, the descriptive diagnostic categories of the DSM are well-suited for meeting the goal of providing a common point of reference that can coordinate among the various practices of mental health professionals.[8]

While I have argued that the DSM should be reconceived as an epistemic hub that facilitates communication, to be a *useful epistemic hub* that genuinely facilitates the exchange of information and integration of explanations, the DSM needs to provide definitions for *valid diagnostic categories.* If this requirement is met, then the DSM could more effectively meet its goals of providing a manual that is useful for guiding treatment and facilitating research. Without the provision of validated diagnostic categories, a high cost of the DSM is the potential reification of artificial kinds (e.g., narcissistic personality disorder, histrionic personality disorder), which have pernicious implications for both treatment and research.[9] As argued in the

[8]For this purpose, the dimensional measures introduced in DSM-5 (APA 2013) may be disadvantageous because they import an unnecessary level of specificity into definitions of mental disorders.

[9]In the human sciences, I assume that artificial kinds are classes that are useful for distinguishing a socially relevant group (e.g., 'liberals,' 'widows,' 'police officers'), *but are not associated with a distinctive biological causal structure* (see Tsou 2013). Some clear examples of artificial

second section of this paper ("Descriptive vs. Theoretical Approaches to Psychiatric Classification"), the DSM ought to individuate and classify natural kinds associated with an identifiable biological basis, which would ensure that the DSM's diagnostic categories are projectable in the sense of yielding successful inductive inferences about members of a kind. In her analysis, Keuck downplays the importance of issues concerning validity, suggesting that epistemic hubs "cannot and should not be considered scientific taxonomies in Hempel's sense" (Kutschenko 2011, p. 585). Against this view, I maintain that in addition to functioning as an epistemic hub that facilitates communication, the DSM should also strive to provide validated diagnostic categories that are useful in contexts of treatment and research.[10] For this goal to be met, the DSM's diagnostic categories ought to incorporate theoretical information about the causes of disorders.

## *The Integration of Causal Information*

If the DSM is understood as an epistemic hub that serves primarily to mediate between different contexts of use in which definitions of mental disorders occur, should the DSM adopt a purely descriptive (atheroretical) or theoretical approach? According to the analysis of this paper, there is no question that the DSM should move towards a theoretical approach. The most problematic aspect of the DSM's neo-Kraepelinian approach is the lack of validity of its categories and adopting a broadly theoretical approach would provide a more effective means for addressing this difficulty. The important question is not *whether*, but *how* theoretical information could be integrated into the DSM and on what basis its diagnostic categories could be revised.

The theoretical approach that I advocate would not necessarily *include* the causes of mental disorders in its diagnostic categories (cf. Hyman 2007); rather, it would *incorporate* information about causes by requiring—as an *ideal*—that the operational symptoms that constitute the diagnostic criteria of a disorder are caused by identifiable biological mechanisms. According to this ideal, in order for a disorder to be included in the DSM, there needs to be evidence that there is a

kinds listed in DSM-5 include histrionic personality disorder, dependent personality disorder, and voyeuristic disorder. There is meager evidence that the characteristic signs of these disorders are caused by identifiable biological mechanisms, and any biological regularities observed for these disorders are better accounted for at a more general level of analysis (e.g., in terms of anxiety or high testosterone levels).

[10]Keuck suggests that epistemic hubs and Hempelian scientific taxonomies are different insofar as "the latter strive for precision, whereas as the former need to uphold a certain degree of fuzziness in their descriptions . . . in order to allow different actors to connect their more restrictive classification systems to the epistemic hub" (Kutschenko 2011, p. 585). However, there is no principled reason why the diagnostic categories of scientific taxonomies cannot be formulated at the level of generality required to serve as an epistemic hub.

*stable and distinctive biological causal structure* for the classified disorder. This stance can accommodate the fact that scientists often do not have knowledge of the biological causes for many DSM disorders (e.g., bipolar disorder), although they have evidence that these disorders possess a distinctive biological causal structure. However, the DSM's diagnostic categories ought to be refined over time such that the characteristic signs that define mental disorders are associated with identifiable biological causes.

As a concrete example of a diagnostic category that possesses some (predictive) validity, consider the DSM's definition of schizophrenia, which requires that two or more of the following symptoms must be present over a 1-month period (APA 2013, pp. 99–100):

1. Delusions
2. Hallucinations
3. Disorganized speech (e.g., frequent derailment or incoherence)
4. Grossly disorganized or catatonic behavior
5. Negative symptoms (i.e., diminished emotional expression or avolition)
   At least one of the symptoms present must be (1), (2) or (3)

The diagnostic criteria for the DSM's schizophrenia classification, while not ideal, are useful precisely because there are compelling theoretical reasons for thinking that these particular symptoms have identifiable causes.[11] For example, neurobiological research indicates that delusions (1) and hallucinations (2)—the "positive symptoms" of schizophrenia—are caused by excessive dopamine activity in the mesolimbic pathway, while "negative symptoms" (5) are caused by deficient dopamine activity in the mesocortical pathway (see Tsou 2012). In this manner, the DSM's classification of schizophrenia represents a *theoretically informed descriptive category*. In connection with the ideal of epistemic hubs defended in this paper, the level of generality that the schizophrenia classification is described is useful because it can be coordinated with more specific descriptions of schizophrenia (e.g., paranoid or catatonic sub-types). Hyman (2007) suggests that a number of disorders listed in the DSM (e.g., schizophrenia, bipolar disorder, major depression, and obsessive-compulsive disorder) possess some validity and pick out natural phenomena because there is evidence for the heritability of these disorders and striking similarities in the symptoms of these disorders as they appear across cultures (p. 726).[12] These disorders provide paradigm examples of disorders that

[11]The DSM's diagnostic criteria for schizophrenia could be improved by including 'cognitive impairments' (e.g., deficits in attention, memory, and executive functioning), which are theorized to be caused by dopamine dysfunction in the prefrontal cortex (Hyman and Fenton 2003). Similarly, I have argued that there are good theoretical reasons for including mental inflexibility ("cognitive rigidity") in the diagnostic criteria for depression because there is evidence that this cognitive trait, which is correlated with both depression and suicide ideation, is caused by deficient serotonin projections to the orbitofrontal cortex (Tsou 2013).

[12]I have elsewhere discussed the importance of cross-cultural research for helping to identify disorders that are natural kinds (Tsou 2007, 2013). On my view, the distinction between natural

should be included in the DSM, and the diagnostic criteria for these disorders should ideally include symptoms that are theorized to be caused by identifiable biological mechanisms.[13]

The hybrid approach to classification outlined herein can accommodate the advantages of purely descriptive and theoretical approaches. Because this approach would retain the DSM's descriptive approach, it would maintain the DSM's ability to facilitate communication and widespread use of the manual. In terms of the ideal of epistemic hubs defended in this paper, diagnostic criteria should be articulated at a level of generality that allows its categories to be compatible with alternative, more specific, descriptions of mental disorders. Integrating theoretical information about the causes of mental disorders into these categories, however, would provide a more effective method for providing validated diagnostic categories, which would be beneficial for guiding clinical treatment and facilitating research.

## *Theoretical Pluralism and the Revision of DSM Categories*

As discussed in the second section of this paper ("Descriptive vs. Theoretical Approaches to Psychiatric Classification"), one of the main disadvantages of theoretical approaches is the difficulty of accommodating different, and sometimes incompatible, theories of psychopathology. Theoretical pluralism is a promising methodological means for overcoming this difficulty. In philosophy of science, proponents of pluralism (e.g., see Feyerabend 1975; Suppes 1978; Dupré 1993; Longino 1990, 2002; Mitchell 2003; Giere 2006; Kellert et al. 2006) argue that—because of the theory-ladenness of scientific observation (Hanson 1958; Kuhn 1962)—a plurality of inconsistent, and sometimes mutually incompatible, scientific theories are *required* to obtain an adequate understanding of scientific phenomena. For example, Paul Feyerabend (1965) contends that:

> Not only is the description of every single fact dependent on *some* theory . . . but there also exist facts that cannot be unearthed except with the help of alternatives to the theory to be tested and that become unavailable as soon as such alternatives are excluded. This suggests that the methodological unit to which we must refer when discussing questions of test and empirical content is constituted by a *whole set of partly overlapping, factually adequate, but mutually inconsistent theories* . . . it suggests a theoretical pluralism as the basis of every test procedure. (p. 175, emphasis in original)

---

and artificial kinds is a distinction of a degree, and the symptoms of 'more natural' disorders (i.e., disorders whose characteristic signs are more directly determined by biological mechanisms) *will exhibit greater uniformity across cultures* (and over time) than artificial kinds.

[13]In this connection, more research efforts should be directed towards to validating proposed criterion sets of disorders by providing evidence that the cluster of signs included in these sets are caused by biological mechanisms (cf. Frances and Widier 2012, pp. 118–120). For this purpose, cross-cultural research is a particularly useful resource for identifying the common cluster of symptoms associated with a disorder (see Kleinman 1988, Chaps. 2–3; Kendler 2009; Tsou 2007; cf. Cooper 2010).

Feyerabend maintains that if the description of scientific facts is theory-dependent, then there are facts relevant to the validity of a theory that *cannot* be revealed *except* with alternatives to the theory being tested. On this view, theoretical pluralism is virtuous because alternative theories can serve as *criticisms* of currently accepted theories that cannot be obtained in other ways (Feyerabend 1965, p. 150).[14] With respect to how the DSM's diagnostic categories should be revised to incorporate theoretical information, the kind of theoretical pluralism advocated by Feyerabend could provide an effective means for avoiding theoretical dogmatism.

In order for the DSM to incorporate theoretical information about its diagnostic categories in a fruitful manner, a plurality of theories should be represented in the DSM revision process. Much disgruntlement surrounding the revision of the DSM in the past resulted from the exclusion of individuals with divergent theoretical views (Sadler 2005, Chap. 3), and the DSM would benefit by broadening the theoretical perspectives represented by experts involved in the DSM revision process (Frances and Widiger 2012). Similarly, the critical review of proposed revisions to the DSM could be improved by soliciting literature reviews from people working outside of the DSM revision process, especially from reviewers who are likely to be critical of proposals (Widiger and Clark 2000; Frances and Widiger 2012).[15] Finally, the ideal of pluralism could be better approached in the field testing of proposed criterion sets, which currently is limited to pilot studies in artificial research settings (Frances and Widiger 2012). At minimum, some field testing should be conducted in clinical settings and criticism of proposed criterion sets should come from practicing clinicians adopting diverse theoretical orientations.

In addition to pluralism in the DSM revision process, the DSM could benefit from pluralism at the taxonomic level (Kutschenko 2011). Given that there are multiple purposes for classifying mental disorders, researchers and clinicians should develop alternative classification systems that can simultaneously serve to inform (and potentially criticize) the DSM's diagnostic categories. One example of such an alternative is the Research Domain Criteria (RDoC) classification system being introduced by the National Institute of Mental Health (NIMH). The RDoC is an explicitly theoretical system of classification intended primarily for facilitating research.[16] While the RDoC was initially formulated as a competing system of classification to the DSM—due to dissatisfaction with the lack of validity of DSM

[14]I have elsewhere discussed the importance of Feyerabend's views on pluralism (Tsou 2003).

[15]While I have argued that disorders included in the DSM should have a clear biological basis, the theories that are employed to criticize existing diagnostic categories ought to represent a wide-spectrum of views, including psychoanalytic, humanistic, and social psychological perspectives. These alternative perspectives could help to clarify which particular symptoms of disorders lack a natural basis and shed light on other mechanisms relevant for the expression of disorders.

[16]As indicated in the Research Domain Criteria Matrix (see Morris and Cuthbert 2012, p. 31), the RDoC seeks to organize research by distinguishing seven units of analysis (viz., genes, molecules, cells, circuits, physiology, behavior, self-reports, and paradigms) and five research domains (viz., negative valence emotional systems, positive valence emotional systems, cognitive systems, systems for social processes, and arousal/ regulatory systems).

diagnostic categories (Insel et al. 2010)—the NIMH and APA have taken a more collaborative stance recently. The NIMH and APA issued a joint press release on May 13, 2013, which stated the following:

> All medical disciplines advance through research progress in characterizing diseases and disorders. DSM-5 and RDoC represent complementary, not competing, frameworks for this goal. DSM-5 . . . reflects the scientific progress seen since the manual's last edition was published in 1994. RDoC is a new, comprehensive effort to redefine the research agenda for mental illness. *As research findings begin to emerge from the RDoC effort, these findings may be incorporated into future DSM revisions* and clinical practice guidelines. But this is a long-term undertaking. It will take years to fulfill the promise that this research effort represents for transforming the diagnosis and treatment of mental disorders. (Insel and Leiberman 2013, emphasis added)

This collaborative effort between the APA and NIMH provides an instructive example of how the DSM could be coordinated with and informed by alternative classification systems, which could help to incorporate theoretical information about the causes of disorders into the DSM's diagnostic categories.[17]

A legitimate worry about adopting a theoretical approach to classification is the possibility of theoretical dogmatism. I have suggested that theoretical pluralism provides a methodological prescription that could alleviate this problem. Whereas theoretical pluralism is the norm within research on psychopathology, the ideal of pluralism is inadequately met in the DSM revision process. The DSM could also benefit by being forced to coordinate itself against a plurality of more specifically formulated classification systems, such as the RDoC, CFTMEA (Misés and Quemada 2002), and OPD (OPD Task Force 2008). Unfortunately, there are few alternative psychiatric classification systems that have been formulated for specific purposes (e.g., treatment manuals); undoubtedly, part of this problem is due to the current hegemonic status of the DSM. However, the development of alternative classification systems might provide the most promising types of interactions that could result in innovative revisions to the DSM.

## Conclusion

In this paper, I argued that the DSM should adopt a theoretical approach to classification that incorporates information about the causes of mental disorders. The particular approach that I favor retains the DSM's descriptive categories, but integrates theoretical information about the causes of disorders into its categories. This hybrid approach would aim to classify mental disorders with a clear natural (biological) basis and ensure that the diagnostic criteria that define mental disorders

[17] Moreover, the efforts of the RDoC may ultimately serve to drastically revise our understanding of what the proper targets of validation are given that the RDoC will explore new ways of classifying mental disorders that do not rely on DSM diagnostic categories (Sanislow et al. 2010; Morris and Cuthbert 2012; Tabb 2015).

have identifiable (or hypothesized) biological causes. Compared to purely descriptive approaches, the main benefits of this approach is that it would: (1) provide a methodological basis for formulating valid and projectable diagnostic categories, (2) offer an explicit basis for deciding when disorders ought to be included or excluded from the DSM, and (3) engage the DSM more closely with a large body of scientific research on psychopathology. On a more general level, I suggested that the DSM could usefully be reconceived of as an epistemic hub that serves to mediate among various contexts of use in which definitions of mental disorders appear. From this perspective, the diagnostic categories of the DSM ought to be defined at a level of generality that permits the manual to be coordinated with alternative descriptions of mental disorders.

While I have characterized the shift towards a theoretical approach to classification as psychiatry's "second revolution," I envisage this shift as a gradual and iterative process. Kendler and First (2010) have urged that the DSM is not ready for a paradigm shift wherein the DSM's purely descriptive approach is *jettisoned* in favor of an etiological and theoretical approach. Others have stressed the importance of conservatism in revising future editions of the DSM (Frances and Widiger 2012). I am sympathetic to these concerns, and the ideal of classification advanced in this paper can accommodate these worries. Because the theoretical approach that I advocate retains the DSM's descriptive approach to classifying mental disorders, theoretical information about the causes of the signs of mental disorders could be incorporated in the DSM's existing diagnostic categories in a gradual and piecemeal fashion. When there is uncertainty regarding the validity or causes of a disorder, my favored approach would recommend retaining the status quo until there are compelling theoretical reasons for making revisions. Regardless of whether the theoretical approach that I advocate would constitute a paradigm-shifting revolution in the classification of mental disorders, it would mark a significant point of scientific progress by supplanting the DSM's antiquated neo-Kraepelinian approach.

**Acknowledgements** I owe thanks to Ian Hacking, Bill Wimsatt, Robert Richards, Trevor Pearce, Lara Keuck, Kathryn Tabb, Doreen Fraser, Tim Kenyon, John Koolage, Judy Segal, Alan Richardson, Sylvia Berryman, Scott Edgar, Robyn Bluhm, Trey Boone, Jacqueline Sullivan, David Stump, Martin Davidson, Sacha Ludgate, Steeves Demazeux, Patrick Singy, and two anonymous reviewers for helpful comments and discussion. Earlier versions of this paper were presented at the Science and Society Speaker Series at the University of British Columbia in January 2009 and the fourth biennial conference of the Society for Philosophy of Science in Practice (SPSP) at the University of Toronto in June 2013; I am grateful for feedback that I received on those occasions.

## References

APA. (1952). *Diagnostic and statistical manual: Mental disorders*. Washington, DC: American Psychiatric Association.

APA. (1968). *Diagnostic and statistical manual of mental disorders, second edition: DSM-II*. Washington, DC: American Psychiatric Association.

APA. (1980). *Diagnostic and statistical manual of mental disorders, third edition: DSM-III*. Washington, DC: American Psychiatric Association.
APA. (2000). *Diagnostic and statistical manual of mental disorders, fourth edition text revision: DSM-IV-TR*. Washington, DC: American Psychiatric Association.
APA. (2013). *Diagnostic and statistical manual of mental disorders, fifth edition: DSM-5*. Washington, DC: American Psychiatric Association.
Bechtel, W., & Richardson, R. C. (1993). *Discovering complexity: Decomposition and localization as strategies in scientific research*. Princeton, NJ: Princeton University Press.
Blashfield, R. K. (1984). *The classification of psychopathology: Neo-Kraepelinian and quantitative approaches*. New York: Plenum Press.
Bolton, D. (2008). *What is mental disorder? An essay in philosophy, science, and values*. Oxford: Oxford University Press.
Boyd, R. (1985). Observations, explanatory power, and simplicity: Toward a non-Humean account. In P. Achinstein & O. Hannaway (Eds.), *Observation, experiment, and hypothesis in modern physical science* (pp. 47–94). Cambridge, MA: MIT Press.
Boyd, R. (1999). Homeostasis, species, and higher taxa. In R. A. Wilson (Ed.), *Species: New interdisciplinary essays* (pp. 141–185). Cambridge, MA: MIT Press.
Boyd, R. (2010). Realism, natural kinds, and philosophical methods. In H. Beebee & N. Sabbarton-Leary (Eds.), *The semantics and metaphysics of natural kinds* (pp. 212–234). New York: Routledge.
Brown, P. (1987). Diagnostic conflict and contradiction in psychiatry. *Journal of Health and Social Behavior, 28*(1), 37–50.
Compton, W. M., & Guze, S. B. (1995). The neo-Kraepelinian revolution in psychiatric diagnosis. *European Archives of Psychiatric and Clinical Neuroscience, 245*(4–5), 196–201.
Cooper, R. (2005). *Classifying madness: A philosophical examination of the Diagnostic and statistical manual of mental disorders*. Dordrecht: Springer.
Cooper, R. (2010). Are culture-bound syndromes as real as universally occurring disorders? *Studies in History and Philosophy of Biological and Biomedical Sciences, 41*(4), 325–332.
Craver, C. F., & Darden, L. (2001). Discovering mechanisms in neurobiology: The case of spatial memory. In P. K. Machamer, R. Grush, & P. McGlaughlin (Eds.), *Theory and method in the neurosciences* (pp. 112–137). Pittsburgh, PA: University of Pittsburgh Press.
Decker, H. S. (2007). How Kraepelinian was Kraepelin? How Kraepelinian are the neo-Kraepelinians?—from Emil Kraepelin to *DSM-III. History of Psychiatry, 18*(3), 337–360.
Decker, H. S. (2013). *The making of DSM-III: A diagnostic manual's conquest of American psychiatry*. Oxford: Oxford University Press.
Dupré, J. (1993). *The disorder of things: Metaphysical foundations of the disunity of science*. Cambridge, MA: Harvard University Press.
Feyerabend, P. K. (1965). Problems of empiricism. In R. G. Colodny (Ed.), *Beyond the edge of certainty: Essays in contemporary science and philosophy* (pp. 145–260). Englewood Cliffs, NJ: Prentice-Hall.
Feyerabend, P. K. (1975). *Against method: Outline of an anarchistic theory of knowledge*. London: New Left Books.
First, M. B., Pincus, H. A., Levine, J. B., Williams, J. B., Ustun, B., & Peele, R. (2004). Clinical utility as a criterion for revising psychiatric diagnoses. *American Journal of Psychiatry, 161*(6), 946–954.
Follette, W. C., & Houts, A. C. (1996). Models of scientific progress and the role of theory in taxonomic development: A case study of the *DSM. Journal of Consulting and Clinical Psychology, 64*(6), 1120–1132.
Frances, A. J., & Widiger, T. (2012). Psychiatric diagnosis: Lessons from the DSM-IV past and cautions for the DSM future. *Annual Review of Clinical Psychology, 8*, 109–130.
Giere, R. N. (2006). *Scientific perspectivism*. Chicago: University of Chicago Press.
Glennan, S. S. (1996). Mechanisms and the nature of causation. *Erkenntnis, 44*(1), 49–71.
Glennan, S. (2002). Rethinking mechanistic explanation. *Philosophy of Science, 69*(S3), S342-S353.

Goodman, N. (1983). *Fact, fiction, and forecast* (4th ed.). Cambridge, MA: Harvard University Press.

Hacking, I. (2013). Lost in the forest. Review of *DSM-5: Diagnostic and statistical manual of mental disorders, fifth edition* by the American Psychiatric Association. *London Review of Books, 35*(15), 7–8.

Hanson, N. R. (1958). *Patterns of discovery*. Cambridge: Cambridge University Press.

Hempel, C. G. (1965). Fundamentals of taxonomy. In C. G. Hempel, *Aspects of scientific explanation and other essays in the philosophy of science* (pp. 137–154). New York: The Free Press.

Horwitz, A. V. (2002). *Creating mental illness*. Chicago: University of Chicago Press.

Houts, A. C. (2001). The diagnostic and statistical manual's new white coat and circularity of plausible dysfunctions: Response to Wakefield, part 1. *Behavior Research and Therapy, 39*(3), 315–345.

Houts, A. C., & Follette, W. C. (1998). Mentalism, mechanisms, and medical analogues: Reply to Wakefield (1998). *Journal of Consulting and Clinical Psychology, 66*(5), 853–855.

Hyman, S. E. (2007). Can neuroscience be integrated into the DSM-V? *Nature Reviews Neuroscience, 8*(9), 725–732.

Hyman, S. E. (2010). The diagnosis of mental disorders: The problem of reification. *Annual Review of Clinical Psychology, 6*, 155–179.

Hyman, S. E., & Fenton, W. S. (2003). What are the right targets for psychopharmacology? *Science, 299*(5605), 350–351.

Insel, T. R., & Lieberman, J. A. (2013, May 13). DSM-5 and RDoC: Shared interests. *Joint press release by the National Institute of Mental Health and the American Psychiatric Association*.http://www.nimh.nih.gov/news/science-news/2013/dsm-5-and-rdoc-shared-interests.shtml

Insel, T., Cuthbert, B., Garvey, M., Heinssen, R., Pine, D. S., Quinn, K., Sanislow, C., & Wang, P. (2010). Research Domain Criteria (RDoC): Toward a new classification framework for research on mental disorders. *American Journal of Psychiatry, 167*(7), 748–751.

Jablenksy, A. (2012). The nosological entity in psychiatry: A historical illusion or a moving target? In K. S. Kendler & J. Parnas (Eds.), *Philosophical issues in psychiatry II: Nosology* (pp. 77-94). Oxford: Oxford University Press.

Kellert, S. H., Longino, H. E., & Waters, C. K. (Eds.). (2006). *Scientific pluralism: Minnesota studies in the philosophy of science* (Vol. 19). Minneapolis, MN: University of Minnesota Press.

Kendell, R. E. (1989). Clinical validity. *Psychological Medicine, 19*(1), 45–55.

Kendell, R., & Jablensky, A. (2003). Distinguishing between the validity and utility of psychiatric diagnoses. *American Journal of Psychiatry, 160*(1), 4–12.

Kendler, K. S. (1990). Toward a scientific psychiatric nosology: Strengths and limitations. *Archives of General Psychiatry, 47*(10), 969–973.

Kendler, K. S. (2009). An historical framework for psychiatric nosology. *Psychological Medicine, 39*(12), 1935–1941.

Kendler, K. S., & First, M. B. (2010). Alternative futures for the DSM revision process: Iteration *v.* paradigm shift. *British Journal of Psychiatry, 197*(4), 263–265.

Kendler, K. S., Zachar, P., & Craver, C. (2011). What kinds of things are psychiatric disorders? *Psychological Medicine, 41*(6), 1143–1150.

Kessler, R. C., Chiu, W. T., Demler, O., & Walters, E. E. (2005). Prevalence, severity, and comorbidity of 12-month DSM-IV disorders in the national comorbidity survey replication. *Archives of General Psychiatry, 62*(7), 617–627.

Khalidi, M. A. (2013). *Natural categories and human kinds: Classification in the natural and social sciences*. New York: University of Cambridge Press.

Kirk, S. A., & Kutchins, H. (1988). Deliberate misdiagnosis in mental health practice. *Social Service Review, 62*(2), 225–237.

Kirk, S. A., & Kutchins, H. (1992). *The selling of DSM: The rhetoric of science in psychiatry*. New York: Aldine de Gruyter.

Kleinman, A. (1988). *Rethinking psychiatry: From cultural category to personal experience*. New York: The Free Press.

Klerman, G. L. (1978). The evolution of scientific nosology. In J. C. Shershow (Ed.), *Schizophrenia: Science and practice* (pp. 99–121). Cambridge, MA: Harvard University Press.

Kuhn, T. S. (1962). *The structure of scientific revolutions*. Chicago: University of Chicago Press.

Kupfer, D. J., & Regier, D. A. (2011). Neuroscience, clinical evidence, and the future of psychiatric classification in *DSM-5. American Journal of Psychiatry, 168*(7), 672–674.

Kupfer, D. J., First, M. B., & Regier, D. A. (2002). Introduction. In D. J. Kupfer, M. B. First, & D. A. Regier (Eds.), *A research agenda for DSM-V* (pp. xv–xxiii). Washington, DC: American Psychiatric Association.

Kupfer, D. J., Kuhl, E. A., & Regier, D. A. (2013). *DSM-5*—the future arrived. *Journal of the American Medical Association, 309*(16), 1691–1692.

Kutschenko, L. K. (2011). How to make sense of broadly applied medical classification systems: Introducing epistemic hubs. *History and Philosophy of the Life Sciences, 33*(4), 583–602.

Longino, H. E. (1990). *Science as social knowledge: Values and objectivity in scientific inquiry*. Princeton, NJ: Princeton University Press.

Longino, H. E. (2002). *The fate of knowledge*. Princeton, NJ: Princeton University Press.

Machamer, P. (2004). Activities and causation: The metaphysics and epistemology of mechanisms. *International Studies in the Philosophy of Science, 18*(1), 27–39.

Machamer, P., Darden, L., & Craver, C. F. (2000). Thinking about mechanisms. *Philosophy of Science, 67*(1), 1–25.

Mayes, R., & Horwitz, A. V. (2005). DSM-III and the revolution in the classification of mental illness. *Journal of the History of the Behavioral Sciences, 41*(3), 249–267.

Meehl, P. E. (1995). Bootstrap taxometrics: Solving the classification problem in psychopathology. *American Psychologist, 50*(4), 266–275.

Misés, R., & Quemada, N. (2002). *Classification Française des troubles mentaux de l'enfant et de l'adolescent: R-2000*. Paris: CTNERHI Éditions.

Mitchell, S. D. (2003). *Biological complexity and integrative pluralism*. Cambridge, UK: Cambridge University Press.

Morris, S. E., & Cuthbert, B. N. (2012). Research domain criteria: Cognitive systems, neural circuits, and dimensions of behavior. *Dialogues in Clinical Neuroscience, 14*(1), 29–37.

Murphy, D. (2006). *Psychiatry in the scientific image*. Cambridge, MA: MIT Press.

Pincus, H. A. (2012). DSM-IV: Context, concepts and controversies. In K. S. Kendler & J. Parnas (Eds.), *Philosophical issues in psychiatry II: Nosology* (pp. 145–160). Oxford: Oxford University Press.

Quine, W. V. (1969). Natural kinds. In W. V. Quine, *Ontological relativity and other essays* (pp. 114-138). New York: Columbia University Press.

Regier, D. A. (2008). Forward: Dimensional approaches to psychiatric classification. In J. E. Helzer, H. C. Kraemer, R. F. Krueger, W. Hans-Ulrich, P. J. Sirovatka, & D. A. Regier (Eds.), *Dimensional approaches in diagnostic classification: Refining the research agenda for DSM-V* (pp. xvii–xxiii). Washington, DC: American Psychiatric Association.

Regier, D. A., Narrow, W. E., Kuhl, E. A., & Kupfer, D. J. (2009). The conceptual development of DSM-V. *American Journal of Psychiatry, 166*(6), 645–650.

Robins, E., & Guze, S. B. (1970). Establishment of diagnostic validity in psychiatric illness: Its application to schizophrenia. *American Journal of Psychiatry, 126*(7), 983–987.

Sadler, J. Z. (2005). *Values and psychiatric diagnosis*. Oxford: Oxford University Press.

Sadler, J. Z. (2013). Considering the economy of DSM alternatives. In J. Paris & J. Phillips (Eds.), *Making the DSM-5: Concepts and controversies* (pp. 21–38). New York: Springer.

Sanislow, C. A., Pine, D. S., Quinn, K. J., Kozak, M. J., Garvey, M. A., Heinssen, R. K., Wang, P. S., & Cuthbert, B. N. (2010). Developing constructs for psychopathology research: Research domain criteria. *Journal of Abnormal Psychology, 119*(4), 631–639.

Schwartz, M. A., & Wiggins, O. P. (1986). Logical empiricism and psychiatric classification. *Comprehensive Psychiatry, 27*(2), 101–114.

Shaffner, K. F. (2012). A philosophical overview of the problems of validity for psychiatric disorders. In K. S. Kendler & J. Parnas (Eds.), *Philosophical issues in psychiatry II: Nosology* (pp. 169–189). Oxford: Oxford University Press.

Spitzer, R. L., & Endicott, J. (1978). Medical and mental disorder: Proposed definition and criteria. In R. L. Spitzer & D. F. Klein (Eds.), *Critical issues in psychiatric diagnosis* (pp. 15–39). New York: Raven.

Spitzer, R. L., Sheehy, M., & Endicott, J. (1977). DSM-III: Guiding principles. In V. M. Rakoff, H. C. Stancer, & H. B. Kedward (Eds.), *Psychiatric diagnosis* (pp. 1–24). New York: Brunner/Mazel.

Suppes, P. (1978). The plurality of science. In P. Asquith & I. Hacking (Eds.), *PSA 1978: Proceedings of the 1978 biennial meeting of the Philosophy of Science Association* (Vol. 2, pp. 3–16). East Lansing, MI: Philosophy of Science Association.

Tabb, K. (2015). Psychiatric progress and the assumption of diagnostic discrimination. *Philosophy of Science, 82*(5).

Tabery, J. G. (2004). Synthesizing activities and interactions in the concept of a mechanism. *Philosophy of Science, 71*(1), 1–15.

Task Force, O. P. D. (Ed.). (2008). *Operationalized psychodynamic diagnosis OPD-2: Manual of diagnosis and treatment*. Cambridge, MA: Hogrefe & Huber Publishers.

Tsou, J. Y. (2003). Reconsidering Feyerabend's 'anarchism'. *Perspectives in Science, 11*(2), 208–235.

Tsou, J. Y. (2007). Hacking on the looping effects of psychiatric classifications: What is an interactive and indifferent kind? *International Studies in the Philosophy of Science, 21*(3), 446–470.

Tsou, J. Y. (2011). The importance of history for philosophy of psychiatry: The case of the DSM and psychiatric classification. *Journal of the Philosophy of History, 5*(3), 446–470.

Tsou, J. Y. (2012). Intervention, causal reasoning, and the neurobiology of mental disorders: Pharmacological drugs as experimental instruments. *Studies in History and Philosophy of Biological and Biomedical Sciences, 43*(2), 542–551.

Tsou, J. Y. (2013). Depression and suicide are natural kinds: Implications for physician-assisted suicide. *International Journal of Law and Psychiatry, 36*(56), 461–470.

Wakefield, J. C. (1998). The *DSM*'s theory-neutral nosology is scientifically progressive: Response to Follette and Houts (1996). *Journal of Consulting and Clinical Psychology, 66*(5), 846–852.

Wakefield, J. C. (1999). Philosophy of science and the progressiveness of the DSM's theory-neutral nosology: Response to Follette and Houts, part 1. *Behavior Research and Therapy, 37*(10), 1001–1027.

Wakefield, J. C. (2001). The myth of DSM's invention of new categories of disorder: Hout's diagnostic discontinuity thesis disconfirmed. *Behavior Research and Therapy, 39*(5), 575–624.

Whooley, O. (2010). Diagnostic ambivalence: Psychiatric workarounds and the Diagnostic and statistical manual of mental disorders. *Sociology of Health & Illness, 32*(3), 452–469.

Widiger, T. A., & Clark, L. A. (2000). Toward DSM-V and the classification of psychopathology. *Psychological Bulletin, 126*(6), 946–963.

Wilson, M. (1993). DSM-III and the transformation of American psychiatry: A history. *American Journal of Psychiatry, 150*(3), 399–410.

# DSM-5: The Delayed Demise of Descriptive Diagnosis

**Stuart A. Kirk, David Cohen, and Tomi Gomory**

**Abstract** In 1980, DSM-III adopted a descriptive approach to psychiatric diagnosis, creating checklists of unwanted behaviors to define and use as required criteria when posing each of several hundred diagnoses. The objective of this novel approach was to validate psychiatry as a scientifically legitimate branch of medicine, by enabling research into hopefully homogeneous groups of patients to pinpoint the implicit hypothesized physiological causes of the disorders the patients were presumably sharing. In each subsequent revision of the DSM including the DSM-5, however, no physiological criteria of any sort are included for any diagnosis, confirming the empirical failure of this attempt to substantiate the medical model of madness. The futile endeavor to validate countless human faults and suffering as medical diseases explains most of the "scientific" conundrums and controversies surrounding the release of DSM-5, including whether to include or exclude diagnoses, where to draw boundaries for each, and why clinicians still fail to agree on which diagnosis they should apply in a given case. Despite DSM-5's insolvency, the essentially moral project of descriptive psychiatric diagnosis has today vast socio-economic ramifications that help to preserve it.

---

To appear in *The DSM-5 in Perspective: Philosophical Reflections on the Psychiatric Babel*, edited by Steeves Demazeux & Patrick Singy, in series *History, Philosophy and Theory of the Life Sciences*, Springer. Some parts of this chapter are adapted with permission from material in *Mad Science: Psychiatric Coercion, Diagnosis, and Drugs*, Transaction Publisher, 2013.

S.A. Kirk (✉) • D. Cohen
University of California, Los Angeles, CA, USA
e-mail: kirk@ucla.edu

T. Gomory
Florida State University, Tallahassee, FL, USA

 
S. Demazeux, P. Singy (eds.), *The DSM-5 in Perspective*, History, Philosophy and Theory of the Life Sciences 10, DOI 10.1007/978-94-017-9765-8_4

## Introduction

In 2013, the American Psychiatric Association released the fifth edition[1] of its best-seller, the *Diagnostic and Statistical Manual of Mental Disorders* (DSM-5). In development for nearly 15 years, the DSM-5 was due for a celebration—as routinely signaled during this time by those spearheading the effort—as another major scientific advance in understanding and diagnosing madness (called mental illness for about a century). The architects of DSM-5 aspired not only to revise the manual, but also to restructure and improve psychiatric diagnosis by perhaps using a dimensional rather than a categorical assessment approach and by incorporating the latest genetic and biochemical advances in identifying mental illness (Kupfer et al. 2002). As it turned out, the release of DSM-5 was accompanied neither by applause nor acknowledgment of its scientific achievements but by a virtual public funeral for the descriptive diagnosis that it represents. In purpose, assumptions and method, DSM-5 shared the genome of its predecessors, an inheritance that should have proved to be fatal. Yet, one should not issue a premature report of the death of biological psychiatry (Torrey 1970), as psychiatric diagnosis has never rested on nor required scientific validation.

The celebrated introduction of descriptive diagnosis came with DSM-III in 1980. That manual launched an approach to diagnosis that supposedly avoided theoretical assumptions about the nature or causes of mental illness[2] by relying on novel checklists of presumably low inference observable behaviors and reported experiences and feelings making up the required criteria to define each of about 200 mental disorders. This so-called atheoretical descriptive diagnosis was invented because the professional credibility of psychiatry historically was profoundly undermined: no convincing scientific evidence existed about the medical causes of, or biological markers associated with, any form of madness. Nor was there evidence that the classification system met any minimal scientific standards. The architects of DSM-III justified descriptive diagnosis by arguing that it would make psychiatric diagnoses more reliable. Increased reliability would provide reassurance that diagnosed people could be placed into homogeneous groups differing from groups of normal people, which would permit researchers to pinpoint eventually the hypothesized biological or psychological causes of madness.

Although atheoretism suggested a justified uncertainty about the causes of madness, the American Psychiatric Association (APA) and legions of psychiatrists and neuropsychiatrists since at least Wilhem Griesinger (1817–1868) and Emil Kraepelin (1856–1926) had proclaimed mental illnesses to be the *manifestations of physiological abnormalities*. The ultimate objective of descriptive diagnosis was presumably to validate mental illness as a family of physical illnesses or at least to appear to be in pursuit of such validation, thereby supporting psychiatry as a

[1]If one counts DSM-III-R (1987) and DSM-IV-TR (2000), DSM-5 is the seventh edition.

[2]This is a term of art with which we disagree and discuss in detail in Kirk et al. (2013).

legitimate branch of modern (biological, that is, organ, cell, and molecule based) allopathic medicine. The scientific effort failed: after countless studies into the possible biological bases of mental disorders diagnosed using DSM-III (APA 1994) and its successors, no revision including DSM-5 could incorporate a single biological criterion as a reliable marker for the existence of, let alone the cause for, any mental disorder.[3] Even the limited goals of establishing diagnostic reliability and grouping clients into meaningfully homogeneous groups have not been achieved. In other research arenas, such a catastrophic failure would mean a turn toward new testable explanatory hypotheses for the phenomena of interest. In the unfalsifiable psychiatric research program, the APA continues to assert, as in the Introduction to DSM-5 (APA 2013), that "DSM is a medical classification of disorders . . . " (p. 10).

Asserting that a vast array of willful, irrational, and banal human behaviors are medical diseases is a scientific conceit,[4] but it also reflects the DSMs' astounding political and economic impact in spearheading the medicalization of ordinary life in the latter part of the twentieth century. The political nature of the DSM-5 is revealed in that virtually all of the controversies, conundrums, and conflicts that were raised in its development—e.g., to include a new or exclude an old diagnosis, to insert a new or remove an old criterion, to modify or leave unchanged an existing phrase—were about where to draw subjective boundaries guiding the allocation of public and private resources to *manage madness* in our exceedingly complex society. The debates erupted because of contemporary psychiatry's unflagging continuation of its historical effort to reformulate human acts, language, cognitions, emotions

[3]As its predecessors, DSM-5 states in several places that probable or definitive biomarkers for mental disorders or illnesses are lacking. Nonetheless, it does includes as "mental disorders" a number of diagnoses—for dementias, for example—whose criteria include evidence for a specified pathophysiological condition known to cause the presenting problem. Despite psychiatry's historical role in managing senile older adults and others with organic psychoses and with degenerative brain diseases, the presence of these diagnoses is scientifically intriguing, as not all physical conditions with spectra of presenting psychological problems (e.g., epilepsy, diabetes) are so treated in the manual. Moreover, throughout the descriptive text that follows the criteria for each diagnostic category, sub-sections titled "Diagnostic Markers" or "Risk and Prognostic Factors" are frequently included. The relevant sub-section for Obsessive-Compulsive Disorder, for example, states the following: "Familial transmission is due in part to genetic factors (e.g., a concordance rate of 0.57 for monozygotic vs. 0.22 for dizygotic twins). Dysfunction in the orbitofrontal cortex, anterior cingulate cortex, and striatum have been most strongly implicated" (APA 2013, p. 240). Since the DSM-5, like its predecessors, lists no references, these assertions (all questionable and publicly questioned in their assumptions, determinations, and interpretations) arguably constitute pseudo-scientific adornments.

[4]The medical illness metaphor is a conceit in that it states as fact that which has not yet been scientifically validated, namely that the behaviors defined as symptoms of mental illnesses are caused by biological dysfunction. The assertion that mental illnesses are physical illnesses has the powerful advantage of borrowing the language and causal model of biological disease to buttress the reputation of psychiatry, while obscuring the absence of any accepted scientific evidence for that claim. The "common vocabulary" that the DSM provides is that of the assumption and statement of illness, backed by the biomedical-industrial complex. The justification that "the DSM at least provides a common vocabulary" ignores (or exploits) the power of language to conquer rhetorically what cannot be validated empirically.

and other bio-psychological adaptations into disease entities that can be described impersonally, without reference to individual human contexts and circumstances. This has always been and remains a moral project, and a morally flawed one at that (Jacobs and Cohen 2012), but there is no clear sign that the public or most experts recognize it as such.

## Converting Moral Judgments into Disease

The psychiatric literature contains no coherent conception of madness as a medical disease, nor, for that matter, an explicit conception of normal behavior against which abnormality could be contrasted (Bolton 2008; Gomory et al. 2013; McNally 2011; Szasz 1994). The confusions are manifest in most recent official attempts at synthesis. The first APA definition of mental disorder appeared in DSM-III as a "syndrome that occurs in an individual" and represents some sort of "dysfunction" (APA 1980, p. 6), but without sufficient clarity about the nature of dysfunction except that it is "associated with ... significant impairments." Twenty years later, psychiatric researchers (Rounsaville et al. 2002) working under APA auspices in preparation for DSM-5 concluded that the 146 word definition of *mental disorder* offered in DSM-IV-TR is not "cast in a way that allows it to be used as a criterion for deciding what is and is not a mental disorder" (p. 3). They noted the existence of a "contentious" debate among the academic experts about "whether *disease, illness, and disorder* are scientific biomedical terms or are sociopolitical terms ..." (p. 3, emphasis in original). They admitted that the contemporaneous DSM "is based on a falsely optimistic assumption: that psychiatric disorders are discrete biomedical entities ..." (p. 8). Another 8 years later, leading psychiatrists working on DSM-5 were compelled to repeat these conclusions for the current long list of disorders. Stein and colleagues (2010) conceded that the "clinically significant" DSM criterion for mental disorders is tautological, that boundaries between normal and pathological are contentious, and that the concept of dysfunction involves speculative assumptions about the purpose of human evolution. Allen Frances, who as chair of the DSM-IV Task Force presided over the largest expansion of the mental disorder vocabulary and population before he became the most vocal denouncer of DSM-5, plainly summarized the definitional difficulties: "There is no definition of a mental disorder. It's bullshit. I mean, you just can't define it" (cited in Greenberg 2011). For its part, the National Institute of Mental Health (NIMH 2008) continued to declare that "Mental disorders are brain disorders."

Psychiatry's inability to define mental disorder coherently continues in the DSM-5, where each disorder is still defined by its "diagnostic criteria," the checklists of behaviors, emotions and cognitions—devoid of any contextual elements, as if they were objective signs of bodily dysfunctions (see Jacobs and Cohen 2010)—that must be observed or reported to determine if a mental disorder is present. Here are some examples among hundreds: restlessness or being on edge; goal directed behavior; irritability and muscle tension; problems with concentration;

difficulty discarding possessions; delayed or premature ejaculation for men and reduced interest in sexual activity or infrequent orgasm for women; lack of interest in eating or, conversely, eating too much food; preoccupation with having or acquiring a serious illness; fear of dying. DSM-5 contains over 60 pages describing problems with not sleeping well or appropriately. Among all editions of DSM, only DSM-III-R (APA 1987) included a 28-page "symptom index" of disorders sharing the same symptom. For example, insomnia is listed as a symptom of 33 and profuse sweating of 18 different disorders. There is little doubt that these thoughts, feelings and behaviors are "real" and that many people commonly experience them as unpleasant, disappointing, or distressing and turn to physicians or others for counsel or advice to manage them or make them disappear. The issue is whether the behaviors or complaints are best defined as "symptoms" of underlying mental disorders, as implied in a "medical classification."

What is the likelihood that such behaviors, singly or in small combinations, constitute "symptoms" logically belonging to a conceptual category "mental disorder," implying that they share a common pathological origin in "brain disease," "internal dysfunction," or "emotional disturbance"? The foundational, neo-Kraepelinian claim of DSM-5 and its predecessors is that all the diagnostic criteria point to underlying medical conditions (Klerman 1978). That is why they are categorized together in a manual of mental disorders. But how likely is it that all of the huge number of diverse behaviors that DSM-5 declares are "symptoms" of medical pathology are not better explained or partially explained by normal human variations, variations in social contexts, patterns of learning, individual motivations and so forth—the province of the social, not medical, sciences? To make the medical pathology argument, as the DSM and much of psychiatry does, in the absence of at least strongly compelling evidence, is psychiatric hubris. Little wonder, then, that definitions of mental disorder are confused and tautological. The semantic category of mental illness and its clone mental disorder is a conceptual black hole, a linguistic sleight of hand, a reification of a medical metaphor as first described by Thomas Szasz (1961). The notion funnels an exceedingly heterogeneous group of people, behaviors, and feelings into a "disjunctive category" (Bruner et al. 1986) of people deemed to require attention from mental health professionals. As we discuss in Kirk et al. (2013):

> [B]efore anything else, madness is a word. Like all words, it is a human artifact stitched together to represent or echo something (abstract or concrete) related to human behavior or one's perception of this behavior. As a linguistic sign, *madness* becomes available for our critical manipulation, but like all linguistic signs it need not be anchored to particular aspects of the material world. The word *trees,* for example, does not directly locate any specific tree in the observable world, nor does it specify what the particular limiting attributes of the categorical term trees are. It serves as a label for collecting many disparate types of plants judged to belong under the category 'trees,' though these plants may share few common elements other than their category name (for example banana trees and oak trees). Such a category might best be called 'disjunctive.' (p. 39)

Since the DSM is unable to document scientifically that these diverse behaviors represent any medical or mental disease, its endurance may be reasonably attributed to other powerful social purposes.

## Incoherent Definitions Produced Scientific Failures

The incoherence of DSM-5's attempts to define madness as disease is revealed, first, in the continuing failure of the diagnostic criteria to create reliable categories. Descriptive diagnosis was intended to settle the contested meaning of madness by positing necessary and sufficient criteria for each category of disorder. The primary distinguishing characteristic of all the DSM editions since DSM-III was to offer lists of explicit criteria, which were expected to greatly improve the scientific and practical problem of inter-rater reliability. The lists generally have failed to do so for 30 years, despite grand pronouncements of "far greater" reliability (Kirk and Kutchins 1992; Kirk et al. 2013). The makers of DSM-5 also sought to improve reliability and conducted field trials to measure it. Just prior to the release of their reliability data at the 2012 APA Annual Meeting, and apparently aware that the data would be damaging (Kraemer et al. 2012; Spitzer et al. 2012), they issued a confusing and convoluted statement arguing that reliability standards for DSM-5 should be *lower* than those established for DSM-III (Brauser 2012; see Kirk et al. 2013 Chap. 5). Only a few diagnostic categories obtained a kappa score[5] above 0.50 (considered in the barely acceptable range by the pre-DSM-III neo-Kraepelinian researchers). Some popular categories, major depressive disorder included, obtained kappas in the 0.20–0.30 range. According to Jacobs and Cohen (2012), "The [DSM-IV] 'Limitations' section admits that discrete entities are fictions necessitated by the medical framework; absent validated conditions, inter-clinician agreement resembles indoctrination more than scientific progress" (p. 89).

If the diagnostic criteria do not establish valid, or even reliable, categories of disease that "cut nature at its joints," DSM-5 fails as a manual of scientific classification. The literature is replete with evidence for the questionable validity of the diagnostic criteria for many mental disorder categories, ranging from conduct disorder (Hsieh and Kirk 2003), to PTSD (Barglow 2012), major depressive disorder (Horwitz and Wakefield 2007), ADHD (Cohen 2006), general anxiety disorder (Horwitz and Wakefield 2012), and many more. Even if no questions were raised about the lists of troubled behaviors and emotions claimed as symptoms of pathology, one cannot establish the validity of a medical illness unless they are linked empirically to a demonstrable physiological dysfunction that can be reasonably believed to *cause* the behaviors and emotions. For example, clinicians say to parents that their "child is inattentive because he has ADHD," implying that some underlying disease (ADHD) causes the child's inattention. But if asked for the evidence of the existence of the presumed disease, the clinician will offer the

[5]Kappa is a chance-corrected measure of reliability, that is, inter-judge agreement or consistency. Kappa coefficients may range from 0.0 to 1.0, from no agreement to perfect agreement, respectively. The developers of DSM-III were the first to use kappa as a measure of diagnostic consistency. They considered kappa scores of .70 or higher to be satisfactory or good and those around .50 or lower to be no better than fair or poor (see Kirk and Kutchins 1992).

child's inattention as evidence. Thus, the child is inattentive because of ADHD and he has ADHD because he is inattentive. Descriptive diagnosis offers tautology in the guise of scientific explanation. A genuine medical diagnosis, by contrast, would offer a non-tautological statement such as the "child is inattentive because he has an identified brain lesion," a form of causal explanation. Unfortunately, DSM offers behavioral diagnostic criteria as if they confirm the existence of a valid medical condition. This is just as true for behaviors that largely overlap with normal or unproblematic behaviors if the social context is accommodating (e.g., ADHD) as it is for behaviors suggesting a "human breakdown" or a rupture with consensual reality given the behaviors' unacceptability and inexcusability across a wide variety of contexts (e.g., schizophrenia).

Even a cursory examination reveals that the diagnostic criteria themselves (independent of their implied causal connections to the "disorder") are flawed (for details, see Kirk et al. Chap. 5). They are ambiguous (e.g., what exactly is "excessive anxiety" in general anxiety disorder?). They are often redundant (e.g., "difficulty sustaining attention" and "easily distracted" are two supposedly distinct criteria of ADHD). They ignore the interpersonal, familial, social and cultural context of a person's behavior. Arbitrary and shifting numbers of criteria for every disorder and expanded age ranges for the onset for some, such as ADHD, increase the proportion of people who can be labeled as mentally disordered. As the DSM-III to DSM-IV-TR editions recognized in their Limitations section, no category could be assumed to group people who share even the key criteria used to define the category (Jacobs and Cohen 2004). Co-morbidity among categories, with as many as up to half the people diagnosed qualifying for multiple disorders, undermine the assumption of discrete categories of illness (Mirowsky and Ross 1990).

## DSM-5 and Its Controversies

With so many fundamental and therefore persistent problems with descriptive diagnosis through each new edition, it was predictable that public controversies would emerge during the development and launch of DSM-5. Almost all revolved around the science-mimicking definitions of mental disease embedded in the diagnostic criteria. Prior editions of DSM spawned well-known public disputes over the validity of diagnostic categories, including the gay community pressuring the APA to remove homosexuality as a mental disorder (Bayer 1981; Kutchins and Kirk 1997) and women fighting against masochistic (aka self-defeating) personality disorder (Caplan 1995) which appeared to define some expected gender role behaviors as symptoms of mental disorder. In both these instances, people who might have been subject to being labeled as mentally ill politically fought against having their behaviors defined as illnesses and add to their social or legal oppression. They won, not simply because the makers of the DSM had no convincing evidence that these were medical illnesses, since that rationale could be extended across the diagnostic manual, but because people mustered enough social and political capital among increasingly acceptable constituencies (gays, women, and their supporters) to influence what are essentially non-medical decisions about the artificial boundaries of mental disorders.

The struggles over DSM-5 again exposed the lack of scientific evidence underlying psychiatric diagnoses, but the disputes now were broader and more critical of the assumptions underlying the DSM. They attacked direct and indirect financial conflicts of interest of people and organizations developing the manual. They recognized that the DSM could be viewed mainly as an insurance codebook—offering entitlements, legitimacy and revenues to interests groups—rather than a scientific or medical compendium. Finally, some of those potentially diagnosable and their families argued to be *included* in, not excluded from, the manual.

## Attacks from Outsiders and Insiders

Attacks on the DSM-5 came from prominent insiders, newspaper columnists, professional organizations, psychiatric experts and concerned families. One early stinging criticism, 5 years before the manual's release, came from Robert Spitzer (2008), the godfather and architect of DSM-III and of descriptive diagnosis. Spitzer posted an open letter accusing the DSM-5 Task Force of revising the manual in enforced secrecy, which, he argued correctly, undercut the very "foundation of science." Naturally, his rebuke garnered wide media coverage (Carey 2008; Grossman 2008; Lane 2008). As early drafts of the manual were eventually released online, mental health professional organizations questioned the entire medical orientation of the DSM. The Society for Humanistic Psychology (2012) of the American Psychological Association circulated online a ten-page detailed critique and petition against the DSM-5. The American Counseling Association (2012) publicized its own concerns. The British Psychological Society (2011) distributed a detailed document questioning the "medicalization of . . . natural and normal responses" of people, "which do not reflect illnesses so much as normal individual variations." Further, it argued that DSM-5's putative diagnoses are based largely on social norms, with "symptoms" that are subjective judgments, with no confirmatory physical "signs" or evidence of biological causation. The British Psychological Society was here clearly espousing a Szaszian stance, something extremely few if any professional organizations had ever done up to this point. It noted that the entire manual is plagued by problems of reliability, validity, prognostic value, and co-morbidity (p. 2). Even the prestigious medical journal *Lancet* (2012) joined the fray, publishing an editorial calling the medicalization of grief in the forthcoming DSM-5 "not only dangerously simplistic, but also flawed."

Few critical voices were as unexpected or public as that of Allen Frances, the former leader of the Task Force for DSM-IV. Frances (2009), after a surprising personal mea culpa for being part of the DSM-IV efforts that loosened diagnostic criteria and created "false epidemics of mental illnesses," began a persistent attack on nearly every central aspect of DSM-5, its goals, methods and products, its weak methodology, its claims to a "paradigm shift." He reminded his colleagues that "there can be no dramatic improvements in psychiatric diagnosis until we make a fundamental leap in our understanding of what causes mental disorders" (2009). He clearly stated the "disappointing fact that not even one biological test is ready for inclusion in the criteria sets of DSM-5" (2009). In regular blogs appearing in the online *Psychiatric Times* (available at www.psychiatrytimes.com) and another series entitled "DSM5 in Distress" for *Psychology Today* (available at www.psychologytoday.com) he relentlessly criticized the DSM-5 Task Force, dozens of proposals for changes in the manual, and the APA for ignoring or mishandling the rising chorus of criticisms. In March 2012, he wrote: "DSM-5 has suffered from a fatal combination of excessive ambition, sloppy method, and closed process. It fully deserves the concerted opposition it has generated from 47 professional organizations, the world press, the Society of Biological Psychiatry, *The Lancet,* and the general public" (Frances 2012). As DSM-5 was released, he published a summary of his arguments in book form (Frances 2013a).

Frances' goals in his critique of DSM-5 have remained unclear, given that his stance is not free of contradictions. He has questioned virtually every technical aspect of DSM-5, but, while recognizing the weak validity of categories like *schizophrenia*, has affirmed the power of psychiatrists *qua* physicians to diagnose it and to incarcerate "dangerous" persons (Francis 2012a).[6] Also, apparently relying on professional status over priority, Frances has not generally credited any of the numerous critics of the DSM approach who, since the late 1970s, trailblazed his path.

## Financial Conflicts of Interest

In prior editions of DSM, the only question raised about financial conflicts of interest was whether the APA's rapid revisions of the manual occurred mainly to generate revenues for its publishing arm. With DSM-5, however, more serious conflicts were exposed (Cosgrove and Krimsky 2012; Cosgrove et al. 2006; Kirk et al. 2013). Investigations found that a higher proportion of psychiatrists than other medical specialists received gifts from drug companies (Harris 2007; Graham 2010). Congressional investigations and the media exposed leading psychiatrists, but not necessarily those directly involved with DSM-5, as having hidden ties to Big Pharma—such as Charles Nemeroff, P. Trey Sunderland, Frederick Goodwin, and Joseph Biederman (Harris 2008a, b; Wilson 2006; Harris and Carey 2008). Medical and psychiatric journals had failed to prevent the publishing of articles

[6]Frances (2012a) writes that "mental disorders are constructs, nothing more but also nothing less. Schizophrenia is certainly not a disease; but equally it is not a myth. As a construct, schizophrenia is useful for purposes of communication and helpful in prediction and decision making — even if ... the term has only descriptive, and not explanatory, power."

ghostwritten by drug company employees (Wilson 2009). There was widespread concern that drug companies had corrupted medical research (Angell 2004; Barber 2008; Whitaker 2010; Healy 2012), particularly in psychiatry.

Against this backdrop, the revelation that most psychiatrists on the DSM-IV and DSM-5 task forces (Cosgrove and Krimsky 2012; Cosgrove et al. 2006) had ties to drug companies exacerbated the concern. Conflicts of interest are troubling in this case precisely because without objective markers for a mental disorder, the process to define it consists ultimately of hand-picked researchers and clinicians tinkering with arbitrary diagnostic criteria. Upjohn Pharmaceuticals was closely involved in virtually every step leading to the creation of "panic disorder" in the late 1970s to promote sales of their new drug alprazolam, including funding the first psychiatric conference to discuss this creation (Lane 2008). The former editor of the *New England Journal of Medicine*, Marcia Angell, has written (Angell 2009) "drug companies have perfected a new and highly effective method to expand their markets. Instead of promoting drugs to treat diseases, they have begun to promote diseases to fit their drugs," a process that is known as disease mongering (for examples, see Angell 2009, 2011; Healy and Le Noury 2007). This confluence of drug industry influence on key decision-makers, arbitrary and invalid definitions of mental illness, and enormous profits at stake from sales of drugs advertised as disorder-specific, heightened outside scrutiny. With each edition of DSM, it became more evident to more people that what had evolved from a small doctor's informational codebook had become a financial instrument for distributing taxpayer funds to numerous corporations, professions and agencies and providing mental health services. But while the scientific status of descriptive diagnosis was increasingly being questioned by critics, its social and economic influence was expanding among providers and consumers. Some consumers were now fighting to be included in the manual. These conflicting trends mirror similar divergences in other fields, such as energy, transportation, or food production, where growing recognition that a dominant approach has severe limitations coexists with renewed efforts by some interested parties to exploit it to the end.

## Seeking and Expanding Medical Entitlements

As with previous editions, various proposals for DSM-5 would expand the number of people who could be labeled as having mental disorders by either loosening the existing criteria or creating new categories of illness. These definitional proposals would qualify more people for insurance coverage or other entitlement programs (e.g. disability, Supplemental Security Income). In the professional literature, on websites, and in the traditional media there were debates about the shifting boundaries for the paraphilias, PTSD, psychotic risk disorder, ADHD, eating disorders and others.

One DSM-5 expansion allowing people grieving for loved ones to receive a diagnosis was particularly contentious, since it so clearly incorporated a completely normal reaction to loss. In prior editions, criteria for Major Depressive Episode (APA 2000, p. 356) acknowledged that "symptoms" of depression could result from the loss of a loved one but that normal bereavement should be excluded from mental disorders—a rare acknowledgment by the developers of the manual that an impairment had to be contextualized. The elimination of the "bereavement exclusion" conveniently justified a new market for the pharmaceutical industry. The media and some experts immediately challenged this proposal on various grounds (Carey 2012a; Frances 2010; Wakefield and First 2012).

Some patient groups and their families also argued against a DSM-5 proposal *to narrow* the criteria for Asperger's syndrome, a "disorder" that DSM-IV had defined as a mild form of autism. Like childhood bipolar disorder, the likelihood that children were diagnosed with autism, defined as a pervasive developmental disorder, had been increasing very rapidly for at least a decade, without clarifying evidence about whether this was due to increasing incidence, better surveillance by clinicians, or the inclusion of formerly normal children in the diagnosis due to DSM's broadened definition (Carey 2012c; New York Times Editorial 2012; Steinberg 2012; Zarembo 2011). The DSM-5 Task Force (2012) proposed that the definition of *autism* be narrowed to avoid defining socially awkward children as autistic. A battle erupted, that played out in blogs and newspapers across the country (e.g. Nugent 2012), pitting those questioning why social awkwardness, whatever its disadvantages, should be considered as a serious mental disorder against those concerned that restricting boundaries of autism would disqualify some children for state-funded education and support. The DSM-5 Task Force replied that the new definition would not exclude many people. This controversy highlighted that defining any behaviors as symptoms of a mental disorder provides a medical rationale for providing *services*. Drawing the arbitrary boundaries for mental disorders controls the allocation of billions of dollars in health, educational, and social services (Carey 2012b). In these controversies, and in many others, what we witness are not debates over whether sufficient scientific evidence validates certain "symptoms" of medical pathology, but DSM's mystification of moral judgments about unwanted behaviors.

## Grappling with Failure

Critics of the DSM have pointed out these serious problems for decades.[7] Even Allen Frances conceded that "diagnosis and the use of psychotropic drugs have both gotten out of hand . . . " and that drug companies exploited the DSM by selling "... the idea that problems of everyday living are really mental disorders, caused

[7] A sample of these critics from various disciplines would include Boyle (2002), Dumont (1984), Eysenck (1986), Kirk and Kutchins (1992), Maj (2005), Mirowsky and Ross (1989), Mishara (1994), Sadler et al. (1994), Sarbin (1997), Szasz (1994), Tucker (1998), and Wakefield (1992).

by a chemical imbalance and cured with a pill" (Frances 2013b). Although the psychiatric establishment grappled with these concerns, at the issuance of every edition of the manual they tout that the DSM criteria "are the best available description of how mental disorders are expressed and can be recognized..." and that the manual is "...an essential educational resource for students and practitioners, and a reference for researchers in the field" (APA 2013, p. xii). As DSM-5 was being released, however, the cheery news of psychiatric progress suffered a major setback.

On April 29, 2013 in a blog post, the Director of NIMH, Thomas Insel, rejected the DSM-5 for its lack of validity and indicated that the NIMH will "re-orient its research away from DSM categories." Insel indicated that DSM-5 was "at best, a dictionary," that it fell far short of normal medical science diagnostic standards by being based merely on clusters of symptoms and that its categories cannot be the "gold standard" for diagnosis. Insel stated that "[a] diagnostic approach based on the biology as well as the symptoms must not be constrained by the current DSM categories" (Insel 2013). This public renunciation by a main DSM benefactor and chief federal funder of mental health research, just days before its release, also made national news (Belluck and Carey 2013). Insel's rebuke, however, was actually a promotion of the approach to diagnosis that he had been floating for several years, the Research Domain Criteria (RDoC), based on the rephrased ancient notion that "mental disorders are biological disorders involving brain circuits" (Insel 2013). Insel did admit that his approach to "transform diagnosis" was not yet usable because, he said, "we lack the data." Of course, lacking relevant data was precisely the stated rationale 30 years ago to create descriptive diagnosis and the goal of the DSM in part to gather. Although acknowledging the lack of data, Insel appears certain that madness is a medical disorder, and he is intent on mobilizing the resources of the NIMH to try to document his conviction.

The response to Insel's broadside came the following week in a public statement from the Chair of the DSM-5 Task Force, David Kupfer (2013). First, he candidly admitted failure:

> In the future, we hope to be able to identify disorders using biological and genetic markers that provide precise diagnoses that can be delivered with complete reliability and validity. Yet this promise, which we have anticipated since the 1970s, remains disappointingly distant. We've been telling patients for several decades that we are waiting for biomarkers. We're still waiting.

Nevertheless, Kupfer defended the new DSM-5 as a "guidebook" with a "common language" for clinicians. He noted that Insel's RDoC was, itself, a wish and a prayer that cannot supplant DSM-5:

> RDoC is a complementary endeavor to move us forward, and its results may someday culminate in the genetic and neuroscience breakthroughs that will revolutionize our field. In the meantime, should we merely hand patients another promissory note that something may happen sometime? *Every day, we are dealing with impairment or tangible suffering, and we must respond.* Our patients deserve no less. (Kupfer, released statement May 6, 2013, our italics)

Most telling in these acknowledgements of failure is the unblinking assumption held by both Insel and Kupfer that undefined impairment and suffering are medical entities. Kupfer adds that medicine should preside over them simply because *that is what psychiatry does, every day*. Apparently, for him, scientific validation for the professional responses of psychiatry is worth pursuing, but unnecessary.

Following this public dispute between eminent members of the psychiatric establishment accusing each other of championing an invalid or clinically ineffectual approach to classification, the release of DSM-5 (which, incidentally, retailed for $199 a copy from the APA) was met with derision. The *New York Times* (Garner 2013), *Wall Street Journal* (Tavris 2013), and *USA Today* (Jayson 2013) introduced DSM-5 by focusing on new books described as "pre-emptive strikes . . . Like Patriot antimissile systems, these volumes propose to knock the forthcoming manual out of the sky" (Garner 2013). Their titles suggested the lines of attack: *The Book of Woe*, *Saving Normal*, and *Mad Science*. A sample of media headlines from May 2013 framed the issue for the public: "Psychiatry's new guide falls short, experts say" (Belluck and Carey 2013); "Why the fuss over the D.S.M.-5?" (Satel 2013); "Shortcomings of a psychiatric bible" (New York Times 2013); "Another go-around in the saga of the psychiatric bible" (Jayson 2013); "How psychiatry went crazy" (Tavris 2013); "Psychiatric manual's role to narrow" (Wang 2013). Each article recounted the flaws of the DSM and descriptive diagnosis that have been known, but widely ignored, for over 30 years.

These popular discussions, however, fail to recognize the fundamental problem of the DSM approach. The many flaws of descriptive diagnosis, including the chief failure to produce any measurable or usable advance in the biomedical diagnosis of "mental illness," which, to some, might herald the demise of DSM, in fact do not seem to matter to the business of psychiatry. An example is an article about the DSM-5 appearing in the *New York Times*, by Richard Friedman (2013), a professor of psychiatry who writes regularly for that newspaper on psychiatric topics. Friedman appears to offer a balanced review of the uses and excesses of the DSM. He notes, for example, that psychiatry's "holy grail" of finding a neural basis of mental disorders is elusive and that perhaps the new edition of the bible is the best we can expect at this time, a book that provides "a common language" for clinicians. As do almost all commentators, however, Friedman—perhaps not surprisingly, since as a psychiatrist his authority rests on being a medical doctor—is immersed in the language of medicine. He acknowledges that the DSM "doesn't really describe *diseases* in the way that pneumoncoccal pneumonia . . . is a disease. Instead, it is a practical way of describing *psychiatric syndromes* that *patients* experience and that often *predict a course of illness* and *responses to various treatments*." He confidently asserts without evidence that these patients "almost certainly suffer from a *biologically heterogeneous group of disorders* . . . " (our italics).

It is precisely the "common language" that should be at issue, but, unfortunately, it is not. In Friedman's emblematic essay, misbehaviors are *symptoms*, suffering and anguish are *diseases*, people with troubles are *patients*, providing assistance is *treatment* or *therapy*, and all manner of distress and misbehavior are *disorders* or *illnesses*. This medical language is powerful, pervasive, but misleading, because admitted in many of these same articles is that there is no known psychiatric medical disease. The "bible" remains an appropriate moniker for the modern DSMs, suggesting beliefs that need no roots in rationality or science. Bio-psychiatry grips its bible tightly, "on faith alone" more so in times of trouble, inspiring renewed resolve among the psychiatric priesthood and acquiescence among the public.

The profound economic troubles experienced in the United States since 2008 suggest an analogy to better grasp important aspects of the psychiatric scene, as those troubles were also about the power of language to deceive. Banks and other lenders made fortunes by easily giving home loans to those who would eventually have trouble making their monthly payments. This was a market of "liar loans," a cooperative charade in which borrowers lied about their income and ability to repay the loan and lenders pretended that the borrowers' claims were valid. Creditworthiness was not grounded in financial facts or scrutiny. Lenders would quickly profit by selling the flawed mortgages to other financial institutions, which bundled them up into seemingly autonomous financial products and resold them in massive quantities to still other investors lured by promises of high returns. The products were exchanged as "securitized debt," "mortgaged-back securities," "collateralized debt obligations," and a host of arcane financial terms that relabeled these liabilities as "assets," a ruse that duped even financial experts. For a while, many institutions made billions of dollars in buying and selling these massive bundles of bad mortgages. Only during the nation's financial collapse were these assets accurately relabeled as "toxic debt."

The DSM-5 is the bundled liar loans of American psychiatry. It is a suspect package of shaky assumptions about human distress and misbehavior, and arbitrary labels passed off as the names of genuine diseases. On it stand the nation's mental health professions, "grass roots" support groups[8] and services, the lucrative selling of psychiatric drugs, and thick layers of policies, laws and regulations—all backed with taxpayer money. When will psychiatric diagnoses also be appraised as uncollectable "toxic debt" in the business of responding to impairment and tangible suffering? At the very least, the DSM-5 should come with a warning about its nature and its possible harms. In conclusion and summary, we offer such an alternative introduction to DSM-5.

## An Alternative Introduction to DSM-5

This DSM-5 again embodies the fiction that madness should be defined as mental disorder, a species of medical illness. Since DSM-III in 1980, creating artificial categories of mental disorder around impersonal, context-free symptoms of presumed

[8]Grass roots' groups are not always genuinely grass roots. For example, the National Alliance on Mental Illness (NAMI), the principal "grass roots" mental health organization in America with a commitment to biological views regarding mental illness, receives between 60 % and 75 % of its annual budget from drug companies (see Kirk et al. 2013, pp. 16–17).

diseases has been a scientific failure: not one mental disorder has been discovered to be a "general medical condition." A major DSM patron, the National Institute of Mental Health, has turned its back on our manual because the DSM categories show no correspondence with any biological signs. The head of the DSM-5 Task Force was compelled to admit that making DSM more reliable and valid—the ostensible purpose of all its editions—"remains disappointingly distant." But as editors of the DSM-5 we had to treat these crucial scientific problems as minor matters, because by turning many normal human reactions to difficult circumstances into treatable mental disorders, the DSMs have been a resounding political, cultural and economic success.

Fuzzy disorder categories could only increase the size and reach of an industrial complex whose main feature is the mass marketing of psychoactive drugs. We acknowledge that the DSMs have by themselves shed no particular light on any causes of the human distress and misbehavior that the manuals catalogue. In fact, inventing many new labels for human disturbances may simply have obscured the truth that anger, fear, shame, loss, stressful circumstances, deprivations, injustices, and innate or acquired dispositions all contribute to serious impairment and disability. The DSM distracts as well from the age-old realization that any human suffering is eased by community support, by companionship, and by legal or illegal psychoactive drugs voluntarily ingested.

DSM-5 is an American political document. As we state elsewhere, "DSM is a medical classification of disorders and as such serves as a historically determined cognitive schema imposed on clinical and scientific information..." (APA 2013, p. 10). The nature and boundaries of the myriad complaints and signs of distress are decided and drawn arbitrarily as mental disorders by interest-conflicted players uncertain of the future and focused on short-term gains for themselves and their constituents. The United States spends over $100 billion a year on so-called mental health services, most of which are dispensed only if the recipients are categorized as medically impaired—which almost everyone believes is essential to helping them in an enlightened manner. With so much money at stake, as the APA Task Force developed the DSM-5 the players in the mental health movement fought over the power to define madness, the legitimacy to manage whatever and whoever was defined, and the privilege to qualify for third-party funding to do so. They negotiated about adding, deleting, or changing words making up a criterion for a disorder category because this has huge financial repercussions on the families, professionals and institutions that deal with people who are believed to need drugs, shelter, or counseling.

If you're a consumer, the diagnoses in DSM-5 are your tickets for most of the available, conventional mental health services. Be warned, however, that the help you want or need may not be what you will receive, and what you will receive could harm you. Be warned also that although we, the editors of DSM-5, acknowledge that DSM diagnoses are invalid, you will never be able to remove such a diagnosis from your health records.

If you're a service provider, buying the DSM-5 registers your willing or unwilling endorsement of the reigning medicalization of human distress and misbehavior.

## References


American Counseling Association. (2012). Retrieved on April 30, 2012, from http://www.counseling.org/PressRoom/NewsReleases.aspx?AGuid=315a280b-4d0b-48af-9421-1f7d3f01b4b7

American Psychiatric Association. (1980). *Diagnostic and statistical manual of mental disorders, third edition (DSM-III)*. Washington, DC: American Psychiatric Association.

American Psychiatric Association. (1987). *Diagnostic and statistical manual of mental disorders, third edition-revised (DSM-III-R)*. Washington, DC: American Psychiatric Association.

American Psychiatric Association. (1994). *Diagnostic and statistical manual of mental disorders, fourth edition (DSM-IV)*. Washington, DC: American Psychiatric Association.

American Psychiatric Association. (2000). *Diagnostic and statistical manual of mental disorders, fourth edition (DSM-IV-TR)*. Washington, DC: American Psychiatric Association.

American Psychiatric Association. (2013). *Diagnostic and statistical manual of mental disorders, fifth edition (DSM-5)*. Washington, DC: American Psychiatric Association.

Angell, M. (2004). *The truth about the drug companies: How they deceive us and what to do about it*. New York: Random House.

Angell, M. (2009). Drug companies & doctors: A story of corruption. *The New York Review of Books, 56*(1), 8–12.

Angell, M. (2011). The epidemic of mental illness: Why? *The New York Review of Books*, 23 June.

Barber, C. (2008). *Comfortably numb: How psychiatry is medicating a nation*. New York: Pantheon Books.

Barglow, P. (2012, May/June). We can't treat soldiers' PTSD without a better diagnosis. *Skeptical Inquirer*, 36.3. Retrieved from http://www.csicop.org/si/show/we_cant_treat_soldiers_ptsd_without_abetterdiagnosis

Bayer, R. (1981). *Homosexuality and American psychiatry: The politics of diagnosis*. Princeton, NJ: Princeton University Press.

Belluck, P. & Carey, B. (2013, May 7–13). Psychiatry's new guide falls short, experts say. *New York Times*, p. A12.

Bolton, D. (2008). *What is mental disorder? An essay in philosophy, science, and values*. New York: Oxford University Press.

Boyle, M. (2002). *Schizophrenia: A scientific delusion?* (2nd ed.). East Sussex: Routledge.

Brauser, D. (2012). DSM-5 field trials generate mixed results. *Medscape Medical News Psychiatry*. Retrieved from www.medscape.com/viewarticle/763519?src=mpnews&spon=12

British Psychological Society (2011). Response to the American Psychiatric Association: DSM-5. Retrieved from http://www.bps.org.uk/news/psychologists-petition-against-dsm-5

Bruner, J. S., Goodnow, J. J., & Austin, G. A. (1986). *A study of thinking*. New Brunswick, NJ: Transaction.

Caplan, P. (1995). *They say you're crazy: How the world's most powerful psychiatrists decide who's normal*. Reading, MA: Addison-Wesley.

Carey, B. (2008, December 18). Psychiatrists revise the book of human troubles. *New York Times*.

Carey, B. (2012a, January 24). Grief could join list of disorders. *New York Times*.

Carey, B. (2012b, January 20). New definition of autism may exclude many, study suggests. *New York Times*.

Carey, B. (2012c, March 29). Rate of autism diagnoses has climbed, study finds. *New York Times*.

Cohen, D. (2006). Critiques of the ADHD enterprise. In G. Lloyd, J. Stead, & D. Cohen (Eds.), *Critical new perspectives on ADHD* (pp. 137–155). London/New York: Routledge.

Cosgrove, L., & Krimsky, S. (2012). A comparison of DSM-IV and DSM-5 panel members' financial association with industry: A pernicious problem persists. *PLoS Medicine, 9*(3), e1001190.

Cosgrove, L., Krimsky, S., Vijayaraghavan, M., & Schneider, L. (2006). Financial ties between DSM-IV panel members and the pharmaceutical industry. *Psychotherapy and Psychosomatics, 75*, 154–160.

DSM Task Force. (2012). Retrieved at http://www.DSM5.org/ProposedRevision/Pages/proposedrevision.aspx?rid=94. 30 Apr 2012.

Dumont, M. P. (1984). The non-specificity of mental illness. *American Journal of Orthopsychiatry, 54*, 326–335.

Eysenck, H. J. (1986). A critique of contemporary classification and diagnosis. In T. Millon & G. L. Klerman (Eds.), *Contemporary directions in psychopathology: Toward DSM-IV* (pp. 73–98). New York: Guilford Press.

Frances, A. (2009, June 26). A warning sign on the road to *DSM-V*: Beware of its unintended consequences. *Psychiatric Times*.http://www.psychiatrictimes.com/display/article/10168/1425378

Frances, A. (2010, June 2). Psychiatric fads and overdiagnosis. *Psychology Today: DSM5 in Distress*. http://www.psychologytoday.com/blog/dsm5-in-distress/201006/psychiatric-fads-and-overdiagnosis

Frances, A. (2012, March 16). Am I a dangerous man? *Psychology Today: DSM5 in Distress*. http://www.psychologytoday.com/blog/dsm5-in-distress/201203/am-i-dangerous-man

Frances, A. (2013a). *Saving normal: An insider's revolt against out-of-control psychiatric diagnosis, DSM-5, big pharma, and the medicalization of ordinary life*. New York: HarperCollins.

Frances, A. (2013b, March 24). Letter to the *New York Times*, Sunday Review, p. 2.

Francis, A. (2012a, August 8). A clinical reality check. *Cato Unbound*.http://www.cato-unbound.org/2012/08/08/allen-frances/clinical-reality-check

Friedman, R. (2013, May 21). The book stops here. *New York Times*, p. D3.

Garner, D. (2013, May 2). Two pleas for sanity in judging saneness. *New York Times*, p. C1.

Gomory, T., Cohen, D., & Kirk, S. (2013). Madness or mental illness? Revisiting historians of psychiatry. *Current Psychology, 32*(2), 119–135.

Graham, J. (2010, November 13). Doctors' ties to drug industry still strong. *Los Angeles Times*.

Greenberg, G. (2011, January). Inside the battle to define mental illness. *Wired*. http://www.wired.com/magazine/2010/12/ff_dsmv/

Grossman, R. (2008, December 29). Psychiatry manual's secrecy criticized. *Los Angeles Times*.

Harris, G. (2007, June 27). Psychiatrists top list in drug maker gifts. *New York Times*. http://www.nytimes.com/2007/06/27/health/psychology/27doctors.html

Harris, G. (2008a, November 21). Radio host has drug company ties. *New York Times*. http://www.nytimes.com/2008/11/22/health/22radio.html?em

Harris, G. (2008b, October 3). Top psychiatrist didn't report drug makers' pay. *New York Times*. http://www.nytimes.com/2008/10/04/health/policy/04drug.html?_r=1&oref=slogin

Harris, G., & Carey, B. (2008, June 8). Child experts fail to reveal full drug pay. *New York Times*. http://www.nytimes.com/2008/06/08/us/08conflict.html?pagewanted=all

Healy, D. (2012). *Pharmageddon*. Berkeley: University of California Press.

Healy, D., & Le Noury, J. (2007). Pediatric bipolar disorder: An object of study in the creation of an illness. *International Journal of Risk and Safety in Medicine, 19*, 209–221.

Horwitz, A. V., & Wakefield, J. C. (2007). *The loss of sadness: How psychiatry transformed normal sorrow into depressive disorder*. New York: Oxford University Press.

Horwitz, A. V., & Wakefield, J. C. (2012). *All we have to fear: Psychiatry's transformation of natural anxieties into mental disorders*. New York: Oxford University Press.

Hsieh, D. K., & Kirk, S. A. (2003). The effect of social context on psychiatrists' judgments of adolescent antisocial behavior. *Journal of Child Psychology and Psychiatry, 44*(6), 877–887.

Insel, T. (2013). Director's blog: Transforming diagnosis. Retrieved from http://www.nimh.nih.gov/about/director/2013/transforming-diagnosis.shtml

Jacobs, D. H., & Cohen, D. (2004). Hidden in plain sight: DSM-IV's rejection of the categorical approach to diagnosis. *Review of Existential Psychology and Psychiatry, 26*, 81–96.

Jacobs, D. H., & Cohen, D. (2010). Does "psychological dysfunction" mean anything? A critical essay on pathology versus agency. *Journal of Humanistic Psychology, 50*, 312–334.

Jacobs, D. H., & Cohen, D. (2012). The end of neo-Kraepelinism. *Ethical Human Psychology and Psychiatry, 14*, 87–90.

Jayson, S. (2013, May 13). Another go-around in the saga of psychiatry's bible. *USA Today*, p. D4.

Kirk, S. A., & Kutchins, H. (1992). *The selling of DSM: The rhetoric of science in psychiatry.* Hawthorne: Aldine de Gruyter.

Kirk, S., Gomory, T., & Cohen, D. (2013). *Mad science: Psychiatric coercion, diagnosis, and drugs.* New Brunswick, NJ: Transaction Publishers.

Klerman, G. (1978). The evolution of a scientific nosology. In J. C. Shershow (Ed.), *Schizophrenia: Science and practice* (pp. 99–119). Cambridge, MA: Harvard University Press.

Kraemer, H. C., Kupfer, D. J., Clarke, D. E., Narrow, W. E., & Regier, D. A. (2012). DSM-5: How reliable is reliable enough? *American Journal of Psychiatry, 169*(1), 13–15.

Kupfer, D. (2013). Chair of DSM-5 task force discusses future of mental health research. Public statement, released May 6.

Kupfer, D. J., First, M. B., & Regier, D. A. (2002). Introduction. In D. J. Kupfer, M. B. First, & D. A. Regier (Eds.), *A research agenda for DSM-V* (pp. xv–xxiii). Washington, DC: American Psychiatric Association.

Kutchins, H., & Kirk, S. A. (1997). *Making us crazy: DSM: The psychiatric bible and the creation of mental disorders.* New York: Free Press.

Lancet. (2012). Living with grief. Retrieved at: http://www.thelancet.com/journals/lancet/article/PIIS0140-6736(12)60248-7/fulltext

Lane, C. (2008, November 16). Wrangling over psychiatry's bible. *Los Angeles Times.*

Maj, M. (2005). Psychiatric comorbidity: An artifact of current diagnostic systems? *British Journal of Psychiatry, 186*, 182–184.

McNally, R. J. (2011). *What is mental illness?* Cambridge: Harvard University Press.

Mirowsky, J., & Ross, C. (1989). Psychiatric diagnosis as reified measurement. *Journal of Health and Social Behavior, 30*, 11–25.

Mirowsky, J., & Ross, C. (1990). Subjective boundaries and combinations in psychiatric diagnosis. *Journal of Mind and Behavior, 11*, 407–420.

Mishara, A. L. (1994). A phenomenological critique of commonsensical assumptions in DSM-III-R: The avoidance of the patient's subjectivity. In J. Z. Sadler, M. A. Schwartz, & O. P. Wiggins (Eds.), *Philosophical perspectives on psychiatric diagnostic classification.* Baltimore: The Johns Hopkins University Press.

National Institute of Mental Health. (2008). *Strategic planning reports.* NIMH website, 15 Feb 2008.

New York Times (Editorial). (2012, April 1). The puzzle of more autism cases.

New York Times (Editorial). (2013, May 12). Shortcomings of a psychiatric bible. Sunday Review, p. 10.

Nugent, B. (2012, January 31). I had Asperger syndrome. Briefly. *New York Times.*

Rounsaville, B., Alcarcon, R., Andrews, G., Jackson, J. S., Kendell, R. E., & Kendler, K. (2002). Basic nomenclature issues for *DSM-V*. In D. J. Kupfer, M. B. First, & D. A. Regier (Eds.), *A research agenda for DSM-V* (pp. 1–29). Washington, DC: American Psychiatric Association.

Sadler, J. Z., Hulgus, Y. F., & Agich, G. J. (1994). On values in recent American psychiatric classification. *Journal of Medicine and Philosophy, 19*, 261–277.

Sarbin, T. R. (1997). On the futility of psychiatric diagnostic manuals (DSMs) and the return of personal agency. *Applied and Preventive Psychology, 6*, 233–243.

Satel, S. (2013, May 12). Why the fuss over the D.S.M.-5? *New York Times*, Sunday Review, p. 5.

Society for Humanistic Psychology (2012). Retrieved April 30 2012 http://www.ipetitions.com/petition/DSM5/].

Spitzer, R. L. (2008). Psychiatrists revise diagnostic manual—In secret. Retrieved on Sept 25 2008 and posted at http://taxa.epi.umn.edu/~mbmiller/sscpnet/20080909Spitzer/

Spitzer, R. L., Williams, J., & Endicott, J. (2012). Standards for DSM-5 reliability. *American Journal of Psychiatry, 169*(5), 537.

Stein, D. J., Phillips, K. A., Bolton, D., Fulford, K. W. M., Sadler, J. Z., & Kendler, K. S. (2010). What is a mental/psychiatric disorder? From DSM-IV to DSM-V. *Psychological Medicine, 40*(11), 1759–1765.

Steinberg, P. (2012, January 31). Asperger's history of over-diagnosis. *New York Times.*

Szasz, T. (1961). *The myth of mental illness: Foundations of a theory of personal conduct*. New York: Hoeber-Harper.
Szasz, T. (1994). *Insanity: The idea and its consequences*. New York: Wiley.
Tavris, C. (2013). How psychiatry went crazy. *Wall Street Journal*, May 18–19, pp. C5–C7.
Torrey. (1970). *The death of psychiatry*. New York: Penguin.
Tucker, G. J. (1998). Putting DSM-IV in perspective. *American Journal of Psychiatry, 155*, 159–161.
Wakefield, J. C. (1992). Disorder as harmful dysfunction: A conceptual critique of DSM-III-R's definition of mental disorder. *Psychological Review, 99*, 232–247.
Wakefield, J. C., & First, M. B. (2012). Validity of the bereavement exclusion to major depression: Does the empirical evidence support the proposal to eliminate the exclusion in DSM-5? *World Psychiatry, 11*, 3–10.
Wang, S. (2013). Psychiatric manual's role to narrow. *Wall Street Journal*, May 18–19, p. A3.
Whitaker, R. (2010). *Anatomy of an epidemic: Magic bullets, psychiatric drugs, and the astonishing rise of mental illness in America*. New York: Crown.
Wilson, D. (2006, December 5). NIH scientist charged with conflict. *Los Angeles Times*.
Wilson, D. (2009, November 18). Medical schools quizzed on ghostwriting. *New York Times*. http://query.nytimes.com/gst/fullpage.html?res=9404EFDA173AF93BA25752C1A96F9C8B63&scp=1&sq=medical+schools+quizzed+on+ghostingwriting&st=nyt
Zarembo, A. (2011, December 11–16). Discovering autism (4 part series). *Los Angeles Times*.

# Must Disorders Cause Harm? The Changing Stance of the DSM

**Rachel Cooper**

**Abstract** Are mental disorders harmful as a matter of definition, or are they simply conditions that quite often cause problems? Should someone who has "symptoms" but who suffers no harm be diagnosed with a mental disorder? In this chapter I shall show that these are crucial questions but have not been given the attention they deserve. The idea that disorders are linked to "distress or impairment" was first explicitly introduced into the DSM in the DSM-III. This criterion originally came to be introduced following the removal of homosexuality from the DSM in 1973. In the DSM-IV both the general definition of mental disorder contained in the introduction and many of the individual diagnostic criteria sets required that symptoms cause (or at least increase the risk of) distress or impairment before a disorder could be diagnosed. In DSM-5 the requirement that disorders cause harm, or an increased risk of harm, has been dropped from the definition of mental disorder. This matters because many people have the "symptoms" of mental disorder but suffer no distress or impairment, and are not plausibly at an increased risk of suffering distress or impairment in the future. Under DSM-IV such people could not be diagnosed, now they may be. This chapter offers a philosophical history. I show how the conceptualisation of the link between harm and disorder has shifted through the different editions of the DSM, and make it clear why this matters.

All are agreed that there is some sort of link between mental disorder and harm. Mental disorders are, at least in general, bad; that's why the mental health professionals exist to treat them. But what exactly is the link between mental disorder and harm? Are mental disorders harmful as a matter of definition, or are they simply conditions that quite often cause problems? In this chapter I shall show that this is a crucial question, but one on which the DSM wavers.

The reason that it matters whether mental disorders are necessarily harmful (i.e. by definition), or are merely conditions that quite often cause harm, is that it makes a difference to who will meet diagnostic criteria. Some people have many of the "symptoms" of mental disorder but find these unproblematic. Consider Asperger's

R. Cooper (✉)
Lancaster University, Lancaster, UK
e-mail: r.v.cooper@lancaster.ac.uk



S. Demazeux, P. Singy (eds.), *The DSM-5 in Perspective*, History, Philosophy and Theory of the Life Sciences 10, DOI 10.1007/978-94-017-9765-8_5

Disorder (high-functioning Autistic Spectrum Disorder under DSM-5), for example. There are some people who meet all the diagnostic criteria – they don't "get" the nuances of social interaction, their style of speech is odd, their interests unusual, and so on – but they are not unhappy or impaired. Should such people be diagnosed as having a mental disorder, or should they just be considered different? The answer hinges crucially on whether one thinks that disorders necessarily cause harm, or are merely "often associated" with harm. If mental disorder by definition causes harm, then a flourishing individual cannot be diagnosed. They are simply different. In contrast, if mental disorder is merely "typically associated" with harm, then the happy person with Asperger's can be given a diagnosis. Many other cases raise similar issues: consider happy and high-functioning voice-hearers[1]; people who have a low I.Q. on tests, but who have no problems in their everyday life[2]; those who have a very low sex-drive but don't care.

My aims in this chapter are modest. I offer a philosophical history. I shall show that the ways in which the DSM has conceptualised links between disorder and harm have shifted over time, and trace the rationale for the changes. I have spent some time in the archives of the American Psychiatric Association (APA) looking at documents relating to the construction of DSM-III and DSM-IV (materials relating to DSM-5 are not currently available to researchers). I am very grateful to the APA for allowing me to visit, and this chapter will draw on materials from the archives. I shall show that often changes to the DSM relating to the conceptualisation of harm have come about in ways that are almost accidental; the committees revising the DSM have often failed to appreciate the importance of conceptual issues. In this chapter I don't aim to fully resolve the issues; my main aim is to highlight that there are important issues that need to be addressed.

## Harm in the DSM-III

The DSM-III was published in 1980. The date is crucial. During the 1970s, the APA reeled under attack from both gay activists and anti-psychiatrists. Defining "mental disorder" came to be rhetorically useful for the APA in both these battles.

In line with psychiatric tradition, homosexuality had been included in the DSM-II. In the late 1960s, gay rights groups began to campaign for the removal of homosexuality from the classification (Bayer 1981). Protesters mobbed the 1970

[1]For a review of studies examining the community prevalence of voice hearing see Beavan et al. (2011). These studies suggest that voice-hearing occurs relatively commonly, and that many who hear voices are not distressed by them and have no other psychiatric symptoms.

[2]One of the rare occasions on which "normal" adults have their I.Q. tested is during medical checks for military service and here surprises emerge. Granat and Granat (1973) studied Swedish males signing up for military service. Those with I.Q.s less than 70 who had been identified during school made up 0.7 % of the population, but tests found a further 1.5 % of the recruits to have I.Q.s less than 70. Apart from their test-scores, these previously unidentified low-I.Q. men differed little from their higher-I.Q. peers – many had jobs and families.

APA annual meeting in San Francisco, shouting down speakers with whom they disagreed, and disrupting much of the meeting. Protests continued throughout 1971 and 1972. Robert Spitzer, who would later become chairman of the DSM-III committee, became involved in the debates. Spitzer sought to find middle ground between those who considered homosexuality to be a mental disorder and those who considered it a normal variant of human sexuality. He suggested that homosexuality in itself is not a disorder but that a diagnosis should be included for homosexual people who experience distress concerning their sexual orientation (Spitzer 1973, 1981). He defended his stance via defining mental disorder. Spitzer claimed that conditions are only considered mental disorders if they cause distress or disability. As many gay people experience no distress or disability, this means that homosexuality in and of itself cannot be a disorder. However, those people who are distressed about their sexual orientation *can* be considered to suffer from a disorder and are appropriately treated by psychiatrists. Spitzer's position on homosexuality came to be adopted by the APA in 1973, when homosexuality was removed as a diagnosis from the DSM-II and "Sexual Orientation Disorder," a diagnosis for those who are unhappy about being gay, was added.

In the early 1970s, the anti-psychiatry movement was also in full swing. The anti-psychiatrists were a motley group, united only in their distrust of psychiatry. They argued on various grounds that mental illness is profoundly unlike physical illness, and that therefore psychiatry is only dubiously a branch of medicine.[3] Thomas Szasz, in particular, was a serious irritation to American psychiatry at this time. Szasz claimed that there is no such thing as mental illness; psychiatric patients are not sick but rather malingerers, social misfits, and people with problems in living (Szasz 1960). Correspondingly, psychiatrists should not be classed with medical doctors but are rather, at best, expensive agony aunts or, at worst, agents of social control.

Against this background, during the 1970s the APA felt under pressure to define mental disorder. In 1973, at the same time as setting up a task force to revise the DSM, the APA set up a task force to define the terms "mental illness" and "psychiatrist" (Barton 1973). Originally it was hoped that these definitions could be used in the preamble to the DSM-III. In the event, the task force to define "mental illness" and "psychiatrist" seems not to have influenced the DSM efforts (in their histories of development of the DSM-III definition, Millon (1983), and Spitzer and Williams (1982), make no mention of the work of this task force). Instead, the work of the task force to define "mental illness" and "psychiatrist" eventually resulted in a pamphlet "What is a psychiatrist?", produced by the APA for psychiatrists to

[3]David Rosenhan (1973) claimed that psychiatrists cannot distinguish the sane from the mentally ill. Thomas Szasz (1960) claimed that psychiatrists wrongly label social misfits. Laing and Esterson (1970) argued that schizophrenics are fundamentally normal individuals who are scape-goated by their families. Michel Foucault (1971) argued that "mental illness" is a historically contingent category. For a philosophical analysis of the arguments of the antipsychiatrists see Cooper (2007a, Chap. 2). For a historical discussion see Decker (2013, pp. 10–33).

distribute to their patients (APA 1978). Still, the efforts of the task force indicate that the idea that mental disorder needed defining had become widely accepted within the APA.

On becoming chairman of the DSM-III Task Force in 1975, Spitzer set about refining his own definition of mental disorder. Together with Jean Endicott, he produced a draft definition for the DSM-III, which claimed that mental disorder could be defined as:

> A relatively distinct condition resulting from an organismic dysfunction which in its fully developed or extreme form is directly and intrinsically associated with distress, disability, or certain other types of disadvantage. The disadvantage may be of a physical, perceptual, sexual or interpersonal nature. (Spitzer and Endicott 1978, p. 18)

We can see that at this point the idea that a mental disorder is a condition that causes harm, or at least one that may develop into a condition that will cause harm, remained from the definitions that Spitzer developed during the debates over homosexuality. However, during the development of the DSM-III the claimed link between disorder and harm came to be weakened. Eventually, the link between disorder and harm came to be down-graded from being regarded as definitional to a "typical association." The downgrading came about due to concern that in some of the sexual perversions (such as necrophilia, or fetishism) the patient might suffer no harm or impairment, and yet should still be diagnosed (Spitzer and Williams 1982, p. 20). Eventually, the published DSM-III definition came to claim that

> each of the mental disorders is conceptualized as a clinically significant behavioural or psychological syndrome or pattern that occurs in an individual and that is typically associated with either a painful symptom (distress) or impairment in one or more areas of functioning (disability). In addition there is an inference that there is a behavioural, psychological, or biological dysfunction, and that the disturbance is not only in the relationship between the individual and society. (APA 1980, p. 6)

## Harm in the DSM-III-R

The DSM-III-R was published in 1987. Robert Spitzer remained in charge. Many of the revisions made in the DSM-III-R sought to iron out ambiguities in the text. The introductory definition of mental disorder was revised. Now it was claimed that

> each of the mental disorders is conceptualised as a clinically significant behavioural or psychological syndrome or pattern that occurs in a person and that is associated with present distress (a painful symptom) or disability (impairment in one or more important areas of functioning) or with significantly increased risk of suffering death, pain, disability, or an important loss of freedom. In addition, this syndrome or pattern must not be merely an expectable response to a particular event, e.g., the death of a loved one. Whatever its original cause, it must currently be considered a manifestation of a behavioural, psychological, or biological dysfunction in the person. Neither deviant behaviour, e.g. political, religious, or sexual, nor conflicts that re primarily between the individual and society are mental disorders unless the deviance or conflict is a symptom of a dysfunction in the person as described above. (APA 1987, p. xxii)

There was very little discussion about the changes to the definition. The revisions were presented as merely clarifying the DSM-III definition rather than as substantially changing it. The DSM-III-R text says "it is useful to present a definition of mental disorder that has influenced the decision to include certain conditions in DSM-III and DSM-III-R" (p. xxii), making it sound as if the definition presented in DSM-III-R and DSM-III is the same. However, there is an important difference. While in the DSM-III mental disorder is said to be "typically associated" with harm, in DSM-III-R the link between disorder and harm (or at least risk of harm) is presented as being necessary and conceptual. Although the definition in the DSM-III was a descendent of definitions produced to provide a rationale for excluding homosexuality from the classification, strictly speaking it would be compatible with the DSM-III definition to diagnose homosexuality; if mental disorder is "typically associated" with harm this means that it need not be associated in every case. The DSM-III-R definition rules out this possibility. When the link between disorder and harm is understood to be definitional, then those who are not distressed or impaired (or at least at risk of future impairment) should not be diagnosed.[4]

In DSM-III-R, the rationale for including a requirement that disorder be linked to harm also began to shift. When it originated in the debates over homosexuality, the harm criterion was intended to distinguish those who are merely different from those who are disordered. In DSM-III-R, a new concern becomes apparent. The idea that disorders cause harm starts to be used also to reduce the risk of "false positives" that might otherwise arise by diagnosing cases that are too mild to be properly considered disorders. In DSM-III-R the diagnostic criteria for simple phobia, social phobia, and obsessive-compulsive disorder include a criterion that the symptom "interferes with occupational functioning or with usual social activities or relationships with others, or there is marked distress about having the fear" (APA 1987, p. 243). In Spitzer and Wakefield (1999), Spitzer explains that the aim

---

[4]Many think that the claim that disorder must be linked to harm (or risk of harm) to the patient can become implausible in some cases, most notably the personality disorders and paraphilias. In the case of the personality disorders, DSM-III-R instructs clinicians that "It is only when personality traits are inflexible and maladaptive and cause either significant functional impairment or subjective distress that they constitute Personality Disorders (p. 335, though this note is in the text rather than the diagnostic criteria and may well pass unread). DSM-III included the same instruction (p. 305). In DSM-III-R the paraphilias (exhibitionism, fetishism, frotteurism, paedophila, sexual masochism, sexual sadism, transvestic fetishism, voyeurism) include as a part of the diagnostic criteria that "The person has acted on these urges, or is markedly distressed by them" (p. 282, 283, 284, 285, 287, 288, 289, 290). This requirement was not included in DSM-III, in which non-distressing fantasies alone were sufficient for diagnosis. Note that the DSM-III-R stance on paraphilias seems inconsistent. The general definition of disorder in the introduction required that all disorders cause harm (or risk of harm), but in DSM-III-R paraphilias can be diagnosed merely on the basis of harmless but unusual activities (eg masturbation with shoes) (Gert 1992). There are other disorders in the DSM-III-R where the link with harm is unclear. Bipolar Disorder Not Otherwise Specified, for example, is diagnosable purely on the basis of a patient experiencing hypomanic episodes (such episodes can be pleasant, although arguably they might indicate that a patient is at increased risk of later depressive episodes and so might be indicative of an increased risk of harm).

of this criterion was to distinguish these disorders from common and mild versions of the symptoms that cause no impairment. The use of the "harm criterion" to avoid the diagnosis of mild cases became more widespread in the DSM-IV, and will be discussed further in the next section.

## Harm in the DSM-IV and IV-TR

The DSM-IV and DSM-IV-TR repeat the DSM-III-R definition of "disorder" (APA 1994, p. xxi; 2000, p. xxxi), but with little enthusiasm, saying "the definition of *mental disorder* that was included in DSM-III and DSM-III-R is presented here because it is as useful as any other available definition" (APA 1994, p. xxi).

The DSM-IV archives reveal that the committees paid little attention to the definition of disorder. Dedicated workgroups considered fairly much every aspect of the classification in very great detail, but no group worked on the definition of disorder, or on the possible importance of requiring that disorder cause distress or impairment. The only discussion of the definition that I found in the archives occurred in a brief exchange of letters between Allen Frances (who chaired the DSM-IV committees) and Jerome Wakefield (a professor of social work and philosopher well known for his work on the concept of disorder – see his contribution in this volume). Frances had come across a paper by Wakefield that he found "interesting" and asked Wakefield for a preprint of another work (Frances 1992; Wakefield 1992).[5] Notably, the letters are dated from late 1992; too late to have any substantial impact on the contents of DSM-IV (DSM-IV workgroups started work around 1988, and a fairly complete draft was in circulation by early 1991 (APA 1991)).

Robert Spitzer chaired the committees for DSM-III and DSM-III-R, and Allen Frances chaired the DSM-IV committees. Both DSM chairs had a good deal of discretion in how they managed the process. While Robert Spitzer found it useful to define disorder, Allen Frances did not.[6] The lack of interest in defining

---

[5]Frances had originally read Wakefield's (1992). The concept of mental disorder: On the boundary between biological facts and social values. *American Psychologist,* 47, 373–388 and requested Wakefield (1992) Disorder as harmful dysfunction: A conceptual critique of D.S.M-III-R's definition of mental disorder, *Psychological Review* 99, 232–247. Wakefield sent the requested paper and also a copy of Wakefield (1993) Limits of operationalization: A critique of Spitzer and Endicott's (1978) proposed operational criteria for mental disorder. *Journal of Abnormal Psychology*, 102, 160–172.

[6]Writing in 2013 Frances states "I have reviewed dozens of definitions of mental disorder (and have written one myself in DSM-IV) and find none of them the slightest bit helpful either in determining which conditions should be considered mental disorders and which not, or in deciding who is sick and who is not" (Frances 2013, p. 17).

disorder can also be explained by a shift in the political background against which the DSM-IV was developed. I have suggested that the APA became interested in defining "disorder" during the 1970s and early 1980s in order to defend against the claims of the anti-psychiatry movement, and because of the need to determine whether homosexuality should be considered a disorder. These concerns were peculiar to a specific time in American history and by the late 1980s had largely disappeared. Those with an interest in the philosophy of psychiatry often imagine that conceptual issues are of perennial significance. In general, however, only philosophy-aficionados have any interest in defining disorder. When others (such as APA committees) become involved their concern is seldom disinterested. While defining disorder seemed useful in the 1970s and early 1980s, by the time the DSM-IV was under development, official interest in conceptual issues had faded.

The DSM-IV retained the DSM-III-R definition of disorder. The rationale for thinking that disorders cause harm continued to move in the direction initiated by DSM-III-R. Rather than primarily being about the distinction of disorder from conditions like homosexuality (that are better understood as mere difference), during the development of the DSM-IV the harm criterion came to be understood as a means of determining the threshold between normality and disorder. Many disorders merge into normal states; the distinction between depression and normal sadness is vague, so too is that between alcoholism and heavy drinking. This makes it difficult to draw a line between disorder and the normal states that they merge into. The DSM-IV committees came to think of the requirement that a disorder cause "clinically significant distress or impairment" as a means of drawing the threshold (APA 2000, p. 7).

In addition to appearing in the introductory definition of mental disorder, a requirement that symptoms cause harm was added to many sets of diagnostic criteria (usually worded "... causes clinically significant distress or impairment in social, occupational, or other important areas of functioning"). The DSM-IV Task Force took it that the general definition of disorder already implied that disorders cause harm First et al. (1995). Adding a requirement that there be distress or impairment to some sets of diagnostic criteria was thought of as merely serving to emphasise a point that was already implicit. As such, some such additions were made late on in the editing process and with little consultation (the criterion was added to tic disorders sometime after the 1993 draft, p. E.13- for example).

However, on publication, it turned out that some mental health researchers did not share the view that it was already implicit that mental disorder necessarily causes harm. The addition of the criterion to Tourette's syndrome and related tic disorders caused particular complaint (Freeman et al. 1995; Comings 1995). The problem is that many young children who tic are not at all distressed by their ticcing, nor do they suffer impairment. Critics thought such children should count as having tic disorders, but they did not meet DSM-IV diagnostic criteria. In response to these complaints the criterion was removed from the tic disorders in the DSM-IV-TR. It is hard to know what to make of this revision, since the DSM-IV-TR retains the DSM-IV general definition of mental disorder. Presumably this definition applies to all disorders in the classification. If so, then removing the "distress or impairment" criterion from tic disorders should make no difference to who is diagnosed. When diagnostic criteria include a criterion that there is "distress or impairment" this is supposed to merely emphasise a point that is already implicit. On the other hand, in so far as critics have been appeased by the revision, it must have been expected to make some difference in practice. Plausibly, the explanation for the fact that the revisions led to an inconsistency, and that this inconsistency seemed tolerable to both critics and those revising the DSM, is that it is tacitly accepted that in practice only the sets of diagnostic criteria included in the DSM really matter; no one is expected to read the general introductory definition of disorder and the sets of diagnostic criteria together.

## Harm in the DSM-5

The DSM-5 was published in May 2013, just prior to this paper being written. At time of writing, documents relating to the DSM-5 have not been deposited in the APA archives. However, multiple drafts were published online over a number of years, giving some insight into the committees changing thinking about the relationship between harm and disorder.[7] The drafts suggest that the definition of mental disorder finally included in the DSM-5 was produced late, with little discussion, and with less consensus.

Early drafts of the DSM-5 (from at least February 2010 to July 2010) included a slightly revised version of the definition of mental disorder included in DSM-IV-TR (detailed in Stein et al. 2010). However, in the draft available December 2011, two competing draft definitions of mental disorder, developed by different groups, were presented, and a note explained that a decision would be made later which to adopt. The first definition continued to be that presented in the earlier draft; a revision of the DSM-IV-TR definition (Stein et al. 2010). The second definition was developed by the Impairment and Disability Assessment Study Group, chaired by Jane Paulsen, which was charged with reviewing the links between the diagnosis of disorder and notions of distress and disability in the DSM (APA 2012). The definition proposed by the Impairment and Disability Assessment Study Group is very similar to the DSM-IV-TR definition except that, crucially, all reference to distress and impairment has been removed. The Impairment and Disability Assessment Study Group also advocated seeking to remove the criterion that symptoms cause "distress or impairment" from as many sets of diagnostic criteria as possible.

[7]Unfortunately since publishing the DSM-5, the APA has removed all draft versions from the website.

The aim of the Impairment and Disability Assessment Study Group was to bring the DSM definition of disorder closer to that implicit in the International Classification of Diseases (ICD), published by the World Health Organization (WHO). The ICD covers all disorders, not just mental disorders. The WHO also publishes a distinct classification, The International Classification of Functioning, Disability and Health, which supplies codes for all forms of disability. In the ICD system, disorder and disability are thought of as being quite distinct. The thinking here will already be familiar to those who have had some exposure to disability studies, where the social model of disability conceptualizes impairment and disability separately; impairment refers to the biological difference (e.g. having no legs), disability refers to problems in everyday living that are conceived of as arising from the social response to the impairment (Oliver 1996). The line is analogous to that taken in gender studies, where sex and gender are similarly held to be distinct (e.g. Stone 2007). In the eyes of the Impairment and Disability Assessment Study Group someone who for instance hears voices but is not bothered by them and has a good life, should be said to have the disorder of schizophrenia (supposing that criteria for duration, etc., are met),[8] but not to be disabled.

Although there was widespread debate about many of the changes proposed for the DSM-5 (in journal papers, blogs, etc.) the proposed changes to the definition of disorder attracted very little attention. Eventually, from June 2012, something very close to the definition actually published in the DSM-5 was included in drafts. The new DSM-5 definition requires that:

> A mental disorder is a syndrome characterized by clinically significant disturbance in an individual's cognition, emotion regulation, or behaviour that reflects a dysfunction in the psychological, biological, or developmental processes underling mental functioning. Mental disorders are usually associated with significant distress or disability in social, occupational, or other important activities. An expectable or culturally approved response to a common stressor or loss, such as the death of a loved one, is not a mental disorder. Socially deviant behavior (e.g. political, religious, or sexual) and conflicts that are primarily between the individual and society are not mental disorders unless the deviance or conflict results from a dysfunction in the individual, as described above. (APA 2013, p. 20)

The idea that disorder is "usually associated" with harm is reminiscent of the DSM-III definition. In the DSM-5, the published definition appears to be an uncomfortable compromise between the definition developed by Stein et al. and by the Impairment and Disability Assessment Study Group. To a philosopher, the published definition looks horrible. A definition that says that disorders are "usually associated" with harm seems to be of little use. The definition gives no guidance as to whether any particular "symptomatic" but flourishing individual should be diagnosed.

---

[8]In contrast to the DSM-IV position, the ICD has always allowed for schizophrenia to be diagnosed in the absence of disability or distress.

The definition of mental disorder included in the DSM-5 was in any event developed far too late to have had any impact on decisions as to which conditions should be included in the DSM. The online drafts showed that fairly much everything had been fixed long before the definition of disorder was agreed. Darrel Regier, who was the co-chair for DSM-5, reports that the committees felt little need to consider conceptual issues: "almost none of the DSM-5 Task Force or Work Group meetings struggled with these definitional issues as they evaluated the research literature to determine the evidential basis for revisions" (Regier, 2012, pp. 292–293).[9]

In DSM-5 the link between disorder and harm in the introductory definition has been downgraded (from definitional to common). The Impairment and Disability Assessment Study Group had also advocated removing the criterion that symptoms cause distress or impairment from as many sets of diagnostic criteria as possible.[10] This hasn't happened. Rather, in DSM-5 more sets of diagnostic criteria than ever before include a criterion that the symptoms cause distress or impairment. Such criteria have been added to dissociative identity disorder, substance/medication induced psychotic disorder, psychotic disorder due to another medical condition, catatonia, autism spectrum disorder, and agoraphobia.

Still, although criteria that symptoms cause distress or impairment have been added to many diagnostic criteria, the DSM-5 makes these additions with a sense of embarrassment. The introduction explains:

> There have been substantial efforts by the DSM-5 Task Force ... to separate the concepts of mental disorder and disability (impairment in social, occupational, or other important areas of functioning)... However, in the absence of clear biological markers or clinically useful measurements for severity for many mental disorders, it has not been possible to completely separate normal and pathological symptom expressions contained in diagnostic criteria. ... Therefore, a generic diagnostic criterion requiring distress or disability has been used to establish disorder thresholds, usually worded "the disturbance causes clinically significant distress or impairment in social, occupational, or important areas of functioning." (APA 2013, p. 21)

[9]There were a few exceptions: The literature review that justified the addition of Hoarding Disorder to DSM-5, for example, makes use of the DSM-IV definition of disorder to argue that hoarding should be included (Mataix-Cols et al. 2010).

[10]Very little has been written about the notion that symptoms should be required to cause distress or impairment, but a notable attack on the criteria has come from an unanticipated source. As argued here, the idea that mental disorders cause harm was first introduced to the DSM largely through the work of Robert Spitzer. Philosophical papers by Jerome Wakefield were also influential in defending the idea that disorders must be harmful. However, writing in the American Journal of Psychiatry, Spitzer and Wakefield (1999) attack the DSM-IV inclusion of "clinical significance" criteria on the basis that it can lead to false negatives, amongst other cases they discuss that of a young child with the symptoms of Tourette's who suffers no distress. In this paper, Spitzer and Wakefield argue that the child should receive a diagnosis (although such a child is not diagnosable under DSM-IV).

In other words, in DSM-5 the idea that certain disorders cause distress or impairment is included in diagnostic criteria only pending the development of some better means of drawing the line between the normal and the pathological. Rather than harm being seen as essential for diagnosis, the DSM-5 thinks of references to harm as only being necessary because science is as yet underdeveloped.

## Summary and Analysis

Since DSM-III (1980) each edition of the DSM has conceptualised disorder as being linked in some way to "clinically significant distress or impairment." The idea first appears in DSM-III where the claim that disorders are linked to harm is employed as the rationale for excluding homosexuality from the classification. Here harm is used to distinguish those with disorders from those who are just different. In DSM-III-R, IV, and IV-TR, the link between disorder and harm (or risk of harm) is presented as being necessary and conceptual. Gradually, however, the thinking behind the idea that disorder is linked to harm shifts. Rather than acting to distinguish between those who are different and those who are disordered, the harm criterion comes to be used as a means of drawing the threshold in cases where disorders merge into normality. In the DSM-5 the link between disorder and harm is downgraded. Rather than it being a definitional truth that disorders cause harm (or risk of harm), now disorders are merely claimed to be "generally associated" with harm. Where the idea that a disorder causes harm remains in sets of diagnostic criteria, the DSM-5 presents this as a place-holder, and looks forward to the day when references to harm can be replaced by "more scientific" criteria.

One of the most striking facts about the changes that have been made to the conceptualisation of the link between disorder and harm in the DSMs is how little thought has gone into these changes and how little debate they have attracted. The DSM-IV archives reveal that no group worked on the definition of disorder, or on the role that the requirement that disorder cause distress or impairment might play in the classification. Jerome Wakefield was approached, but only in late 1992, when most of the work on the DSM-IV was already complete. Similarly, with the DSM-5, competing definitions of disorder remained in drafts until after the rest of the classification was fairly much complete.

Despite this neglect, being clear on the link between disorder and harm is very important. In the introduction I pointed out that whether a happy voice-hearer, or a high functioning but low I.Q. individual, should be diagnosed depends on whether disorder is conceived of as always causing harm (or risk of harm), or as only sometimes being linked to harm. If the link is conceptual, then those who flourish with symptoms should not be diagnosed. If the link is weaker, then they may be. Huge numbers of people are potentially affected by this decision. The Epidemiological Catchment Area study used DSM-III diagnostic criteria to estimate the prevalence of mental disorder in a community sample (Robins and Regier 1991).

In this study, adding the requirement that symptoms cause distress or disability reduced the 1-year prevalence rate of mental disorder from 28 % to 18 % (Regier 2012, p. 288). In other words, in any 1 year, about an extra 10 % of the population is potentially diagnosable if the DSM states that disorder need not be linked to harm. One explanation for why this difference is sometimes neglected is that the effect of the conceptualisation of harm on rates of diagnosis only shows up in certain contexts. In many clinical contexts, it is fair to assume that those who request therapy are unhappy. Those who flourish with symptoms show up in epidemiological studies rather than in clinics. Still, it is important that the DSM be clear whether those who meet diagnostic criteria but do not suffer (and are not at risk of suffering) should be eligible for diagnosis or not. Prevalence rates gained from epidemiological studies are sometimes used in planning mental health studies. In addition, the DSM plays a huge cultural role in shaping lay perceptions of the normal and the pathological. If the DSM says that one can be normal and have "symptoms" then this helps to legitimise the claims of those who are psychologically different but claim they are fine. This is why when homosexuality was included in the DSM this quite rightly bothered gay people, even though in practice those who kept clear of mental health services were at little actual risk of being diagnosed.

Changes to the way in which the DSM conceptualises the link between disorder and harm have not been well considered. What's more I suggest that the shifts over time have been unfortunate. Originally, the notion that disorders cause harm was introduced to help distinguish those who are merely different (in this case homosexual people) from disordered people. I suggest that this is a job that the harm criterion can do well (for a fuller defense see Cooper 2013). Gradually, however, the DSM committees came to think of the criterion that symptoms cause distress or impairment as being necessary only when there is no other means of drawing a threshold where a disorder shades into normality. This has led to there being a widespread view that the harm criterion can be replaced as soon as some biomarker, or reliable behavioural measure, becomes available. This is a mistake. The criterion that disorder causes harm is essential for distinguishing those who are merely different from those who are disordered and should be maintained.

To claim that disorders necessarily cause harm is to invite many questions: what exactly does it mean to say that a disorder causes harm? Must the harm be to the patient, or can harm to others suffice for diagnosis? How should we conceptualise conditions that cause no harm now, but that increase the risk for the harm in the future? Such issues cannot be explored further in this chapter (although I start to address them in Cooper 2002, 2007b, forthcoming). My principle aim in this chapter isn't to put forward a full account of the harmfulness of disorder, but rather to show that getting straight about the conceptual links between disorder and harm is important, and that the DSM approach to harm has shifted over time.

## References

### Unpublished Materials from the American Psychiatric Association Archives[11]

Barton, W. (1973). Letter from Walter Barton to Sidney Malitz, dated 20 Mar 1973. Archives of the American Psychiatric Association. Professional Affairs. Box 2. Folder 74.

Frances, F. (1992). Letter dated 17 Aug 1992. To Jerome Wakefield. Archives of the American Psychiatric Association. Box labelled "Proofs Full Text," Folder "DSM-IV Introduction."

Spitzer, R. (1973). Progress report of the task force on nomenclature and statistics to the council on research and development. Sept 1975. Archives of the American Psychiatric Association. Professional affairs. Box 2. Folder 76. Nomenclature and statistics 1976.

Spitzer, R., Sheehy, M., & Endicott, J. (1977). DSM-III guiding principles. In the 1977 draft of the DSM-III. Archives of the American Psychiatric Association. Professional affairs. Box 16. Folder 177, Chap. 4, p. 3.

Wakefield, J. (1992). Letter dated 22 Oct 1992. To Allen Frances. Archives of the American Psychiatric Association. Box labelled "Proofs Full Text," Folder "DSM-IV Introduction."

### Published References

American Psychiatric Association. (1978). Report from council on emerging issues. *American Journal of Psychiatry, 135*, 278.

American Psychiatric Association. (1980). *Diagnostic and statistical manual of mental disorders* (3rd ed.). Washington, DC: American Psychiatric Association.

American Psychiatric Association. (1987). *Diagnostic and statistical manual of mental disorders* (3rd ed. – revised). Washington, DC: American Psychiatric Association.

American Psychiatric Association. (1991). *DSM-IV options book*. Washington, DC: American Psychiatric Association.

American Psychiatric Association. (1993). *DSM-IV draft criteria*. Washington, DC: American Psychiatric Association.

American Psychiatric Association. (1994). *Diagnostic and statistical manual of mental disorders* (4th ed.). Washington, DC: American Psychiatric Association.

American Psychiatric Association. (2000). *Diagnostic and statistical manual of mental disorders* (4th ed. text revision). Washington, DC: American Psychiatric Association.

American Psychiatric Association. (2012). Impairment and disability. Available at http://www.dsm5.org/MeetUs/Pages/ImpairmentandDisability.aspx. Accessed 21 Jan 2015.

American Psychiatric Association. (2013). *Diagnostic and statistical manual of mental disorders* (5th ed.). Washington, DC: American Psychiatric Association.

Bayer, R. (1981). *Homosexuality and American psychiatry*. New York: Basic Books.

Beavan, V., Read, J., & Cartwright, C. (2011). The prevalence of voice-hearers in the general population: A literature review. *Journal of Mental Health, 20*, 281–292.

Comings, D. (1995). DSM-IV criteria for Tourette's. *Journal of the American Academy of Child and Adolescent Psychiatry, 34*, 401–402.

[11]At the time of my visit (February 2013) the DSM-IV archives had only just become available to researchers and had not yet been catalogued. Boxes and folders thus had no numbers and can be referred to here only descriptively. Materials relating to DSM-III were visited on an earlier visit, in 1998.

Cooper, R. (2002). Disease. *Studies in History and Philosophy of Biological and Biomedical Sciences, 33*, 263–282.
Cooper, R. (2007a). *Psychiatry and philosophy of science*. Stocksfield: Acumen.
Cooper, R. (2007b). Can it be a good thing to be deaf? *Journal of Medicine and Philosophy, 32*, 563–583.
Cooper, R. (2013). Avoiding false positives; the role of zones of rarity, the threshold problem, and the DSM clinical significance criterion. *Canadian Journal of Psychiatry, 58*, 606–611.
Cooper, R. (forthcoming). On harm. In L. Faucher, D. Forest (Eds.). *Defining mental disorders: Jerome Wakefield and his critics*. M.I.T. Press: Forthcoming.
Decker, H. (2013). *The making of DSM-III*. New York: Oxford University Press.
First, M., Frances, A., & Pincus, H. (1995). DSM-IV editors reply. *Journal of the American Academy of Child and Adolescent Psychiatry, 34*, 402.
Foucault, M. (1971). *Madness and civilisation*. London: Routledge.
Frances, A. (2013). *Saving normal*. New York: Harper Collins.
Freeman, R., Fast, D., & Kent, M. (1995). DSM-IV criteria for Tourette's. *Journal of the America Academy of Child and Adolescent Psychiatry, 34*, 400–401.
Gert, B. (1992). A sex caused inconsistency in DSM-III-R: The definition of mental disorder and the definition of paraphilias. *The Journal of Medicine and Philosophy, 17*, 155–171.
Granat, K., & Granat, S. (1973). Below average intelligence and mental retardation. *American Journal of Mental Deficiency, 78*, 27–32.
Laing, R. D., & Esterson, A. (1970). *Sanity, madness and the family*. Harmondsworth: Penguin.
Mataix-Cols, D., Frost, R. O., Pertusa, A., Clark, L. A., Saxena, S., Leckman, J. F., et al. (2010). Hoarding disorder: A new diagnosis for DSM-V? *Depression and Anxiety, 27*(6), 556–572.
Millon, T. (1983). The DSM-III: An insider's perspective. *American Psychologist, 38*, 804–814.
Oliver, M. (1996). Defining impairment and disability; issues at stake. In C. Barnes & G. Mercer (Eds.), *Exploring the divide* (pp. 29–54). Leeds: The Disability Press.
Regier, D. (2012). Diagnostic threshold considerations for DSM-5. In K. Kendler & J. Parnas (Eds.), *Philosophical issues in psychiatry II: Nosology* (pp. 285–297). Oxford: Oxford Universiy Press.
Robins, L., & Regier, D. (Eds.). (1991). *Psychiatric disorder in america: The epidemiological catchment area study*. New York: The Free Press.
Rosenhan, D. (1973). On being sane in insane places. *Science, 179*, 250–258.
Spitzer, R. (1973). A proposal about homosexuality and the APA nomenclature: Homosexuality as an irregular form of sexual behaviour and sexual orientation disturbance as a psychiatric disorder. *American Journal of Psychiatry, 130*, 1214–1216.
Spitzer, R. (1981). The diagnostic status of homosexuality in D.S.M-III: A reformulation of the issues. *American Journal of Psychiatry, 138*, 210–215.
Spitzer, R., & Endicott, J. (1978). Medical and mental disorder: Proposed definition and criteria. In R. Spitzer & D. Klein (Eds.), *Critical issues in psychiatric diagnosis* (pp. 15–39). New York: Raven.
Spitzer, R., & Wakefield, J. (1999). DSM-IV diagnostic criterion for clinical significance: Does it help solve the false positives problem? *American Journal of Psychiatry, 156*, 1856–1864.
Spitzer, R., & Williams, J. (1982). The definition and diagnosis of mental disorder. In W. Gove (Ed.), *Deviance and mental disorder* (pp. 15–31). Beverly Hills: Sage.
Stein, D., Phillips, K., Bolton, D., Fulford, K., Sadler, J., & Kendler, K. (2010). What is a mental/psychiatric disorder? From DSM-IV to DSM-V. *Psychological Medicine, 40*, 1759–1765.
Stone, A. (2007). *An introduction to feminist philosophy*. Cambridge: Polity Press.
Szasz, T. (1960). The myth of mental illness. Reprinted in Szsaz (1970) *Ideology and insanity* (pp. 12–24). Harmondsworth: Penguin.
Wakefield, J. (1993). Limits of operationalization: A critique of Spitzer and Endicott's (1978) proposed operational criteria for mental disorder. *Journal of Abnormal Psychology, 102*(1), 160–172.

# "Deviant Deviance": Cultural Diversity in DSM-5

**Dominic Murphy**

**Abstract** The DSM-IV-TR dealt with cultural diversity in mental illness by adopting the category of culture-bound syndromes (American Psychiatric Association, Diagnostic and statistical manual of mental disorders, 4th edn, text revision [DSM-IV-TR]. American Psychiatric Association, Washington, DC, 2000). DSM-5 makes some reforms including a structured program for assessing cultural influence on mental illness (American Psychiatric Association, Diagnostic and statistical manual of mental disorders, 5th edn [DSM-5]. American Psychiatric Association, Washington, DC, 2013). It also drops the category of culture-bound syndrome. These are improvements, but much of the underlying logic of DSM-IV-TR remains. In particular, the DSM-5 is still prone to see Western psychology as the human norm. These changes are discussed and situated in a wider intellectual context. Recent work in the cognitive sciences and cognitive anthropology has drawn attention to the dangers inherent in taking Western psychology as the norm. However, strong claims of cultural difference may rest on questionable philosophical assumptions about the semantics of kind terms. The relevant literature is surveyed and suggestions are made for developing psychiatry along lines that take its lessons into account.

## Introduction

An old joke has it that the behavioural sciences employ two model organisms, the rat and the American undergraduate. Rats are perfectly adequate model organisms, but Americans may not be. The problem is that Americans are WEIRD (Henrich et al. 2010). That is, they're Western, Educated, Industrialised, Rich and Democratic. In fact, most Westerners are WEIRD, but Americans are the WEIRDest of all. People in Western countries have values and minds that are not like those of the rest of humanity (that's not to say that the rest of humanity are all the same, just that Westerners seem to be distinctive along several dimensions). These differences should not be overstated; claims that there are no human universals are probably

D. Murphy (✉)
Unit for History and Philosophy of Science, University of Sydney, Sydney, Australia
e-mail: dominic.murphy@sydney.edu.au

 
S. Demazeux, P. Singy (eds.), *The DSM-5 in Perspective*, History, Philosophy and Theory of the Life Sciences 10, DOI 10.1007/978-94-017-9765-8_6

insupportable. However, the differences are real and they have implications for the sciences of the mind and brain that we are only just beginning to explore. From the perspective that takes WEIRD minds as normal, cultural variation is a deviation from the human norm, and cultural variation in mental illness looks like, as Hughes (1986, p. 3) noted, "deviant deviance": ways of being mentally ill that depart from the expectations taught in Western medical schools.

This essay looks at the implications of cultural diversity for psychiatry, in the context of DSM-5's changes. I will discuss the shifts in theory from DSM-IV-TR to DSM-5 and try to locate the specific issue of psychiatric cultural diversity in the wider context of the sciences of the mind. I will try to sketch the broader issues and advocate a conceptual shift that does not take Western minds as the default setting in psychiatry. It is here that DSM-5 really falls down. We need to take seriously the idea that WEIRD people might be an exception to the rule, or that there may be no rule at all.

## Culture and Psychiatry in DSM-IV and DSM-5

DSM-5 follows its predecessor in coping with cross-cultural variety in two ways. First, some conditions, such as depression and anxiety, are treated as universal – found in every culture, despite local differences in character. DSM-IV (p. xxiv) introduced the notion of culture-bound syndromes to express the idea that some cultures exhibited distinct forms of mental illness, but was especially concerned with conditions that western psychiatrists might see, for example among immigrant groups. This reflects a sensitivity to cross cultural differences in psychopathology that was growing at the time as medical anthropology began to influence psychiatry. Arthur Kleinman, a very influential psychiatrist and medical anthropologist, made the case for cultural variation with great authority for depression. He (1987) argued that depression takes different forms around the world. In Chinese subjects, it is likely to manifest as physical aches and pains. The experience of lower back pain and the experience of depression in the West, he said, are different enough that Chinese and American subjects might as well have different diagnoses. I will come back to this claim of Kleinman's, because it acknowledges unity but emphasises that diversity is what counts, and raises in abbreviated form most of the issues we need to discuss.

A related example is *Wacinko*, which is found only among the Oglala Sioux in the northwestern United States. *Wacinko* is a state of withdrawn, mute anger, directed at someone else, which may last for years. The Oglala live in the United States, but their culture is not shared with most Americans, and neither is *wacinko. Wacinko* is often a reaction to disappointment or interpersonal problems. Some Lakota writers see it as a form of strategic behaviour, aimed at manipulating another person's will. But because in extreme cases some subjects may commit suicide, this condition has been seen by Western-trained psychiatrists as a form of reactive depression (Lewis 1975).

If one adopts this unificationist route and sees different diagnoses as manifestations of the same underlying condition, the challenge is to widen the scope of the diagnosis and produce a more nuanced picture of the typical symptoms. It has been suggested that anxiety disorders, for example, should have their "worry domains" expanded because people from other cultures may worry about things that Euro-Americans don't fret over (Lewis-Fernandez et al. 2010). The assumption here is that anxiety disorders are common enough in all human cultures to merit treating as a universal, but that this can be missed if clinicians do not appreciate the different triggers, reactions and conceptions that distinguish the anxiety disorders in different populations. We need to recognise the greater variety of symptoms among sufferers from anxiety who live in Western cultures but don't show the typical symptoms of a Westerner. Lewis-Fernandez and his team (2010) argue that a majority of Latino sufferers from *ataque de nervios* would meet panic disorder criteria, for example, if the DSM amended the requirement that a panic attack lasts only a few minutes. *Ataque de nervios* was listed as a culture-bound syndrome in DSM-IV-TR. Its symptoms commonly include a mix of classic panic attack symptoms like palpitations and sensations of bodily heat and loss of control. But it may involve uncontrollable shouting and threats of violence, together with other symptoms less soften found in panic attacks. Lewis-Fernandez is a lumper; he is at least open to the idea that *ataque de nervios* can be assimilated to existing anxiety diagnoses provided we take the variation in symptomatology more seriously.

So the unificationist project involves seeing some non-Western conditions as expressions of one underlying process, which gives rise to different syndromes when it is active in different cultures. This fits well with a view of psychiatry as a branch of cognitive neuroscience: in (Murphy 2006) I alluded to the idea that different mental representations, transmitted from mind to mind within a culture, might interact with neurological mechanisms to give rise to specific local forms of a universal disorder. Theorists will disagree over whether the local or general forms should command more attention when building a theory, as we will see later on, but the strategy assumes that human minds have enough in common to build general models that permit local variation.

The second tactic used by American psychiatry to deal with the problem of cultural diversity has been the concept of "culture-bound syndromes." Many culture-bound syndromes have been recognised by psychiatrists. DSM-IV-TR defined culture-bound syndromes as "recurrent, locality-specific patterns of aberrant behavior and troubling experience that may or may not be linked to a particular DSM-IV diagnostic category" (p. 898). The distinction between different manifestations of a diagnosis and a culture-bound syndrome is drawn clearly:

> Although presentations conforming to the major DSM-IV categories can be found throughout the world, the particular symptoms, course, and social response are very often influenced by local cultural factors. In contrast, culture-bound syndromes are generally limited to specific societies or culture areas and are localized, folk, diagnostic categories that frame coherent meanings for certain repetitive, patterned, and troubling sets of experiences and observations.

DSM-IV-TR treats *ataque de nervios* as a culture-bound syndrome and not a form of panic disorder on the grounds that it is often associated with a precipitating event (panic attacks are supposed to come out of the blue) and is not marked by sufficiently acute feelings of fear and apprehension. Therefore, it does not conform to the diagnostic exemplar closely enough to count as panic disorder. DSM-5 did not adopt the amendments to the panic disorder criteria that were urged by Lewis-Fernandez and his team. Its treatment of *ataque de nervios* continues to see it as a distinct psychiatric kind.

Another example (not in DSM-IV-TR) is *Ashanti psychosis* (from West Africa) which comes in two main subtypes. In the "Fear-Guilt-Frenzy" subtype, subjects believe that they are being punished for some offense, often witchcraft. They become withdrawn, then paranoid and then hyperactive. The individual may sing, dance, rip their own clothes off, and consume faeces. The other frequent subtype is depressive. This looks to most people like psychotic behaviour, but it does not map well onto existing Western categories. (Edgerton 1966, in East Africa, found considerable local agreement on what counts as psychotic among the tribal societies he studied, and regular patterns of behaviour in which nudity was common, but hallucinations rare.) Interestingly, despite its insistence on the specific localities of culture-bound syndromes, DSM-IV-TR recognises similarities across them in some cases. *Amok*, an aggressive dissociative behavioural pattern from Malaya that gave us the English phrase 'running amok' is noted as similar to several other conditions in Southeast Asia, the Pacific and North America. This raises the question whether amok and its relatives should be seen as a cross-cultural condition. That cannot be admitted by the American Psychiatric Association, though, because in their view only DSM categories can be cross-cultural. In other words, American mental illness is universal, and other cultures have specific conditions. The American (indeed, the American of Non-Hispanic European descent) has humanity's factory model psychology, and other cultures just provide noisy local variation.

In theory, culture-bound syndromes are those folk illnesses in which alterations of behavior and experience figure prominently, relative to local norms. Culture-bound syndromes are conceived of as limited: they exist among specific societies. Standard psychiatric diagnoses are not thought of in this way, even if they are culturally limited. Multiple Personality Disorder/Dissociative Identity Disorder, for example, was widely diagnosed in the USA in the 1980s, but seldom found elsewhere. That does not lead psychiatrists to treat it as a culture-bound syndrome, even though it seems to fit the definition. (In fact DSM-5 has introduced the 'possession form' of Dissociative Identity Disorder, to include experiences of possession as well as those of internal takeover of the personality by an alter. I will come back to this.) In such cases, scepticism is likely to focus on the idea that there is a genuine condition here at all, but culture-bound syndromes are not seen as unreal or artifactual, just not standard. The assumption in mainstream psychiatry is that Western conditions are not culture-bound; they represent abnormalities in a universal human endowment. DSM-5 (pp. 243–244) for instance, treats muscle dysmorphia as a regular diagnosis, not a culture-bound syndrome. This is despite prevalence data for the wider condition of body dysmorphic disorder being noted only for Germany and the USA, and the characteristic behaviours being weightlifting and steroid use, which largely exist in specific western contexts. Avoidant/Restrictive Food Intake Disorder is noted as occurring in the USA, Canada, Australia and Europe (DSM-5, p. 336). Although these countries are not identical to each other, one would expect a wider range of cultures before it can be seen as universal. The point is that DSM-5 has main text for diagnoses that occur in the West and a special appendix for the rest if the world. It is hard not to see this as the expression of a view that Western minds are the norm. In this picture, culture-bound syndromes can come to be seen not as distinctive local forms of illness or distress, but as different expressions of the same underlying abnormality that Western subjects exhibit. There should be dysmorphia everywhere, but it won't always look American. It will however be the expression of a condition that is best understood by modelling the minds of people from NATO countries and then treating other populations as deviant. Before going on to a fuller philosophical discussion of these issues, let me discuss the changes introduced in DSM-5. They represent some improvements over the DSM-IV treatment of culture-bound syndromes, but the basic problematic assumption that Western minds are the default model is still there. This would be less of a problem if the ambitions of the DSM were limited, with only North America (or perhaps, the NATO countries) being seen as the context within which DSM categories are to be applied. However, DSM categories are seen as corresponding to universal conditions, as noted above. And certainly psychiatrists trained in DSM categories are found all over the world. Even if we anticipated that DSM would be used in just a few western countries, the presence of immigrants from other cultures in those countries would raise questions about whether the manual was handling cross-cultural issues adroitly. As it is, though, the DSM is being exported around the world, which makes the issues we are looking at most pressing.

DSM-5 has heavily qualified concept of a culture-bound syndrome. It has reformed psychiatry's approach to what it now calls *cultural formulation*. DSM-5 has moved to a different model, that of "cultural concepts of distress" (p. 758) which are the ways that "cultural groups experience, understand and communicate suffering, behavioural problems or troublesome thoughts and emotions." DSM-5 goes on to distinguish three main types of cultural concepts: syndromes, idioms (ways of expressing distress), and local explanations. The last is particularly interesting: a standard picture of mental illness in both non-Western and premodern societies represents it as explained by supernatural forces like ghost possession. But Edgerton (1966) reported that the East African tribes he studied were divided over the explanation of psychosis. Some tribes did understand it in terms of supernatural agency, but others thought of it as having an organic cause ("brain worms"). DSM-5 makes room for incorporating such findings as part of the integration of local explanations into a theory of mental illness.

DSM-5 supplements its conceptual reform with a thorough overhaul of the "Outline for Cultural Formulation" that was introduced in DSM-IV-TR, in part in response to pleas from cross cultural psychiatrists dissatisfied with the old ways (e.g. Canino and Alegria 2008). It reflects an unprecedented investment in applying anthropology to psychiatry by the DSM-5 Cultural Issues Subgroup, who ran field trials of the new Cultural Formulation Interview (CFI) in 6 sites around the world (Aggarwal 2013). The CFI calls for systematically assessing the cultural or ethnic groups that the subject belongs to, together with the ways in which they understand the problem and affect the subject's experience of it and quest for help.

Everyone has a culture, and the DSM-5 is officially committed to a much less WEIRD centric picture of mental illness than its predecessor. This is presumably linked to the fact that, historically, the editions of the DSM following DSM-III were aimed at a broader audience than merely American psychiatrists. DSM IV became, alongside the ICD (a general medical classification issued by the World Health Organisation), the international reference for the classification of mental disorders. The vast majority of "national" psychiatric classifications through the world – and even the Chinese classification – have been influenced by the DSM.

The CFI is a structured interview, in two versions. One is used for dealing with the patient and the other for informants such as family members who may be informative about the patients. It comprises a series of questions designed to gather information from the point of view of the patient and his or her cultural peers, including specific questions that tell the patient not to repeat what the doctor has told them but to express their problems as they see them and describe which aspects of their identity are most important to them. It could be used with any subject anywhere, and the manual acknowledges that all mental illness, including the DSM diagnoses, is shaped by the surrounding culture. On the other hand, the CFI and its associated conceptual material are poorly integrated into the DSM-5 – many diagnoses pay no attention to cultural formulation at all – and the culture-bound syndromes do in fact survive. They have been renamed "Cultural Concepts of Distress" but the appendix listing them survives, so one may wonder how thorough the overhaul has been.

Despite the rhetoric, it is unclear how the new appreciation for the cultural shaping of all psychopathology is going to influence the field, given that the treatment of these issues by the manual is still cursory (less than 20 pages out of 900) and poorly integrated with the diagnostic material. And if one thinks of one's own culture as just the way things are, or ought to be, then the problem of seeing cultural influence as interesting only if it marks a departure from one's own ways will persist.

Despite some hopeful signs in DSM-5, then, I do not think it has moved away from the traditional picture in psychiatry with its two approaches: the idea that cultural variation is either the distinctive cultural shaping of a universal condition or the idea that that culture-bound syndromes are peculiar local forms of distress with no claims to generality.

That is, the understanding of (e.g.) depression that derives from Western psychiatry is seen as the basic case, and other conditions are viewed as falling under the relevant diagnosis depending on how far they resemble the Western condition. Other conditions are local matters of no real psychiatric import, for they do not make us rethink our categories. Of course, nothing I have said so far means any of this is wrong, so now let me draw some morals via a discussion of the wider issues.

I have mentioned two ways that psychiatry deals with cultural variation: assimilation to the Western norm, or marginalization as culture bound syndromes. It is interesting that these occur together, because they can be seen as defining two opposite strategies: one insists on the psychic unity of human kind, despite local noise, and the other argues that there is no sense in which psychology is shared cross culturally. The latter strategy sees every culture-bound syndrome as a distinctive mental kind that cannot be understood other than in context, and cannot be assimilated to a wider condition.

## A Digression on Normativity

In abnormal psychology and psychiatry, most existing diagnoses have emerged from the tradition of asylum based, broadly biological psychiatry that began in late nineteenth-century Europe. This tradition validates diagnoses by formulating them in the context of a clinical population and then taking the resulting predictive instruments into the community, typically in the form of questionnaires. Answers to questionnaires determine whether subjects in the community will be diagnosed with the same condition as the clinical population. For this method to work cultural variation cannot make any relevant difference to the population to whom the instrument is being applied. This could be either because they share a cultural background with the initial clinical population, or because their cultural background, though different, does not mask the underlying similarities. As more and more psychiatrist are trained using Western methods and concepts these diagnoses are spreading into developing countries and traditional cultures. This first happened in remote areas of North America, such as the Spanish speaking villages of the New Mexico *altiplano*, and is occurring with increasing frequency around the world. The American journalist Watters, in a popular but rigorous treatment based on extensive interviews (2011) argues on the basis of several case studies that American psychiatry is launched on a campaign of world domination This is abetted by the pharmaceutical industry and the advertising it pays for and by social workers and carers with Western assumptions who try to counsel the suffering in developing countries by redescribing their experiences using Western categories. As cultures all over the world have their mental lives distorted and traditional ways of coping eroded by the spread of DSM categories, and the conceptual and clinical repertoire that comes with it, we will increasingly face important normative issues. In general, if we are inclined to think about curing, or mitigating, disorder as the goal of psychiatry, we face the question of whether curing would involve imposing a new form of life in members of a traditional culture. We would, in effect, be speeding up the arrival of Westernised modernity. Is eradicating a culture-bound syndrome akin to eradicating a traditional language (which is both a human tragedy and a loss of data) or more like just getting rid of smallpox? Smallpox has few champions, but if you incline to the view that culture-bound syndromes are not mental illnesses at all, just local forms of life, then trying to medicalize them and treat them as problems to be cured looks like cultural imperialism of the worst sort. All psychiatric conditions involve a judgement about whether someone's life is going badly. Applying these judgements to subjects in very different cultural contexts will always be even more fraught than making them in one's own cultural neighborhood. I will focus on the epistemological issues. These are hard enough, but probably easier than the moral ones. In the rest of the essay, then, I will try to situate the DSM-5 approach within the wider sciences of the mind.

## The Epistemology of Human Universality and Difference

First, then, the problem of WEIRD people. Henrich et al. survey a number of experimental results that suggest that people of European descent are not representative of human beings generally. This does not just include performance on high level cognitive tasks. It also involves systematic differences at lower levels. Take the famous Muller-Lyer illusion, for example, in which parallel straight lines of the same length are tipped with arrowheads pointing either inwards or outwards and respondents must judge which is longer. Philosophers and psychologists of perception teach their students that we always judge the line to be longer that has outward pointing fins.

But this is not an illusion at all as far as the San foragers of the Kalahari are concerned; they see no difference (Henrich et al. 2010, p. 65) whereas South Africans of European descent require the two lines to differ in length by about 20 % before the illusion goes away. Among American undergraduates an even greater difference in length is required before the two lines are seen as equal.

Results like this are significant for psychiatry because they suggest that cultural differences can run very deep within our psychology. Some experimental differences in cross-cultural cognitive psychology, such as those detected by Richard Nisbet (2003) between people of European and East Asian descent, implicate high-level processing systems, and one might think that these are of limited psychiatric interest, perhaps of relevance only to the content of delusions or breakdowns in thought disorder. You might think that mental disorders are more often the product of low-level neurological pathologies. If so, the Muller-Lyer results should give you pause, because we are talking here about the effects of culture on very low-level visual processing. Indeed, visual illusions were cited by Fodor (1983) as examples of the way in which early vision is insulated from higher cognitive control.

Another lesson from the Muller-Lyer results concerns the mechanism that the experimenters whose work Henrich et al. report (Segall et al. 1966) suggested for its explanation. Their idea is that the visual system in Western subjects adapts early in life to an environment full of "carpentered corners." People from Western countries grow up in a world of right angles and this biases their visual system to make specific inferences about distance from the edges and contours in its visual field. But this early environment is not the norm in human history, so it may be that WEIRD people are the exception to the human rule. Conversely, Segall et al.'s results suggest that children experience the Muller-Lyer illusion more than their elders, including African children. That may suggest that Africans grow up among building with rounded edges, to which they adapt. Perhaps Africans are exceptions to a pattern of development in which the visual system is set up to expect a world of edges (edge detection is a very important part of early vision and visual development). Or perhaps there is no default system and nobody is an exception, and the visual system is just highly malleable. We don't know, though the issue seems amenable to empirical proof. The point is that we have no grounds for thinking either that very early, low-level processes are exempt from culture, nor that Westerners have minds that are the human norm.

Even basic aspects of the mind can be cross-culturally distinguished. Geurts (2002) argues that the senses are culturally constructed: so that her subjects, the Anlo-Ewe in Ghana, have a different folk model of the sensorium, which privileges balance and reception of sound as the most important senses, rather than sight as in the Western "five-sense" model. She argues that this has significant implications for moral development and conceptions of health and disease. But, in fact, the idea that there are more than five senses has been recently canvassed among Western philosophers, and actually enumerating the senses turns out to be very difficult; the debate kicked off by Keeley (2002) turns out to be hard to resolve, especially once we widen its focus to include non-human animals. Is your sense of balance a distinct, sixth sense? Keely proposed a materialist program that individuates senses in terms of specific organs linked to the physical transduction of energy. This is very different from an approach like Geurts', which essentially says the senses are whatever the local culture takes them to be. This tension between the scientific approach and the local ways of knowing gets to the heart of the issues under discussion, for the natural way of following a universalist program is to look for, and privilege, the neuropsychology of a presumed universal diagnosis. The skeptical response is that even basic neuropsychology can be shaped by culture.

So far I have talked about cognitive processes. When it comes to affective psychology, claims about human universals are heard less often, and cross-cultural variety is widely taken for granted. In this case, too, a little bit of skepticism is in order, but in the reverse direction, since the appearances may mask a lesser degree of diversity. Whereas basic cognitive systems like vision can seem impenetrable by culture, emotions can seem to be nothing but culture. The anthropologist Catherine Lutz (1988, p. 10), on the basis of her work in Melanesia, argues that to "understand the meaning of an emotion word is to be able to envisage (and perhaps to find oneself able to participate in) a complicated scene with actors, actions, interpersonal relationships in a particular state of repair, moral points of view, facial expressions, personal and social goals, and sequences of events." Because of this interweaving of emotional vocabulary and cultural forms, Lutz is skeptical that there could be any genuinely cross-cultural emotions. For example, she compares (148–9) the Ifaluk emotion of *fago* to sadness, grief and depression in Western contexts. *Fago* most commonly arises because of the death of someone close, or their removal from the local group in another way (e.g. people may leave the island to work elsewhere).

Lutz sees sadness and depression in the West as essentially bound up with negative self-attributions and lowered self-esteem, and these are absent in *fago*. She rejects "sadness" as an appropriate translation of *fago* because it does not capture the way the concept works in Ifaluk discourse.

Mallon and Stich (2000) regard Lutz as arguing on the basis of an (unacknowledged) philosophical theory of how terms get their meaning. On the approach they attribute to Lutz, we can think of mental state terms as akin to theoretical terms in the sciences, and the meaning of terms as implicitly defined by the theory. That is one way of answering the question, in virtue of what do scientific terms refer to the things they talk about? Answering this question is part of general philosophical project that aims to understand the semantics of names, and in general terms there are two solutions. Going back to Frege (1892/1948) and up to Jackson (1998) and beyond we find the descriptive theory, or the idea that each name has associated with it a description that fixes its reference. In scientific terms, we can think of this as a definition provided by a theory, so "gene" would be defined as the entity that obeys Mendel's laws and is responsible for the replication and development of inherited traits. On the other hand, the causal-historical theory associated with Kripke (1972/80) argues that names, including scientific terms, are introduced into a linguistic community in order to refer to an object. Its continued reference to that object is preserved as long as its uses are linked to the original object via *a causal chain* of successive users, in which each user of the name learns it from a prior user all the way back to the first user. Crudely, the descriptive theory sees the meaning of a theoretical term as a set of concepts and the causal-historical theory sees the meaning of a term as the thing in the world it refers to. Mallon and Stich argues that Lutz adopts a descriptive theory, because she defines *fago* in Ifaluk terms as an emotion which is characteristically evoked by circumstances of loss, separation, etc., and which causes a set of typical behaviours and reactions.

This semantic digression does have a point: if you think of a culture's expectations of how mental life normally unfolds as defining its terms, then the meaning of those terms will not translate into a different context, because all the "theoretical" vocabulary shifts across contexts, and without the surrounding theoretical terms that implicitly define it, *fago* can have no meaning. But this is not the only way of defining theoretical terms, and it has well-known drawbacks. (For example, if a theory changes over time rather than across cultures, the new terminology should render the old concepts meaningless, just as cross-cultural change allegedly does for *fago*. But many scientific terms survive changes in theory.)

Now, if you think that there are likely to be cross-cultural emotions or other mental states, but you recognize the significant cross-cultural differences in the situations which provoke them and the behaviours they cause, then you cannot adopt the description theory of meaning that Mallon and Stich find in Lutz. It is quite possible to insist that a theoretical term refers to the same underlying phenomena in several cases. On this view of the meaning of a term, the associated behaviors and concepts are part of a full understanding of the phenomena we wish to explain in each cultural manifestation of the psychological entity in question. But the entity itself is the same in all cases – perhaps some shared underlying process that adapt or responds to different cultural settings to produce different behaviors.

A common response is to argue that the underlying abnormality, even if it exists, is beside the point. Our response to the condition should be determined by its expression, even if there is an underlying psychological kind. Here is Kleinman on depression in the West versus depression in East Asia (1987, 450): "Depression experienced entirely as low back pain and depression experienced entirely as guilt-ridden existential despair are such substantially different forms of illness behavior with different symptoms, patterns of help-seeking, course and treatment responses that though the disease in each instance may be the same, the illness rather than the disease is the determinant factor."

Horwitz and Wakefield (2007, p. 199) opt for universality in response: "we agree with Kleinman's distinction between disease as a universal underlying dysfunction and illness as the culturally shaped expression of a given dysfunction... [but] if there are indeed underlying common dysfunctions, then treatment presumably depends in large part on the science of identifying and intervening in such dysfunctions irrespective of their cultural presentation."

A number of points are raised by this exchange. They are connected to the argument about theoretical terms that I brought up above. We can dispense with two right away: first, it is very unclear exactly how we should view the semantics of theoretical terms, or indeed any others. Although I think the descriptive theory that Mallon and Stich read into Lutz faces important objections, it is certainly reasonable for any theorist to adopt such a view of the concepts she is studying in their context. Getting the hang of a cultural term does involve understanding how it is embedded in the practices and discourses of the people who use it. This leads to the second point we can put aside. Although Kleinman, Horwitz and Wakefield seem very sure of themselves, I am reluctant to tell people what they ought to be working on. It may be that if we are interested in treatment we should look for a common pathway that is always present and allows us to intervene. But it might also be that the best treatments are culturally specific. This seems to be an empirical issue, albeit one complicated by the difficulties that we would meet if we tried to establish what counts as a good treatment outcome in different societies. Still, it is entirely reasonable to opt for either a culturally conditioned investigation or an investigation aimed at uncovering universal processes.

There are two points to take away from this discussion: we have no reason to think that culture cannot penetrate even very low-level psychological processes, and we also have no reason to think that culture penetrates psychology so completely as to make cross-cultural generalizations impossible. Although I have used only a few illustrative examples, this is, I think, the right pair of lessons to draw, at least with respect to matters that psychiatrists will want to think about.

## Some Tentative Morals

We have reached a rather unsatisfactory position: on the one hand, people are probably quite a bit alike, especially if you drop the descriptive theory of reference, which accentuates difference by tying classification to local concepts. On the other hand, people are probably a bit different too, and culture shapes many aspects of our minds. This is probably the right conclusion to draw, since humans are alike and different, but it does not solve many problems. I think it premature to argue that all psychopathology is local, and also premature to announce the success of the univeralist conception of mental illness. But it is possible to make some suggestions about how we might proceed.

First, I have suggested that DSM-5, despite its notionally greater openness to medical anthropology, retains many of the bad habits of earlier editions. In particular, it has not fully broken free of the idea that Western psychiatric categories represent normal deviance, and non-Western ones represent deviant deviance. In the light of the results that Henrich et al. survey, this supposition must be repudiated. It is entirely reasonable to suggest that many disorders are experienced by Westerners in ways that reflect a distinctive cultural heritage that the rest of the world does not share, so that even if there are genuinely universal conditions, Westerners in fact may be the ones with the culture-bound syndrome. Take depression, for instance. The more we look, the more it seems as though the cognitive aspects of depression – the complex psychology of self-blame and immiserating introspection – may be distinctively Western (see also Kitanaka 2012 for a Japanese perspective which links depression to characteristic types of people and to excessive fatigue)

This does not mean that the universalising project cannot be carried on. The extent to which there are universal conditions is an empirical matter, although it is also a matter of finding the right grain at which to construct a family of models of human psychology and biology. I am reluctant to state that there is only diversity. But if we are to pursue the universalizing project we must be open to the idea of genuine culture-bound syndromes (i.e. distinctive conditions that only exist in a small number of related populations) alongside the universal ones. And we must also keep in mind the idea that if there are norms and exceptions, then Western categories may be the exceptions.

Reorienting our approach in this way might cause us to reconceive many accepted diagnoses. The introduction of possession-based forms of dissociative identity disorder may be another example: perhaps the epidemic of multiple personality that swept the USA in the 1970s and 1980s was a strange cultural malformation of a basic human tendency to feel like one's self has been take over? It may also be that once we look in Western cultures, we can find behaviors and experiences that conform to culture-based syndromes. Sumipathala et al. (2004), studying *dhat* (semen-loss anxiety) were able to find discussions of it in nineteenth-century British and Australian records, as well as many treatments of semen-loss in the Western medical tradition.

Second, it is pretty clear that essentialism in psychiatry is untenable, and that this takes simple reductionism down with it. Disorders do not share essences, but it nonetheless appears possible to think of them as natural kinds, using Boyd's cluster conception (Kendler et al. 2010; Boyd 1999). On this account, the characteristic properties of psychiatric kind would not occur in invariable sets, but in overlapping clusters, inhabiting a chunk of feature space because of biological (including evolutionary) and developmental processes that make it likely that if one feature is present, so are others. Mental illnesses depend on the coming together of numerous causal factors, and there is nothing conceptually peculiar about adding cultural features to the mix. Certain cultural forms may be quite widespread and robust and 'socially constructed' phenomena are perfectly real. The dynamics of cultural spread may raise the probability of some forms of mental illness just as surely as do developmental or genetic risks. We should not expect mental illnesses to look the same everywhere, but if we are going to pursue the universalising project we can try to build families of models of related conditions, clustering together with spaces separating them from other clusters. And we must not expect that the centre of the cluster will be a DSM diagnosis: that might be a rare form of a more generic condition.

So the universalising project may be possible, though it is quite likely that some local conditions will depend almost entirely on cultural formations that continue to resist assimilation to broader categories. Whether we call these "mental illness" will then be up for grabs; perhaps they are just local forms of life that occur when people suffer, without indicating that their minds are damaged in pathological ways. Whether the universalising project should be pursued is another matter; it raises ethical questions that I have already touched on, and to which I do not know the answer.

**Acknowledgements** I thanks the editors and two anonymous referees for their comments, and Pieter Adriaens and Ivan Crozier for sharing their work on culture-bound syndromes.

## References

Aggarwal, N. K. (2013). Cultural psychiatry, medical anthropology, and the DSM-5 field trials. *Medical Anthropology, 32*, 393–398.

American Psychiatric Association. (2010). *Diagnostic and statistical manual of mental disorders* (4th ed., text revision [DSM-IV-TR]). Washington, DC: American Psychiatric Association.

American Psychiatric Association. (2013). *Diagnostic and statistical manual of mental disorders* (5th ed. [DSM-5]). Washington, DC: American Psychiatric Association.

Boyd, R. (1999). Homeostasis, Species, and Higher Taxa. In R. A. Wilson (Ed.), *Species: New interdisciplinary essays* (A Bradford book, pp. 141–185). Cambridge, MA: The MIT Press.

Canino, G., & Alegria, M. (2008). Psychiatric diagnosis – is it universal or relative to culture? *Journal of Child Psychology and Psychiatry, 49*, 237–250.

Edgerton, R. B. (1966). Conceptions of psychosis in four East African societies. *American Anthropologist, 68*, 408–425.

Fodor, J. A. (1983). *Modularity of mind*. Cambridge, MA: MIT Press.

Frege, G. (1892/1948). On sense and reference. Translated by M. Black. *The Philosophical Review, 57*, 207–230.

Geurts, K. L. (2002). *Culture and the senses: Bodily ways of knowing in an African community*. Berkeley, CA: University of California Press.

Henrich, J., Heine, S. J., & Norenzayan, A. (2010). The weirdest people in the world. *Behavioral and Brain Sciences, 33*, 61–83.

Horwitz, A., & Wakefield, J. (2007). *The loss of sadness*. New York: Oxford University Press.

Hughes, C. C. (1986). Culture-bound or construct-bound. In R. C. Simons & C. C. Hughes (Eds.), *The culture-bound syndromes* (pp. 3–24). Dordrecht: Dordrecht Reidel.

Jackson, F. (1998). *From metaphysics to ethics*. New York: Oxford University Press.

Keeley, B. L. (2002). Making sense of the senses: Individuating modalities in humans and other animals. *The Journal of Philosophy, 99*, 5–28.

Kendler, K., Zachar, P., & Craver, C. F. (2010). What kinds of things are psychiatric disorders? *Psychological Medicine, 22*, 1–8.

Kitanaka, J. (2012). *Depression in Japan*. Princeton, NJ: Princeton University Press.

Kleinman, A. (1987). Anthropology and psychiatry. The role of culture in cross-cultural research on illness. *British Journal of Psychiatry, 151*, 447–454.

Kripke, S. (1972/1980). *Naming and necessity*. Cambridge, MA: Harvard, UP.

Lewis, T. H. (1975). A syndrome of depression and mutism in the oglala sioux. *American Journal of Psychiatry, 132*, 753–755.

Lewis-Fernández, R., Hinton, D. E., Laria, A. J., Patterson, E. H., Hofmann, S. G., Craske, M. G., Stein, D. J., Asnaani, A., & Liao, B. (2010). Culture and the anxiety disorders: Recommendations for DSM-V. *Depression and Anxiety, 27*, 212–229.

Lutz, C. A. (1998). *Unnatural emotions*. Chicago: University of Chicago Press.

Mallon, R., & Stich, S. (2000). The odd couple: The compatibility of social construction and evolutionary psychology. *Philosophy of Science, 67*, 133–154.

Murphy, D. (2006). *Psychiatry in the scientific image*. Cambridge, MA: MIT Press.

Nisbett, R. (2003). *The geography of thought*. New York: New York Free Press.

Segall, M. H., Campbell, D. T., & Herskovits, M. J. (1966). *The influence of culture on visual perception*. New York: Bobbs-Merrill.

Sumipathala, A., Siribaddana, S. H., & Bhugra, D. (2004). Culture-bound syndromes: The story of dhat syndrome. *British Journal of Psychiatry, 184*, 200–209.

Watters, E. (2011). *Crazy like us: The globalization of the American psyche*. New York: Free Press.

# Part II
# Specific Issues

# Danger and Difference: The Stakes of Hebephilia

**Patrick Singy**

**Abstract** The diagnostic category of "hebephilia" (the erotic preference for pubescent children) was suggested in 2008 for inclusion in the DSM-5. Immediately, a violent debate took place about whether this condition should be considered a disease or not, and the proposal to include hebephilia in the DSM-5 was rejected in 2012. In this paper I argue that the debate about the diagnostic validity of hebephilia was profoundly misguided. I first describe how the diagnosis of hebephilia plays a role in "sexually violent predator" (SVP) laws, which can preventively deprive "dangerous" people of their liberty if they are deemed mentally ill (for instance by suffering from hebephilia). I show that the legal requirement of mental illness for the application of SVP laws is supposed to serve two functions: to identify the most dangerous people, and to define them out of humanity by transforming them into quasi animals, thus safeguarding the constitutionality of SVP laws in a liberal context. I then argue that it fails to accomplish both tasks, and that the debate about hebephilia should have targeted this unsound legal requirement itself. Instead, because it was centered around the issue of diagnostic validity, the hebephilia debate rested on an implicit acceptance of the requirement of mental illness for the application of SVP laws.

## Introduction

In the summer of 2008 the *Archives of Sexual Behavior* published an article that created a great deal of controversy. Entitled "Pedophilia, Hebephilia, and the DSM-V," this article describes the concept of hebephilia, which denotes "the erotic preference for pubescent children," that is, for adolescents who are roughly between the ages of 11 and 15. Hebephilia is to be distinguished from pedophilia, which is an erotic preference for prepubescent children, and from teleiophilia, which is "the erotic preference for persons between the ages of physical maturity and physical decline" (Blanchard et al. 2009, 335–336). The eight authors of the article, led

P. Singy (✉)
Union College, Schenectady, NY, USA
e-mail: pbsingy@gmail.com

 
S. Demazeux, P. Singy (eds.), *The DSM-5 in Perspective*, History, Philosophy and Theory of the Life Sciences 10, DOI 10.1007/978-94-017-9765-8_7

by Ray Blanchard of the Department of Psychiatry at the University of Toronto, argued that hebephilia should be included in the DSM-5. Since Blanchard was a key member of the DSM-5's Sexual and Gender Identity Disorders Work Group, his proposal was bound to provoke some reaction.

From the start, it was met with great resistance. Most notable among Blanchard's critics was Allen Frances, who wrote blog entries, articles, and book chapters against the DSM-5 in general, and against the inclusion of hebephilia in it in particular.[1] More than for the depth of his thought, Frances is an important figure because of his role as chair of the DSM-IV task force.

Many people have supported Blanchard's proposal, and many people have joined Frances in his fight against Blanchard, but in the rest of this chapter I will use these two individuals as mascots of two opposite teams, the pro-hebephilia team of Blanchard (the members of this team think that hebephilia is a valid clinical entity that should be included in the DSM), and the anti-hebephilia team of Frances (the members of this team think that hebephilia is an absurd concept that should not be included in the DSM).

On the side of the anti-hebephilia team, two arguments were repeatedly invoked to demonstrate that hebephilia is not a disease. The first one made use of evolutionary theory: if something has an evolutionary function, then it cannot be a disease. Applied to hebephilia, the reasoning is straightforward: the natural function of sex is reproduction, a pubescent child is able to reproduce, ergo hebephilia is not a disease (Frances 2013, 201–2; Franklin 2009, 319–20).

The second argument against the inclusion of hebephilia in the DSM-5 used the criterion of statistical abnormality, or more loosely of bizarreness: something is not a disease if something is (statistically) "normal." Frances claimed that being attracted to young teenagers is very common for men, so hebephilia cannot be a disease. As proof, Frances and others pointed to some advertisements that use very young people posing in a sexually provocative way (Frances and First 2011, 84; Wakefield 2011, 206).

The pro-hebephilia team offered a rebuttal to each argument. First, Blanchard explained that if we want to use evolutionary theory to decide what is a disease and what is healthy, then we should reinstate homosexuality in the DSM (Blanchard 2013, 677). Since this has never been an appealing option for any of the two teams, Blanchard concludes that the evolutionary argument against hebephilia falls flat.[2]

[1] See in particular several entries in his blog, www.psychologytoday.com/blog/dsm5-in-distress, as well as his recent book against the DSM-5 (Frances 2013).

[2] As far as I know, Blanchard never discusses Jerome Wakefield's influential "harmful dysfunction" analysis of mental disorder. Yet it would be an obvious response to his point. According to Wakefield, two criteria must be met for a condition to qualify as a disease: it must be at the same time evolutionary dysfunctional and harmful. With these criteria, homosexuality is not a disease because while it might be evolutionary dysfunctional (if we assume that the evolutionary function of sex is reproduction), it is not harmful. Hebephilia would also not be a disease, but for the opposite reason: while it is harmful (by virtue of being a crime), it is not dysfunctional (because pubescent children are able to reproduce). See Wakefield (1992) and (2011).

Moreover, according to Blanchard it is in fact not true that hebephilia is evolutionary adaptive. He argued that hebephiles have a reduced reproductive fitness, since, according to a statistical study, they have less biological children than teleiophiles (Blanchard 2010a; Hames and Blanchard 2012).

In response to the second argument, Blanchard explained that his opponents have misunderstood him. He agreed that most men are somewhat attracted to pubescent children. But hebephilia is not a sexual *attraction* to pubescent children, it is a sexual *preference* for pubescent children. While it is statistically normal, and therefore non-pathological, for men to be at least a little bit attracted to pubescent children, it is statistically abnormal, and therefore pathological, to *prefer* pubescent children over adults (Blanchard 2013, 676).

On the surface, the debate between the two teams had the appearance of a serious intellectual debate that brought into play a conceptual clarification, statistical references, and evolutionary theory. Yet it quickly turned into a fairly brutal confrontation (by academic standards). Frances warned that the DSM-5 will be "an embarrassment and a burden to the field" (Frances 2009) plagued by "poor and inconsistent writing" (Frances 2010a). According to him, the Sexual Disorders Work Group in particular "has strayed furthest off the reservation," with suggestions "based on the thinnest of research support"—and among these suggestions was the inclusion of hebephilia (Frances 2010b). In retaliation Frances has been accused of offering "poorly reasoned claims," of firing off "criticisms as quickly as his grandchildren might tweet to their friends," and of having enabled the "logical absurdities" of the DSM-IV (Zucker 2010). The President of the American Psychiatric Association (Alan F. Schatzberg), the CEO and Medical Director of the APA (James H. Scully), the Chair of the DSM-5 Task Force (David J. Kupfer) as well as its Vice Chair (Darrel A. Regier), all accused Frances of being motivated by money in his criticism of the DSM-5: the publication of the latter signals the end of the DSM-IV, and therefore the end of the royalties that Frances received for his work on it (Schatzberg et al. 2009). In order to defend the DSM-5, the APA went so far as to hire a former Pentagon spokesman, who told a reporter that Frances "is a 'dangerous' man trying to undermine an earnest academic endeavor" (Cloud 2012).

In order to understand why the debate about hebephilia became so heated, it is crucial to consider the forensic context without which the concept of hebephilia would certainly not have been suggested in the first place. Since 1990 the USA have passed laws against what is called "sexually violent predators" (SVP). Washington was the first state to pass an SVP law in 1990. Since then many states have adopted similar laws, whose constitutionality has been reaffirmed by the US Supreme Court. SVP laws allow for the indefinite civil commitment of people who are deemed sexually dangerous. People are locked up in psychiatric hospitals for fear of the crimes they *might* commit if we let them loose into society.[3]

[3]For an excellent critical introduction to SVP laws, see Janus (2009). In a recent novel, Russell Banks (2011) offers a gripping first-person account of a sexual offender subjected to SVP laws.

Not every rapist qualifies for civil confinement, however. Only those with a mental illness might. A rapist who is deemed mentally sane will go to jail and be released after completion of his sentence. A rapist who is deemed mentally insane might on the other hand qualify for civil confinement after he has served his jail sentence. As it happens, one of the mental disorders that lawyers sometimes invoke to get someone civilly committed is "hebephilia"—along with "antisocial personality disorder," "pedophilia," and more generally "paraphilia not otherwise specified" (Franklin 2010, 753). Because of the way the law is framed—because a necessary condition for civil commitment is mental illness—it is indeed quite important to know whether hebephilia should be included in the DSM or not.

The role of the DSM in the courtroom is in reality a little bit more ambiguous. On the one hand, while the law requires the presence of a mental illness in the defendant for him to be civilly committed, it does not require expert witnesses to use DSM diagnoses *per se*. In the Supreme Court case of *Kansas v. Hendricks* (521 U.S. 346, 1997), the majority clearly stated that the legal and the psychiatric concepts of mental disorder are not and do not need to be equivalent: "The legal definitions of 'insanity' and 'competency' ... vary substantially from their psychiatric counterparts .... Legal definitions ... need not mirror those advanced by the medical profession" (*Kansas v. Hendricks*, 359). Hebephilia was used in the courtroom before Blanchard's article came out, and Blanchard's intention in introducing hebephilia into the DSM-5 was simply, as he himself explained, to "systematize what is already happening unsystematically" (Blanchard 2010b, 311).[4]

On the other hand, it is also true that a defendant who receives a diagnosis that has been vetted by the APA is more likely to be seen as "really" mentally ill. Again in *Kansas v. Hendricks*, the decision to commit the pedophile Hendricks was upheld by the Supreme Court because "the mental health professionals who evaluated Hendricks diagnosed him as suffering from pedophilia, *a condition the psychiatric profession itself classifies as a serious mental disorder*" (*Kansas v. Hendricks*, 360, my italics; this point is reaffirmed in *Kansas v. Crane*, 534 U.S. 407 (2002)). The DSM, in other words, can give legitimacy to what otherwise could be seen as a dubious diagnosis. In a more recent court decision, civil confinement was in fact rejected on the ground that hebephilia is not an established diagnosis (*United States v. Neuhauser*, U.S.D.C. (E.D. N.C.), Case No. 5:07-HC-2101-BO; 2012 WL 174363 (Jan. 20, 2012); see also Wakefield 2011, 206).

In short, while from a forensic standpoint the inclusion of hebephilia in the DSM-5 would not have been decisive, it would not have been inconsequential either. The

[4]To be fair to Blanchard, it is important to note that he has never "testified in any criminal case in any country, [has] never earned one penny from private practice [, and has] the same discomfort with American SVP civil commitment laws that many others have expressed" (Ray Blanchard, personal communication, August 9, 2013). There is therefore no direct causal link between SVP laws and Blanchard's research. It is however obvious that his research would have no *raison d'être* without a forensic context. It is this context that explains why he studies hebephilia or pedophilia rather than the sexual preference for red-hair people, for instance. As a historical fact, all the main sexual perversions have a forensic origin, and hebephilia is no different in that respect.

debate about hebephilia was so heated because for some people whether hebephilia is officially considered a disease or not might mean the difference between freedom and indefinite civil confinement in a hospital.

On December 1, 2012, the Board of Trustees of the American Psychiatric Association opted for the status quo and decided to reject the inclusion of hebephilia in the DSM-5. Frances's team won, Blanchard's lost.

In the rest of this chapter I will argue that Frances's victory is in reality a Pyrrhic victory. The price to pay has been heavy, since the fight against hebephilia was also indirectly a defense of the legal and political conditions of possibility for this concept. Both teams have accepted much too quickly the rules of the game that the Supreme Court has set up for them. Those rules, as I will now show, are flawed, and instead of arguing about the best way to play the game, our critique should take the game itself as its target. Against both Blanchard and Frances, and ultimately against the Supreme Court, I will argue that whether hebephilia is a disease or not should be irrelevant. What should be at stake in the concept of hebephilia is not the problem of diagnostic validity, but the issues of danger and difference.

## SVP Laws and Liberalism

In order to understand why the debate about the diagnostic validity of hebephilia was misguided, we first need to step back and look more broadly at the strategic function of SVP laws in our modern liberal societies. What is the larger historical and political context that can explain the existence of SVP laws?

Since the Enlightenment, in reaction against absolute monarchies, the political philosophy of liberalism has become increasingly dominant in the Western world. The label "liberalism" can apply to many different approaches to governing, but the common thread is an emphasis on individual liberty and personal happiness. A central tenet of liberalism is that "everyone is the best judge of his/her interests and of the management of his/her life, is left as free as possible as long as he/she does not harm others and does not make an attempt on anyone's vital interests" (Audard 2009, 10).

As Michel Foucault and others have observed, the emphasis on liberty means also, in an apparent paradox, an emphasis on security and all the constraints on liberty that this implies (Foucault 2004, 66–69). One of the main founders of liberal thought, John Stuart Mill, made the indisputable point that without security one cannot enjoy liberty.[5] What liberalism promotes is not liberty per se, but what the opinion of the court in *Kansas v. Hendricks* calls "ordered liberty" (*Kansas v. Hendricks*, 357).

[5]See Mill (1998, 190): "security no human being can possibly do without; on it we depend for all our immunity from evil, and for the whole value of all and every good, beyond the passing moment, since nothing but the gratification of the instant could be of any worth to us, if we could be deprived of everything the next instant by whoever was momentarily stronger than ourselves."

While liberty and security are both necessary for liberalism, they pull in different directions. A very secure society would be a society with very few liberties, and a very free society would be a society with very few securities. We need to balance liberty and security, but where we draw the line has been a constant battle since the early nineteenth century.

SVP laws embody this tension between liberty and security. The following statement from *Kansas v. Hendricks* shows quite clearly how the requirement of the presence of a mental disorder is meant to satisfy the two liberal demands of liberty and security: "the precommitment requirement of 'mental abnormality' or 'personality disorder' ... narrows the class of persons eligible for confinement to those who are unable to control their dangerousness" (*Kansas v. Hendricks*, 358). In other words, the criterion of mental illness is supposed at the same time to identify the most dangerous individuals and to curtail the reach of the law by narrowing its application to a limited number of individuals. By identifying the most dangerous individuals and protecting us against them, SVP laws supposedly increase security; and by limiting their reach to people with a mental illness, SVP laws claim to safeguard the liberty of everybody else. What needs to be determined now is whether the criterion of mental illness can legitimately do these two jobs that are being asked of it.

## Danger

In the first instance, does the criterion of mental disorder help us identify the most dangerous individuals? In the previously quoted sentence from *Kansas v. Hendricks* we discern a chain of connections: very clearly, for the Supreme Court there is a connection between mental disorder and lack of self-control, and more implicitly between lack of self-control and dangerousness. SVP laws target only people who have a mental disorder, because those people are said to lack self-control, which makes them more dangerous.

Everything about this statement is problematic, to say the least. First, is there really a direct link between mental illness and lack of self-control? The writers of the DSM, including of the DSM-5, do not think so: "a diagnosis does not carry any necessary implication regarding ... the individual's degree of control over behaviors that may be associated with the disorder" (American Psychiatric Association 2013, 25). Unlike psychiatrists, the Supreme Court is victim of a supreme confusion between having a disease and suffering from an

---

See also, for instance, the following passage from the 1905 case of *Jacobson v. Massachusetts*, quoted in *Kansas v. Hendricks*, 356–7: "[T]he liberty secured by the Constitution of the United States to every person within its jurisdiction does not import an absolute right in each person to be, at all times and in all circumstances, wholly free from restraint. There are manifold restraints to which every person is necessarily subject for the common good. On any other basis organized society could not exist with safety to its members."

incapacity such as lack of self-control. One can have a disease and be able to control oneself, and one can be sane and unable to control oneself (see Singy 2015).

The fact that psychiatrists and lawyers do not see eye to eye is not surprising. What is harder to understand is that the connection between mental disorder and lack of self-control is undermined by the way SVP laws themselves actually work. As mentioned above, in general the person who is civilly committed with SVP laws first went to jail. The insanity defense was not used for him, he was considered by the law to be a fully responsible individual who freely chose to commit a crime. Yet now, after he has served his sentence, he is considered to be unable to control himself. This is an obvious contradiction. The same person is removed from society twice, the first time because he is judged to have committed his crime in full control of himself, and the second time because he is judged to be unable to control himself. It is as if lawyers did not really believe in the connection between mental disorder and lack of self-control, even though the constitutionality of SVP laws depends on this connection.

Would there be at least a connection between mental disorder and dangerousness – whether or not mentally ill people are able to control themselves? Are mentally ill people the most dangerous people? We might be inclined to answer positively here if we automatically equate the perpetration of evil with mental illness. In this case, Hitler or Bin Laden necessarily suffered from mental illnesses. This is a proposition that has gained some strength since the nineteenth century, but that continues to sit awkwardly with our faith in free will.[6] Yet even if we were to agree that dangerousness is itself a mental illness, mental illness still should not be used as a legal criterion to limit the law. Think of it this way: most sexual predators are men, not women, but it would be discriminatory to write in the law that SVP laws can only target men, and never women. *As a matter of fact*, SVP laws might target mostly or only men, but *as a matter of law* they should target all sexually dangerous individuals, whatever their gender, race, religion, length of their nose, IQ level—or mental health. If the function of SVP laws is to protect us against dangerous sexual criminals, then being sexually dangerous should be the one and only criterion. Being a man or being mentally ill might be correlated with being a dangerous sexual criminal, but this empirical regularity cannot serve as a legal criterion.

The fate of sexual predators is often compared to the one of alleged terrorists, who are pre-emptively sent to places like Guantanamo Bay to protect society. This is not an original argument: SVP laws are part of a broad trend in our society toward greater and greater security (see Hudson 2003; Janus 2009, 94). In our legal reasoning about alleged terrorists, their mental state is irrelevant, and rightfully so: what matters is our perception of the danger they pose. The mental state or lack

[6]On this topic, see the historically important discussion in the *Annales médico-psychologiques* 2 (1863), especially Dally's contribution to the discussion, 260–295. For a more modern discussion of this issue, see Eagleman (2012, 151–192).

of self-control of sexual predators should be just as irrelevant as it is for alleged terrorists.[7] The application of security measures needs to be entirely detached from diagnostic issues.

Needless to say that I am not defending the way we handle presumed terrorists any more than I am advocating extreme security measures for sexual criminals. I am only trying to point out that knowing whether a condition is a disease or not is a very different problem from knowing whether someone is dangerous or not. In order to evaluate the amount of danger someone represents, we need to settle on a threshold of danger beyond which we, as a society, think it is worth taking preventive action, and we need to figure out a way to assess risk in a reliable way. Those are thorny political and technical issues, and I do not have the audacity to try to elucidate them here. My point is only that they are logically independent from any question of nosology.

## Difference

When it comes to the fundamental tension between liberty and security that lies at the core of liberalism, SVP laws clearly draw the line in favor of security, since they enable the confinement of people who *might* commit sexual crimes in the future. But since liberty remains what liberalism is primarily about, lawyers and politicians who deploy preemptive measures such as SVP laws suffer from what Barbara Hudson has called a "bad conscience" (Hudson 2003, 35). In a totalitarian regime SVP laws would be rather unproblematic, but in liberal societies they are always controversial since they deprive people of their liberty preventively.

We have however found an original way to alleviate our bad conscience without sacrificing security: we define sexual predators outside humanity and limit the application of SVP laws to them. There is a meaningful analogy to be made here with our treatment of dangerous animals. Imagine a tiger roaming free in the heart of New York City. We are not going to wait for this tiger to kill someone before we intervene. We will protect ourselves preemptively by putting the tiger in a cage or relocating it to an isolated region of the world. Most importantly, we do not have a bad conscience taking preemptive measures against the tiger, because our society (despite the work of animal rights activists) does not find it unjust to deprive animals of some of the rights it gives to humans. In order to alleviate our bad conscience when we enforce SVP laws against hebephiles, we therefore bring the latter

[7] As Richard Green says in an article against the inclusion of hebephilia in the DSM-5, "Thwarted suicide bombers who continue to pose a public threat can be caged without terrorism entering the DSM"; Green (2010, 586).

ontologically closer to wild beasts than to humans. The word "predator" is in that regard very well chosen: we have given an animal-like nature to the sexual criminal.[8]

The criterion of mental illness required by the Supreme Court has the purpose not only of identifying danger (and we saw how it fails to do that), but also of identifying difference—that is, radical difference, difference not only in some attributes of the person, but in the very nature of his being: an ontological difference. The constitutionality of SVP laws rests on this radical difference between "them" and "us." If hebephiles and other types of sexual predator are different enough from "normal" people, then they can legitimately be deprived of some of the rights that the latter enjoy, without this security measure contradicting a fundamental principle of liberalism. As Eric Janus explains, "the central task for the courts has been to show that there is some *principled* limitation that will prevent uncontrolled spread of what is otherwise an unconstitutional application of government power. In order for this judicial narrative to be successful, it must articulate some *fundamental difference*—some difference in kind—between those who are entitled to the normal protections of the Constitution and those who are not" (Janus 2009, 26).

But how different are hebephiles from "normal" people? Blanchard does make a fairly convincing case that hebephiles are in some sense different. His main tool is phallometric testing. Taking his inspiration from his mentor Kurt Freund, he sits his subjects in a chair, covers their penises with a glass cylinder, seals the cylinder, and plugs the whole thing to a machine that can measure extremely precisely the changes in air pressure in the cylinder caused by the tumescence or detumescence of the penis. He then presents the subjects with different erotically charged images and observes if there is a correlation between certain types of images and the tumescence of the penis. In this way he can show that some people have the strongest erections when they are presented with images of children (he calls them "pedophiles"), that others have the strongest erections when they are presented with images of adult women (he calls them "teleiophiles," what other people call "normal"), and finally that some people fall in between: their erections correlate most strongly with images of pubescent children. This last group is the hebephiles. Blanchard's concern is to establish different categories of sexual preference. Assuming that his research is empirically sound, then it is true that hebephiles are in a sense different from "normal" people: hebephilia is "a discriminable erotic age-preference" (Blanchard et al. 2009, 335).

But this obviously does not answer the question of ontological difference and its legal implications. In and of itself, a psychological or biological difference does not necessarily imply an ontological difference. We are all of different sizes, different

[8]My tiger analogy is not randomly chosen: in the early nineteenth century André Matthey suggested that people who have an extreme propensity to violence without delirium — what Pinel had called "mania without delirium" — should be diagnosed with "tigridomanie" (Matthey 1816, 117).

intelligence, different ages, we have different tastes, we suffer from different diseases, etc., but these differences in themselves do not imply that we need to be subjected to different laws.

The question remains: are hebephiles ontologically different from normal people? As far as I can tell, the only plausible answer to this question is a cynical one. It requires from us to turn upside down our habitual way of thinking about sexual criminals: it is not because sexual criminals are no longer human that we do not need to have a bad conscience when we use SVP laws against them; rather, it is because we want to alleviate our bad conscience that we create the illusion of an ontological difference.

Since the early nineteenth century—since the emergence of liberalism—the manufacture of a criminal/animal has typically been a two-step process. It begins with a simple reaction of horror: a crime has been committed that we feel is particularly disgusting. We demand exceptional legal action. We then proceed to confirm our first feeling of horror as well as to rationalize our extraordinary legal demands by creating a radical difference between the criminal and us. This is the role of the science of criminology, which emerged in the nineteenth century. What criminologists and their associates do is to detail and collect the criminal's biological and psychological attributes, like so many brush strokes delineating a distinctive silhouette. Blanchard and his colleagues, for instance, do not only measure erections: "Studies have compared pedophilic, hebephilic, and teleiophilic men on a variety of dependent measures. The results have shown hebephiles to be intermediate between pedophiles and teleiophiles with regard to IQ . . . , completed education . . . , school grade failure and special education placement . . . , head injuries before age 13 . . . , left-handedness . . . , and stature" (Blanchard et al. 2009, 336). The singular "nature" of the criminal emerges through the correlation of all these measures.

The more the law wants to apply preemptive measures against some people, the more it gives an incentive to science to provide thick descriptions of these people, who at some point will begin to look like they have their own singular nature, at the margins of human nature. The sexual predator understood as a quasi-animal is a product of liberalism, and more precisely a product of the tension within liberalism between security and liberty.

In any case, it is not by deciding whether something is a disease or not that we will know whether having this disease signals such a radically different nature that special laws need to be used. We cannot simply accept as a given that hebephiles are ontologically different from the rest of us, and then debate about the pathological status of hebephilia. Rather, with the help of philosophical and historical tools, we need first and foremost to question the constitution of difference itself.

## Conclusion

In an article published in 2011 as the hebephilia debate was raging on, the most philosophically astute member of the anti-hebephilia team, Jerome Wakefield, claimed that "the first task should be to 'get it right' with respect to diagnostic validity, and then to explore the practical needs in the forensic context within the framework set by validity considerations. Otherwise, the forensic tail is wagging the validity dog" (Wakefield 2011, 208). Unlike Wakefield, I have argued that our first task should be to show that we are dealing here with a case of mistaken implant. The forensic tail should not in fact belong to the validity dog, and we should direct our efforts at cutting the tail off.

Two other scholars, Michael First and Robert Halon, have suggested something a little bit different from Wakefield: they claim, and they are certainly right, that forensic experts who testify in SVP cases make important errors when they diagnose paraphilias. These experts focus for instance exclusively on the behavior of the criminal without taking into consideration his urges and fantasies. As a remedy, First and Halon offer a three-step approach that is meant to raise the level of diagnostic accuracy (First and Halon 2008). Here again, while I applaud the good intentions of the authors, I worry that they only contribute to the problem. Instead of trying to increase diagnostic accuracy, we should show that diagnoses must be irrelevant to SVP cases.

Ironically, getting the validity question right like Wakefield or improving diagnostic accuracy like First and Halon, means reinforcing what is wrong with SVP laws and hebephilia, since it implies that we uncritically accept that the pathological status of hebephilia is legally relevant. Our critique should not target the application of the criteria used by SVP laws if these criteria themselves are fundamentally flawed.

When we reflect about hebephilia, we need to think about danger and difference, not about disease. At what point is someone dangerous enough that we want to cage him preemptively? How do we decide when we are dealing with a radical, ontological difference that justifies the suspension of basic legal rights in a liberal context? Psychiatry cannot help us decide what to do with hebephiles, not because this discipline has a questionable level of scientificity, but because the issues raised by hebephilia are not of a psychiatric nature. They are philosophical, legal, and above all political.

## References


American Psychiatric Association. (2013). *Diagnostic and statistical manual of mental disorders: DSM-5*. Arlington, VA: American Psychiatric Association.

Audard, C. (2009). *Qu'est-ce que le libéralisme? Ethique, politique, société*. Paris: Gallimard.

Banks, R. (2011). *Lost memory of skin*. New York: HarperCollins.

Blanchard, R. (2010a). The fertility of hebephiles and the adaptationist argument against including hebephilia in DSM-5. *Archives of Sexual Behavior, 39*, 817–818.

Blanchard, R. (2010b). The DSM diagnostic criteria for pedophilia. *Archives of Sexual Behavior, 39*, 304–316.

Blanchard, R. (2013). A dissenting opinion on DSM-5 pedophilic disorder. *Archives of Sexual Behavior, 42*, 675–678.

Blanchard, R., Lykins, A. D., Wherrett, D., Kuban, M. E., Cantor, J. M., Blak, T., Dickey, R., & Klassen, P. E. (2009). Pedophilia, hebephilia, and the DSM-V. *Archives of Sexual Behavior, 38*, 335–350.

Cloud, J. (2012). What counts as crazy? *Time*, 19 Mar.

Eagleman, D. (2012). *Incognito: The secret lives of the brain*. New York: Vintage Books.

First, M. B., & Halon, R. L. (2008). Use of DSM paraphilia diagnoses in sexually violent predator commitment cases. *The Journal of the American Academy of Psychiatry and the Law, 36*, 443–454.

Foucault, M. (2004). *Naissance de la biopolitique*. Paris: Seuil/Gallimard.

Frances, A. (2009). DSM-V badly off track. *Psychiatric Times*, 26 June.

Frances, A. (2010a). Opening Pandora's box: The 19 worst suggestions for DSM5. *Psychiatric Times*, 11 Feb.

Frances, A. (2010b). DSM5 and sexual disorders—just say no. *Psychiatric Times*, 17 Mar.

Frances, A. (2013). *Saving normal: An insider's revolt against out-of-control psychiatric diagnosis, DSM-5, big pharma, and the medicalization of ordinary life*. New York: William Morrow.

Frances, A., & First, M. B. (2011). Hebephilia is not a mental disorder in DSM-IV-TR and should not become one in DSM-5. *The Journal of the American Academy of Psychiatry and the Law, 39*, 78–85.

Franklin, K. (2009). The public policy implications of 'Hebephilia': A response to Blanchard et al. (2008). *Archives of Sexual Behavior, 38*, 319–320.

Franklin, K. (2010). Hebephilia: Quintessence of diagnostic pretextuality. *Behavioral Sciences & the Law, 28*, 751–768.

Green, R. (2010). Sexual preference for 14-year-olds as a mental disorder: You can't be serious! *Archives of Sexual Behavior, 39*, 585–586.

Hames, R., & Blanchard, R. (2012). Anthropological data regarding the adaptiveness of Hebephilia. *Archives of Sexual Behavior, 41*, 745–747.

Hudson, B. (2003). *Justice in the risk society*. London: Sage Publications.

Janus, E. S. (2009). *Failure to protect: America's sexual predator laws and the rise of the preventive state*. Ithaca: Cornell University Press.

Matthey, A. (1816). *Nouvelles recherches sur les maladies de l'esprit, précédées de Considérations sur les difficultés de l'art de guérir*. Paris: J. J. Paschoud.

Mill, J. S. (1998). Utilitarianism. In *On liberty and other essays* (pp. 129–201). Oxford: Oxford University Press.

Schatzberg, A. F., Scully, J. H., Jr., Kupfer, D. J., & Regier, D. A. (2009). Setting the record straight: A response to Frances commentary on DSM-V. *Psychiatric Times*, 1 July.

Singy, P. (2015). Power, knowledge and laughter: Forensic psychiatry and the misuse of the DSM. In: G. Lambert, M. Silberstein, & P. Huneman (Eds.). *Classification, disease and evidence: New essays in the philosophy of medicine* (pp. 131–145). Dordrecht: Springer.

Wakefield, J. C. (1992). The concept of mental disorder: On the boundary between biological facts and social values. *American Psychologist, 47*, 373–388.

Wakefield, J. C. (2011). DSM-5 proposed diagnostic criteria for sexual paraphilias: Tensions between diagnostic validity and forensic utility. *International Journal of Law and Psychiatry, 34*, 195–209.

Zucker, K. J. (2010). DSM5: Pandora replies to Dr Frances. *Psychiatric Times*, 22 Feb.

# Sexual Dysfunctions and Asexuality in DSM-5

**Andrew Hinderliter**

**Abstract** In the past 15 years or so, there has been a large increase in the amount of industry funding for sex research, as well as a backlash against the "medicalization" of female sexuality. During the same time, there has been a growing community of people identifying as "asexual," which is often seen as a sexual orientation. This chapter considers the interaction of these in leading up to DSM-5, as well as concerns about clinical significance criteria and difficulties in distinguishing between "sexual dysfunctions" and "normal variation."

## Introduction

In the various controversies, debates, and arguments surrounding conditions that were in DSM-IV's (APA 1994) "Sexual and Gender Identity Disorders" chapter, it has not been uncommon for the declassification of homosexuality in 1973 to be invoked by those arguing for the declassification or serious modification of other diagnoses. This is true of the paraphilias (e.g. Green 2002; Hinderliter 2010; Moser and Kleinplatz 2005) and for gender identity disorder (e.g. Winters 2009). In the past 10–15 years, there has been an increasingly visible group of people identifying as "asexual," and many asexuals have advanced similar arguments concerning what, in DSM-IV was called hypoactive sexual desire disorder (HSDD). Although the precise relationship between asexuality and HSDD is unclear (e.g. Bogaert 2006; Hinderliter 2013), roughly speaking, both are categories involving (certain kinds of) lack of interest in sex, but they frame the matter differently and have rather different histories.

DSM-III (APA 1980) introduced dramatic changes to the DSM, and among these changes was the addition of a section on sexual dysfunctions. Following the publication of *Human Sexual Inadequacy* (Masters and Johnson 1970), many

A. Hinderliter (✉)
University of Illinois, Chicago, IL, USA
e-mail: andrewc.hinderliter@gmail.com

 
S. Demazeux, P. Singy (eds.), *The DSM-5 in Perspective*, History, Philosophy and Theory of the Life Sciences 10, DOI 10.1007/978-94-017-9765-8_8

psychiatrists and psychologists became interested in sex therapy, often using unofficial diagnostic systems (Segraves et al. 2007). Adding sexual dysfunction in DSM-III provided a standardized system.

The section on sexual dysfunctions in each version of the DSM since DSM-III has included at least one diagnosis having something to do with lack of interest in sex. Along with the rest of the sexual dysfunctions, it has lifelong and acquired subtypes, as well as situational and generalized subtypes. In DSM-III, it was called inhibited sexual desire. In DSM-III-R (APA 1987), the name was changed to hypoactive sexual desire disorders as it was felt that the previous name suggested a psychodynamic etiology. DSM-III-R's definition was "Persistently or recurrently deficient (or absent) sexual fantasies and desire for sexual activity. The judgment of deficiency or absence is made by the clinician, taking into account factors that affect sexual functioning, such as age and the context of the person's life" (p. 292). In DSM-IV, clinical a significance criterion was added to each of the sexual dysfunctions, which read "The disturbance cases marked distress of interpersonal difficulty." This was the state of the DSM when asexual communities began to form (for a more detailed overview of the development of HSDD, see Hinderliter 2013; Segraves et al. 2007).

"Asexuality" frames things differently. According to the largest asexual website, the Asexual Visibility and Education Network (AVEN, asexuality.org), an asexual is "a person who does not experience sexual attraction," and "unlike celibacy, which is a choice, asexuality is a sexual orientation." Asexual organizing has only occurred on any significant level with the socially widespread use of the Internet. Whereas HSDD is a diagnostic category, "asexuality" is largely an identity category promoted by people self-identifying as asexual, although "asexuality" (using the same term and a similar definition) sometimes appeared in sexological literature well before the formation of these communities (e.g. Storms 1980). In asexual communities, there is generally opposition to seeing asexuality as pathological (Brotto et al. 2010).

The situation for asexuality is well illustrated by the Wikipedia article on HSDD: Since November 2012, the article's introduction has said that asexual activists have criticized the diagnostic category, comparing the current situation of asexuality in the DSM to homosexuality from 1974 to 1987.[1] Shortly afterwards, it was edited to say a citation was needed, and this remains the situation as of writing—although it is correct that some asexuals made this comparison to criticize DSM-IV on various blogs and forums, as well as in an unpublished report submitted to the chair of the DSM-5 sexual dysfunctions subworkgroup (AVEN DSM Task Force 2008), the status of asexual activism was such that there were no citable sources acceptable under Wikipedia standards.

Despite the rather limited amount of research on—and public awareness of—asexuality, it is discussed in the DSM-5 literature reviews for HSDD in women and in men (Brotto 2010a, b) in the context of whether or not to retain the clinical significance criteria; further, these reviews suggested that an exclusionary criterion

[1]Retrieved 8/30/2013 from http://en.wikipedia.org/wiki/Hypoactive_sexual_desire_disorder

for asexuals be added to whatever form HSDD appeared in DSM-5. In DSM-5 itself, no reference to asexuality is found in the diagnostic criteria of any sexual dysfunctions, but in the supporting text for female sexual interest/arousal disorder, it states, "If a lifelong lack of sexual desire is better explained by one's self-identification as 'asexual,' then a diagnosis of female sexual interest/arousal disorder would not be made," (APA 2013, p. 434). In the online version, a conference presentation (Brotto 2012) is cited in which evidence was presented highlighting a number of differences between asexuals and people with HSDD. A similar statement is made in the supporting text for male HSDD (APA 2013, p. 443).

In this chapter, I will focus on the DSM's sexual dysfunctions,[2] especially for low/no sexual desire, and also on asexuality research and activism, examining the background to the comments about asexuality in the DSM. While the focus is relatively narrow, many of the issues that arise are common among many domains of the DSM.

## How Does One Distinguish Sexual Dysfunctions from Normal Variation?

The question of determining what can and cannot be legitimately regarded as a disorder remains unsettled. Although this is true in medicine as a whole, mental disorders tend to be more controversial. Some approaches to addressing this matter involve trying to define "(mental) disorder" and then implement that definition. I suggest a heuristic for classifying such approaches, inspired by Wakefield's (1992a, b, 1993) "harmful dysfunction" definition of a disorder. For Wakefield, "harmful" means that a condition is negatively valued in the society that regards it as a disorder, and "dysfunction" means that the condition is a failure of an internal mechanism to perform what it was "designed" by evolution to do. Thus, to be a disorder a condition must satisfy both the "harm" prong (which is value-laden) and the "dysfunction" prong (which is an objective criterion).[3]

I propose that attempts to limit the reach of "disorder" via a definition take one or both of two approaches: by limiting the relevant "harms" involved (i.e. what negative values may be used), and by limiting what can be called a "dysfunction" (i.e. purportedly an objective, scientific criterion). For those who think it is possible

[2]DSM-5 has two chapters of sexual disorders. The paraphilias concern "deviant" sexual interests, and the ones currently listed are in the DSM largely because of their forensic history and it is in this group that homosexuality was previously included. The sexual dysfunctions include things like sexual pain disorders, erectile dysfunction, low/no sexual desire and/or arousal, etc.

[3]The term "dysfunction" here is different than in the collocation "sexual dysfunction." The collocation "sexual dysfunctions" generally entails being a disorder. "Sexual disorders" is often a superordinate term for the sexual dysfunctions and paraphilias.

to have an objective component, there are two main options (Bolton 2008): Defining *function* (and thus *dysfunction*) in terms of statistical norms, and defining them in terms of evolved function.

Robert Spitzer was largely responsible for working out the compromise that enabled the declassification of homosexuality in 1973, and an important part of that compromise involved defining *mental disorder* so as to exclude homosexuality. Spitzer was later made chair of the DSM-III Taskforce, and in writing about the status of homosexuality in DSM-III, he explains his 1973 reasoning: "With the exception of homosexuality and perhaps some of the other 'sexual deviations' [the mental disorders in DSM-II] all regularly caused subjective distress or were associated with generalized impairment in social effectiveness" (Spitzer 1981, p. 211). Further, whether a condition is regarded as a disorder should be because of its consequences, rather than its (presumed) causes. He notes that using these criteria, homosexuality "as well as states like inhibited sexual desire, inhibited sexual excitement (frigidity), premature ejaculation and Don Juanism, do not qualify as mental disorders" (p. 211). This approach focuses on (limiting) the harm prong.

Spitzer states that during the course of DSM-III, it was felt that the solution involved in removing homosexuality was unsatisfactory (i.e. it failed to justify retaining the sexual deviations, which were renamed paraphilias, and failed to justify adding the sexual dysfunctions). "Generalized impairment" was replaced with "impairment in one or more important areas of functioning" and it was said that mental disorders are "typically" associated with distress or impairment. Using this, the addition of the sexual dysfunctions in DSM-III was justified via "the assumption that sexuality is an important area of functioning and there is the normative notion of the 'complete sexual repose cycle' such that the absence of any of the component phases justifies the inference of a dysfunction whether or not the individual is distressed by the symptom" (p. 212). This reasoning is plausible, but many are likely to find it unsettling: Although certainly important in the lives of a large part of the population, sex has a sense of optionality to it that many domains of life do not. The approach involves both limiting harms (i.e. "impairment in one of more important areas of functioning"), and requiring that a condition be a dysfunction. Here function (and thus dysfunction) were defined not in terms of a general theory of (dys)function, but with respect to a particular (normative) model: Kaplan's (1983) triphasic model of the human sexual response cycle (desire, excitement, orgasm) which is the basis for DSM-III's organization of sexual dysfunctions.

In DSM-III-R, the DSM's definition of mental disorder was expanded to include a third harm (in addition to distress and disability), and this remained in the definitions in DSM-IV and IV-TR. Specifically, it includes conditions "with a significantly increased risk of suffering death, pain, disability or an important loss of freedom" (APA 2000, p. xxxi).[4] This allows "disorder" to include risk conditions. Outside of psychiatry there are well-accepted disorders of this sort,

[4]This addition largely reflects part of a definition of (mental) disorder proposed by Culver and Gert (1982). Their approach also focuses on limiting relevant harms (see also Culver and Gert 2009),

such as hypertension and osteoporosis, although the attempt to introduce a risk condition in DSM-5 (psychosis risk syndrome) proved extremely controversial, and the proposal was dropped (Maxmen 2012).

## *Clinical Significance Criteria*

In DSM-IV (APA 1994), clinical significance criteria were added to many diagnoses in the DSM, including the sexual dysfunctions. The ones for the sexual dysfunctions require that, to make a diagnosis, "the disturbance causes marked distress or interpersonal difficulty." These two ills correspond to "distress" and "impairment" in DSM-IV's definition of mental disorder.

A consequence of adding this criterion is that it leaves in doubt what the relevant harm for the sexual dysfunctions is: any set of diagnostic criteria whatsoever, if they have a clinical significance criterion added, can meet the harm prong (provided all harms listed are acceptable ones). For example, Something-Or-Other Disorder is a condition that causes clinically significant distress or impairment.

Criterion A: The individual has property P.
Criterion B: The state of affairs described in Criterion A causes clinically significant distress or impairment.

If the mere fact that the individual is distressed about the fact that they have property P is enough to satisfy B, then there are no restrictions from the harm prong on what kinds of properties can be listed in A, other than the requirement that it has construct validity. For people who believe that an objective dysfunction prong can and should do most of the work for excluding inappropriate "disorders," this may not be a serious problem. For those who do not, the problem may be more serious.

Across the DSM, the addition of the clinical significance criteria remains controversial (for discussion, see Spitzer and Wakefield 1999), although present discussion will be limited to the sexual disorders. Some members of the DSM-IV workgroup for the sexual dysfunctions did not even remember discussing the matter at the time or what consequences it might have (Segraves et al. 2007).

An influential consensus document proposing an updated classification of "female sexual dysfunction" (Basson et al. 2000) replaced the DSM's clinical significance criteria with "causes personal distress." Althof's (2001) response to this illustrates a number of recurrent issues pertaining to the clinical significance criteria. He argues for their removal from the sexual dysfunctions, claiming that they "[detract] from the field's attempt to maintain scientific rigor" (p. 124). Realizing that his position might not be "politically correct," he suggests that if a woman is unable to achieve orgasm under any circumstances, she "qualifies for the diagnosis

but has been less influential. Rather than a notion of "dysfunction," they require that the condition be a "nondistinct sustaining cause" (Culver and Gert 1982, p. 72).

of orgasmic disorder whether or not she is distressed by her lack of orgasms. The phenomenon is present. It simply exists" (p. 124).[5] He suggests that the current situation may be because of moral reasons (with which he is sympathetic), but offers an alternative explanation: "It also may be because the field does not have adequate norms on which to base the presence of a dysfunction or to establish cut-off points for delimiting the range of normal function" (p. 124). He also argues via analogy: diagnoses of obsessive-compulsive disorder, substance abuse disorders, and hypertension do not require distress to be diagnosed. Some subsequent authors endorsed this view, some repeating the analogy with hypertension (e.g. Segraves et al. 2007).

I want to draw attention to three issues. The first issue is the analogy with obsessive-compulsive disorder, substance abuse disorders, and hypertension. In the first two cases, distress is not required because they may be diagnosed if there is distress *or impairment*. Hypertension is a (treatable) risk condition, and thus meets the harm prong by the third harm listed in DSM-IV (increased risk of death, disability, or an important loss of freedom). The analogy with these suggests confusion and/or lack of clarity over what, exactly, the relevant "harm" involved with the sexual dysfunctions is. It may be that these are regarded as (sexual) disabilities, which is my reading of Spitzer's (1981) logic, but the prevailing atmosphere discourages people from stating this outright. Alternatively, people may be primarily relying on a general feeling that something in the individual is not working right, without making a clear distinction between "harm" and "dysfunction."

The second major issue is Althof's suggestion that the difficulty in determining what is a "dysfunction" and what is within a "normal range of function" is simply a matter of current lack of data. This view is not uncommon. For example, in the DSM-5's Introduction the discussion of clinical significance criteria suggests that the current difficulty in separating the pathological from the non-pathological is because of a lack of biological markers and clinically useful severity measures (APA 2013, p. 21). Such suggestions make one ask, even if they had as large and high quality a dataset as they could possibly want, what would they do with it? How would they use it to distinguish "normal" and "pathological"? Create an arbitrary cut-off?

Furthermore, this relies on the assumption that statistical norms can be used to distinguish between disorder and non-disorder, even though it is well known that this simply does not work. The classic counter-example is intelligence: both the extreme high and extreme low ends are statistically deviant, but one is considered a disorder, while the other is not, and this seems to derive from a negative valuation of one and a positive valuation of the other.

The third issue is that clinical significance criteria replace a 2-way distinction (the individual has the condition or does not have it) with a 3-way distinction. For simplicity's sake, I will refer to the clinical significance criterion as Criterion B, and to everything else as Criterion A. The three categories are:

[5] Spitzer and Wakefield (1999) had previously made similar comments about the sexual dysfunctions and paraphilias.

1. The individual meets neither Criterion A nor Criterion B
2. The individual meets A but not B.
3. The individual meets both A and B.[6]

What is the status of (2)? On account of the issues involved, in the paraphilias chapter of DSM-5, a distinction between "paraphilias" (2) and "paraphilic disorders" (3) is drawn, and this distinction attracted some controversy (e.g. Hinderliter 2010; Kamens 2011; Moser 2010; O'Donohue 2010; Serano 2009; Singy 2010). And yet the issue is one that pervades much of the DSM system.

Additionally, there have been doubts about the value of the distress criterion for clinical practice, and for various matters of clinical research. Intuitively, it seems likely that (2) is more useful for etiological research than is (3).

However, there are also concerns about excluding such criteria, as this may itself cause distress. Sexuality is an area in which there is considerable variation among people, yet people often have especially strong feelings about what is and is not normal, and many have worries that they might not be "normal." Some variant of "Am I normal?" is often said to be among the most common questions sex educators receive. In many cases, reassurance that there is nothing wrong with the person is effective. My own intuition is that, if telling the person there is nothing wrong with them and giving them information about similarly situated people is an effective "cure," it seems inappropriate to call it a disorder in the first place (even if there is distress), and I have spoken with others who share this intuition. This is primarily a moral argument, which is consistent with the view that the purpose of diagnoses is to help patients. Even if there can be an objective component for "disorder," and a condition meets that criterion, if telling patients that they do not have a disorder is a more effective treatment that trying to treat the condition, a diagnosis seems inappropriate. Further, ruling out such cases from the category of "disorder" would need to be accomplished via the "harm" criterion, interpreted according to moral issues relating to the role of clinicians.

## The Rise of Asexual Communities

In the past 10–15 years, there has been a growing community of people identifying as "asexual," although the necessary conceptual resources have been around for longer. References to "asexuals" or "asexuality" in lists of sexual orientation groups can occasionally be found in US and UK newspapers since at least the late 1960s/early 1970s (although sometimes as a joke), and evidence of people identifying as asexual can be occasionally be found in letters printed in advice columns since the late 1970s (Hinderliter, in preparation). Storms (1980) proposed a model of sexual orientation that included asexuality, and 'asexuals' have appeared

[6]Meeting Criterion B presupposes meeting Criterion A, so "meeting B but not A" can be excluded.

for a while in literature of sexual-orientation based typologies of male-to-female transsexuals.

DSM-III specifically used the term 'asexual' in its description of transsexualism, and DSM-III-R's diagnosis gender identity disorder of adolescence or adulthood non-transsexual type, calls it a 'sexual orientation': "Specify history of sexual orientation: asexual, homosexual, heterosexual, or unspecified" (APA 1987, pp. 66–67, bold removed). Because the use of "homosexual" and "heterosexual" was offensive to many patients and confusing to clinicians (Bradley et al. 1997), this was replaced in DSM-IV with "Sexually Attracted to Males, Sexually Attracted to Females, Sexually Attracted to Both, and Sexually Attracted to Neither" (APA 1994, p. 534). In the introduction to the sexual and gender identity disorders chapter, DSM-IV-TR added a definition of sexual orientation in the part about gender identity disorder, defining it as "erotic attraction to males, females, or both" (APA 2000, p. 535). "Neither" is conspicuously absent, although the existence of such individuals is more or less acknowledged only four pages later, where "lifelong" is listed as a subtype of HSDD. This inconsistency presumably reflects a general lack of interest in/awareness of asexuality.

There seems to have been virtually no asexual organizing until widespread social use of the Internet in the second half of the 1990s. At first there were mostly static content sites and some listservs. In May 2002, David Jay, the founder of AVEN, obtained the domain asexuality.org and set up forums there. AVEN quickly came to overwhelmingly dominate online asexual discourse, at least in terms of volume, for a number of years. Presently, it is one of the two largest online asexual communities, with the other being asexuals on the microblogging site Tumblr. AVEN also hosts online asexual communities in a few languages besides English (mostly Germanic and Romance languages).

Prior to the growing visibility of these online communities, there was a lack of any socially widespread categories describing (roughly) this group. From the materials I have read, the available categories in late twentieth- and early Twenty-first-century Anglo-American countries for making sense of asexuality are fairly limited. The main ones seem to be asexuality as sexual orientation, as (psycho)pathology, as celibacy and/or abstinence, and as moral opposition to sex (antisexuality). In online asexual communities, there has consistently been an opposition to seeing asexuality as pathology, and there has been a tendency to see it as a sexual orientation.

Over time, conceptualizations of asexuality have tended to move away from celibacy. Antisexuality is strongly discouraged in the largest communities, although not uncommon among new members. Consistent with seeing asexuality as a sexual orientation, asexuals have tried to make alliances with LGBT groups and this has had a reasonable amount of success.

A consequence of this is that, through analogical reasoning, many asexuals have reasoned that if asexuality is a sexual minority, and other sexual minorities are oppressed, asexuals must be oppressed as well. The question of whether asexuals are oppressed is presently a controversial one in asexual communities. Very roughly, opposition to this view seems to be based partly on the lack of evidence to support it, and partly on a dislike of the politics that tend to go with such claims. Proponents sometimes maintain their view despite a lack of evidence by saying that, once there is more widespread awareness of asexuality, there will be more oppression.

One piece of evidence sometimes invoked to defend the view that asexuals are oppressed has been the existence of HSDD in the DSM. Further, attempts to "cure" this may be seen as analogous to attempts to "cure" homosexuality, which are widely condemned by health organizations. Many asexuals suspect that existence of this diagnosis harms asexuals, although there is hardly even anecdotal evidence to support this. I have only heard of one case that might be an instance of this (described in Rubio-Sheffrey 2011), which involved extended attempts with therapists and even hormone therapies to generate sexual interest, but to no avail. An asexual friend reading this chapter recalled reading a few possible cases of asexuals being harmed by the diagnosis, although none were from the past few years.

## Developments in Research and Academic Literature

Roughly coinciding with the rise of online asexual communities, the field of "sexual medicine" has emerged. In the 1990s, it was discovered (unintentionally) that sildenafil could induce penile erections, leading to the production and marketing of Viagra, which was found to be incredibly profitable. However, it was only marketable to half of the population. Attempts to develop a (profitable) sexual medicine for women have been ongoing. On account of this, there has been an enormous growth in industry-funded sexuality research, as well as a great deal of publicity about "female sexual dysfunction" (a term popularized in this time), and also a backlash against the medicalization of female sexuality (Moynihan and Mintzes 2010).

Concerns about the role of industry influence are widespread for conditions throughout the DSM, to the point that accusations of the DSM being the tool of Pharma have become stock criticisms of the DSM. The situation is especially severe in the realm of sexuality research because of few other sources of funding (Brotto 2007): at least in the US and Canada, research to help treat sexual problems is not given priority by government funding agencies, and government funding for sexuality research is periodically made the object of political attack.

There has been considerable growth in the amount of academic literature on asexuality, coinciding with the growth in online asexual communities, and this comes primarily from psychology, sociology, and gender studies. Presumably, this is due to a combination of increased social interest and increased ease of recruiting convenience samples. I strongly suspect that part of the interest is because studying asexuality is politically useful for people reacting against the medicalization of low sexual desire (which is a goal that is congruent with asexual interests).

HSDD is commonly mentioned in this literature. Bogaert (2006) considers whether asexuality is pathological, and answers the matter by referring to the sexual dysfunctions' clinical significance criterion: if asexuals were not distressed by their asexuality and if it did not cause (clinically significant) interpersonal difficulty, then asexuality as a whole should not be pathologized. Prause and Graham (2007) note that "Implicit in the debate about what constitutes a 'normal' level of sexual desire is an assumption that *some* level of sexual desire is normative" (p. 343), and asexuality draws this into question and highlights the importance of clinical significance criteria. Further, noting that "If personal distress is primarily due to conflicts with social expectations or worry that a physical problem exists, then a psychiatric diagnosis implying abnormality may exacerbate concerns in an asexual individual" (p. 353). Brotto et al. (2010) found that only about 10 % of asexuals in their sample had clinically significant levels of sexual desire related distress. Brotto and Yule (2011) report results from a study measuring genital and subjective sexual arousal in heterosexual, homosexual, bisexual, and asexual women, and found no significant difference between the groups, and used this combined with lack of distress to claim that this provides evidence against the suggestion that asexuality is a sexual dysfunction.[7]

There is considerable opposition in asexual communities towards seeing asexuality as pathology, although a general sense that there must be something wrong with themselves is a fairly common part of coming to an asexual identity (Carrigan 2011).

In summer 2008, David Jay put together a small group of people to do small-scale lobbying campaign about asexuality and HSDD. Previously, interested asexuals thought this would not be possible due to the relatively small size of the community, its limited political power, and the lack of community-internal academics in relevant fields. However, at the advice of a transgender activist, an effort was started. A total of six people were involved. I had just finished my first year of graduate school, and I was in charge of background research.

It quickly became clear that the situation was complex. Asexuality most closely resembles lifelong HSDD. However, the large majority of patients, it seems, have the acquired subtype. Few studies even use the subtyping system, and there do not appear to have ever been any studies specifically about the lifelong subtype. Further, because of lack of expertise, and because of the potential for backlash if it looked like we were trying to pressure the sexual dysfunctions subworkgroup, the plan was to try to interview people with relevant expertise about their thoughts on the matter. I was told that the goal would be to generate discussion around the issues, rather than to specifically advocate for any particular position. Initially, we were concerned about trying to get the subworkgroup to take us seriously, but this proved to be unexpectedly easy. One of the four people on the sexual dysfunctions subworkgroup, Cynthia Graham, had been the faculty advisor for one of the first studies on asexuality (Prause and Graham 2007), and another, Lori Brotto, is one of the leading researchers on asexuality.

---

[7]In fact, they had difficulty finding asexual women, and so the asexual sample size was smaller than for other groups. The graph they show for physiological arousal (Figure 1, p. 705) suggests that, with increased power from a larger sample, they would likely find that asexual women do have lower levels of physiological sexual arousal in the condition they studied.

The result of AVEN's small-scale advocacy work was briefly mentioned in Brotto's DSM-5 literature review on HSDD in men: "In fact, an AVEN DSM Task Force prepared a 75-page document which included interviews on seven academics with expertise in human sexuality, which concluded that the DSM-5 should explicitly exclude asexual individuals from receiving a diagnosis of HSDD" (Brotto 2010a, p. 2023). Conceptually, I am not sure how much sense an exclusionary clause for asexuals makes, but an exception-clause for self-identified asexuals was included in DSM-5's supporting text.[8] When discussed in online asexual discourse (which is fairly infrequently), responses to this exception-clause are positive.

## Influence of the Pharmaceutical Industry?

The DSM-5 literature reviews for the sexual and gender identity disorders workgroup were published in *Archives of Sexual Behavior*, with the exception of sexual dysfunctions in men, which were published in the *Journal of Sexual Medicine*. There were three main changes proposed for the desire-related disorders: They suggested deleting sexual aversion disorder (which was added in DSM-III-R along with hypoactive desire disorder to replace DSM-III's inhibited sexual desire) because it was rarely used and rarely researched. They proposed splitting HSDD into separate diagnoses for men and women, and they proposed merging female HSDD with female sexual arousal disorder into a category called female sexual interest/arousal disorder.

When DSM-III was published, in accordance with the approach taken by Masters and Johnson's highly influential work, similarities between male and female sexuality were emphasized, while differences were downplayed; the prevailing view now is there are considerable differences between male and female sexuality. The most common sexual complaints among women seeking sex therapy concern low/no sexual desire, while the most common complaints among men relate to erection and ejaculation (Leiblum 2006). Because of this, there has been much more research on HSDD in women than in men, and in absence of much data, the subworkgroup expressed uncertainty about what to do with the male version (Brotto 2010a).

A number of commentaries to the proposed changes were published in the *Journal of Sexual Medicine* (Derogatis et al. 2010) and one (Derogatis et al. 2011) was published in *Archives of Sexual Behavior*. There was no opposition to separating male and female diagnoses. Reactions to merging female HSDD and female sexual arousal disorder were mixed. While ASB required acknowledgment of recent industry connections, JSM did not, but it was hard to miss that the strongest

[8]In a footnote in our report, I indicated that I neither endorsed nor opposed this option. We had not asked about this possibility in our interviews, and in the absence of any feedback from clinicians I did not want to support it.

opposition to the merger all were either consultants for (or employees of) Boehriner Ingelheim in their attempt to get Flibanserin approved by the FDA for female HSDD, or employees of Fabre Kramer Pharmaceuticals, which was also developing a possible treatment.

Their arguments also tended to use a great deal of facts and figures and statistics, while being surprisingly weak if one looked closely at those figures (e.g. Kramer and Smith [in Derogatis et al. 2010] try to demonstrate how low cormorbidity is by describing a dataset in which 56 % of women who met the diagnosis of female sexual arousal disorder also met the criteria for HSDD—but they used the word "only" and only gave the numbers necessary for calculating this figure. Later they show a graph indicating that two drugs likely affect both desire and arousal roughly the same way, but then highlight relatively trivial differences to argue against merging).

Outside of the academic commentaries, in blog posts, forum discussions, etc. during this time period, if someone did not like a particular newly proposed diagnostic category, it seemed that almost invariably there would be a declaration that this new "disorder" was being cooked up by the pharmaceutical industry.

In the professional commentaries, there were no comments on the proposal to exclude self-identified asexuals from the diagnosis.

## Text of the DSM-5 and Its Reception

The relationship between asexuality and the DSM-5 diagnoses sexual arousal/interest disorder in women and hypoactive sexual desire disorder in men is complex. Whatever the futures of these categories will be, it is safe to expect that they will interact in important ways. Asexual organizing and visibility brings increased salience to the question that many were already asking: "What is normal sexuality?" Variations of this question can be found repeatedly in discussions about potential changes for DSM-5's sexual dysfunctions.

In reading the DSM-5's text for the sexual dysfunctions, one thing that strikes me is that, although the very notion presupposes the existence of "normal sexuality" (with "sexual dysfunction" being certain sorts of deviations from that), the text sometimes appears to be at pains to avoid taking any position on what is "normal sexuality," while it makes numerous comments trying to broaden what is deemed normal.[9]

[9]By contrast, the chapter on the paraphilias appears to be quite up front about what is not normal in its definition of 'paraphilia', which it contrasts with 'normophilia': "The term paraphilia denotes any intense and persistent sexual interest other than sexual interest in genital stimulation or preparatory fondling with phenotypically normal, physically mature, consenting human partners" (APA 2013, p. 685). I have no idea what this means, but it does not give the impression of avoiding taking a position on what is normal.

I have found relatively few mentions of DSM-5's comment about asexuals in the low/no desire disorders outside of online asexual discourse, although it is discussed in some media articles about asexuality (Morgensberg 2013), all of which frame the change in a neutral-to-positive way. For several months, most of the discussion of this within online asexual discourse seems to have based things on one person making a brief comment on tumblr. A few months later, someone posted the relevant text. The change is mentioned only a few times on AVEN, and a number of these are discussions of media articles where the change is mentioned.

The Wikipedia page for HSDD lists the proposed changes for DSM-5, but has no mention of the actual changes. Neither the Wikipedia pages on vaginismus and on dyspareunia has any mention of the fact that these were merged in DSM-5. This suggests that these disorder categories are areas of relatively little social interest compared to other parts of the DSM, which is consistent with the fact that the NIMH's website has almost no information about sexual disorders. Using Google to search their site, there are no mentions of paraphilia(s), and the only mentions of sexual dysfunction(s) is in discussion of possible side effects of SSRIs.

## References

Althof, S. E. (2001). My personal distress over the inclusion of personal distress. *Journal of Sex and Marital Therapy, 27*, 123–125.

American Psychiatric Association. (1980). *Diagnostic and statistical manual of mental disorders* (3rd ed.). Washington, DC: American Psychiatric Association.

American Psychiatric Association. (1987). *Diagnostic and statistical manual of mental disorders 3rd edition revised.* Washington, DC: American Psychiatric Association.

American Psychiatric Association. (1994). *Diagnostic and statistical manual of mental disorders* (4th ed.). Washington, DC: American Psychiatric Association.

American Psychiatric Association. (2000). *Diagnostic and statistical manual of mental disorders 4th edition text revision.* Washington DC: American Psychiatric Association.

American Psychiatric Association. (2013). *Diagnostic and statistical manual of mental disorders* (5th ed.). Washington, DC: American Psychiatric Association.

AVEN DSM Task Force. (2008). Asexuality, HSDD, and the DSM: A collaborative report. Unpublished manuscript.

Basson, R., Berman, J., Burnett, A., Derogatis, L., Ferguson, D., Fourcroy, J., Goldstein, I., Graziottin, A., Heiman, J., Laan, E., Leiblum, S., Padma-Nathan, H., Rosen, R., Kathleen, S., Taylor Segraves, R., Shabsigh, R., Sipski, M., Wagner, G., & Whipple, B. (2000). Report of the international consensus development conference of female sexual dysfunction: Definitions and classification. *The Journal of Urology, 163*, 888–893.

Bogaert, A. F. (2006). Towards a conceptual understanding of asexuality. *Review of General Psychology, 10*, 241–250.

Bolton, D. (2008). *What is Mental Disorder? An Essay in philosophy, science, and values.* Oxford: Oxford University Press.

Bradley, S. J., Blanchard, R., Coates, S., Green, R., Levine, S., Meyer-Bahlburg, H., Pauly, I. B., & Zucker, K. J. (1997). Gender identity disorder. In T. A. Widiger, A. J. Frances, H. A. Pincus, R. Ross, M. B. First, & W. Davis (Eds.), *DSM-IV sourcebook volume 3* (pp. 317–325). Washington, DC: American Psychiatric Association.

Brotto, L. A. (2007). More nonindustry funding is needed for sexological sciences to endure. *Journal of Sex and Marital Therapy, 33*, 411–413.

Brotto, L. A. (2010a). The DSM diagnostic criteria for hypoactive sexual desire disorder in men. *Journal of Sexual Medicine, 7*, 2015–2030.

Brotto, L. A. (2010b). The DSM diagnostic criteria for hypoactive sexual desire disorder in women. *Archives of Sexual Behavior, 39*, 221–239.

Brotto, L. A. (2012). *Diagnostic and clinical issues in the distinction between low desire and asexuality*. Paper presented at the annual meeting of the International Academy of Sex Research, Estoril, Portugal, July 2012.

Brotto, L. A., & Yule, M. A. (2011). Physiological and subjective sexual arousal in self-identified asexual women. *Archives of Sexual Behavior, 40*, 699–712.

Brotto, L. A., Knudson, G., Inskip, J., Rhodes, K., & Erskine, Y. (2010). Asexuality: A mixed methods approach. *Archives of Sexual Behavior, 39*, 599–618.

Carrigan, M. (2011). There's more to life than sex? Difference and commonality within the asexual community. *Sexualities, 14*, 462–478.

Culver, C. M., & Gert, B. (1982). *Philosophy in medicine*. New York: Oxford University Press.

Culver, C. M., & Gert, B. (2009). Sex, immorality, and mental disorders. *Journal of Medicine and Philosophy, 34*, 487–495.

Derogatis, L. R., Laan, E., Brauer, M., Van Lunsen, R. H. W., Jannini, E. A., Davis, S. R., Fabre, L., Smith, L. C., Basson, R., Guay, A. T., Rubio-Aurioles, E., Goldstein, A., Pukall, C., Kellogg, S., Burrows, L., Morrison, P., Krychman, M., Goldstein, S. W., & Goldstein, I. (2010). Responses to the proposed DSM-V changes. *Journal of Sexual Medicine, 7*, 1998–2014.

DeRogatis, L. R., Clayton, A. H., Rosen, R. C., Sand, M., & Pyke, R. E. (2011). Should sexual desire and sexual arousal disorders in women be merged? *Archives of Sexual Behavior, 40*, 217–219.

Green, R. (2002). Is pedophilia a mental disorder? *Archives of Sexual Behavior, 31*, 467–471.

Hinderliter, A. C. (2010). Defining paraphilia: Excluding exclusion. *Open Access Journal of Forensic Psychology, 2*, 241–272.

Hinderliter, A. C. (In preparation). *The evolution of online asexual language*. Ph.D. dissertation University of Illinois-Urbana Champaign.

Kamens, S. R. (2011). On the proposed sexual and gender identity diagnoses for DSM-5: History and controversies. *The Humanistic Psychologist, 39*, 37–59.

Kaplan, H. S. (1983). *The evaluation of sexual disorders: Psychological and medical aspects*. Levittown, PA: Taylor & Francis.

Leiblum, S. R. (2006). Sex therapy today: Current issues and future perspectives. In S. R. Leiblum (Ed.), *Principles and practice of sex therapy* (4th ed., pp. 3–22). New York: The Guilford Press.

Master, W. H., & Johnson, V. E. (1970). *Human sexual inadequacy*. Toronto/New York: Bantam Books.

Maxmen, A. (2012). Psychosis risk syndrome excluded from DSM-5. *Nature*. doi:10.1038/nature.2012.10610.

Morgensberg, D. (2013). Asexual disorder? The search for ace identity is part recognition, part definition. *Huffington Post*. Retrieved from http://www.huffingtonpost.com/2013/06/18/asexual-disorder_n_3361472.html. 30 Oct 2013.

Moser, C. (2010). Problems with ascertainment. *Archives of Sexual Behavior, 39*, 1225–1227.

Moser, C., & Kleinplatz, P. K. (2005). DSM-IV-TR and the paraphilias: An argument for removal. *Journal of Psychology and Human Sexuality, 17*, 91–109.

Moynihan, R., & Mintzes, B. (2010). *Sex, lies and pharmaceuticals*. Vancouver: Greystone Books.

O'Donohue, W. (2010). A critique of the proposed DSM-V diagnosis of pedophilia. *Archives of Sexual Behavior, 39*, 587–590.

Prause, N., & Graham, C. A. (2007). Asexuality: Classification and clarification. *Archives of Sexual Behavior, 36*, 335–341.

Rubio-Sheffrey, E. (February 1, 2011). Lambda archives honoree Sarah Beth brooks is a true leader for the next generation of LGBT activists. Retrieved from http://www.sdgln.com/social/2011/02/01/lambda-archive-honoree-sara-beth-brooks-true-leader-next-generation-lgbt-activist. 3 Dec 2013.

Segraves, R., Balon, R., & Clayton, A. (2007). Proposals for changes in diagnostic criteria for sexual dysfunctions. *Journal of Sexual Medicine, 4*, 647–580.

Serano, J. (2009). Psychology, sexualization and trans-invalidations. Keynote lecture presented at the 8th Annual Philadelphia Trans-Health Conference

Singy, P. (2010). What's wrong with sex? *Archives of Sexual Behavior, 39*, 1231–1233.

Spitzer, R. L. (1981). The diagnostic status of homosexuality in DSM-III: A reformulation of the issues. *American Journal of Psychiatry, 138*, 210–215.

Spitzer, R. L., & Wakefield, J. C. (1999). DSM-IV diagnostic criterion for clinical significance: Does it help solve the false positive problem? *American Journal of Psychiatry, 156*, 1856–1864.

Storms, M. D. (1980). Theories of sexual orientation. *Journal or Personality and Social Psychology, 38*, 783–792.

Wakefield, J. C. (1992a). The concept of mental disorder: On the boundary between biological facts and social values. *American Psychologist, 47*, 373–388.

Wakefield, J. C. (1992b). Disorder as harmful dysfunction: A conceptual critique of *DSM-III-R*'s definition of mental disorder. *Psychological Review, 99*, 232–247.

Wakefield, J. C. (1993). Limits of operationalization: A critique of Spitzer and Endicott's (1978) proposed operational criteria for mental disorder. *Journal of Abnormal Psychology, 102*, 160–172.

Winters, K. (2009). *Gender madness in American psychiatry: Essays from the struggle for dignity*. Dillon, CO: GID Reform Advocates.

# The Crippling Legacy of Monomanias in DSM-5

**John Z. Sadler**

**Abstract** Since the disjunctive change of DSM-III, architects of the DSMs, including the recent DSM-5, have tended to ignore the problematic status of disorder categories limited by single symptoms or single symptom clusters. These disorders include kleptomania, pyromania, intermittent explosive disorder, paraphilic disorders, factitious disorder, among others. Single symptom/cluster disorders resemble Esquirol's nineteenth-century French concept of monomania, which is defined as an "*idée fixe*, a single pathological preoccupation in an otherwise sound mind" (Goldstein 1998, p. 388). Monomania is also a variation on the early idea of partial insanity, where the patient's reasoning remained unaffected, while insanity was circumscribed by either a single delusional belief or overwhelming irrational impulses. In this chapter I briefly review the history of monomania and why it came to be rejected by Anglo-European alienists. These historical critiques are then related to contemporary DSM-5 single symptom/cluster disorders. I then supplement this historical-to-contemporary critique with an original conceptual critique of single-symptom/cluster disorders. The chapter concludes with some methodological approaches to classification which address the crippling legacy of monomanias in today's DSMs.

## Introduction: The Monomania Concept

The term 'monomania' has a complex international history, pedigree, and multiple meanings. Used today in ordinary language, www.dictionary.com lists "monomania" with the following definitions:

1. (no longer in technical use) a psychosis characterized by thoughts confined to one idea or group of ideas.
2. An inordinate or obsessive zeal for or interest in a single thing, idea, subject, or the like. (http://dictionary.reference.com/browse/monomania?s=t Last accessed 8/5/13)

J.Z. Sadler (✉)
University of Texas Southwestern Medical Center, Dallas, TX, USA
e-mail: John.Sadler@UTSouthwestern.edu

 
S. Demazeux, P. Singy (eds.), *The DSM-5 in Perspective*, History, Philosophy and Theory of the Life Sciences 10, DOI 10.1007/978-94-017-9765-8_9

The term has been adjectivized into 'monomaniacal' and personalized into 'monomaniac' and often appears in popular discourse, from the rhetorical tarnishing of political rivals to the authoritative defining of comic-book supervillains.

Such was not always the case, and in its origins, 'monomania' was a psychiatric diagnostic term attributed to Jean-Etienne-Dominique Esquirol (1772–1840) (Goldstein 1998; Goldstein 2001; Bynum 2003). An ambitious French psychiatrist, initially the student of and later the rival of Phillipe Pinel (1745–1826), Esquirol was not only a major diagnostic theorist of his time, but one of the founding fathers of forensic psychiatry.

The monomania concept as articulated by Esquirol was arcane from the beginning, and its complexity and consequent ambiguity likely contributed to the semantic twists and turns it was to undergo at the hands of European and American "alienists" (the nineteenth-century term for psychiatrists), who adapted the term in a *laissez-faire* manner. Central to the monomania concept was the French symptom of the *idée fixe*, which was, as Goldstein puts it, "a single pathological preoccupation in an otherwise sound mind" (Goldstein 2001, pp. 154–5). Monomania was a variety of the 'partial insanities', which nineteenth-century Anglo-American psychiatrists viewed as separate from traditional insanities like melancholia and delirium: the crucial distinction being that insanity affected the entire mental life of the patient, the latter being, using a DSM convention today, a 'pervasive' disorder. Partial insanities involved syndromes that produced profound symptoms in some mental domains, leaving others functionally intact.

In his *Mental Maladies: A Treatise on Insanity* (Esquirol 1845), Esquirol provides a general discussion of monomania, and breaks the numerous varieties of monomanias into three general groupings: intellectual, affective, and instinctive.

Esquirol describes intellectual monomanias:

> At one time, the intellectual disorder is confined to a single object, or a limited number of objects. The patients seize upon a false principle, which they pursue without deviating from logical reasonings, and from which they deduce legitimate consequences, which modify their afflictions, and the acts of their will. Aside from this partial delirium, they think, reason, and act, like other men. Illusions, hallucinations, vicious associations of ideas, false and strange convictions, are the basis of this delirium, which I would denominate *intellectual monomania*. (Esquirol 1845, p. 320)

For affective monomania, Esquirol notes:

> At another [type]; monomaniacs are not deprived of the use of their reason, but their affections and dispositions are perverted. By plausible motives, by very reasonable explanations, they justify the actual condition of their sentiments, and excuse the strangeness and inconsistency of their conduct. It is this, which authors have called *reasoning mania*, but which I would name *affective* monomania. (Esquirol 1845, p. 320)

Finally, for instinctive forms:

> In a third class of cases a lesion of the will exists. The patient is drawn away from his accustomed course, to the commission of acts, to which neither reason nor sentiment determine, which conscience rebukes, and which the will has no longer the power to restrain. The actions are involuntary, instinctive, irresistible. This is *monomania without delirium*, or *instinctive monomania*. (Esquirol 1845, p. 320)

In his *Treatise*, Esquirol discusses multiple examples of four major varieties of monomania: erotic monomania (erotomania), monomania resulting from drunkenness (elsewhere called dipsomania), incendiary monomania (pyromania), and homicidal monomania. Esquirol's interest in these four categories should be no surprise, given the forensic overlap of these syndromes with crimes of adultery, intoxication, arson, and murder. These conditions were of premier interest in his forensic testimony practice, as they expanded the early domains of criminal excuse due to 'insanity' – a use that was ultimately to prove quite controversial on both sides of the Atlantic.

With this single intellectual template, it was possible for subsequent alienists to identify dozens of specific mental disorders with names that specified the particular preoccupation or behavioral extremity. The story of the international proliferation of the monomania concept, along with the manifold particular monomanias, is fascinating but too complex to recount in this brief chapter. However, a few highlights should be noted. Bynum links monomania to contemporary concepts of nymphomania and kleptomania (2003) and Hunter and MacAlpine (1963) as well as Goldstein (2001) note the relationship to James Prichard's concept of 'moral insanity' as a partial insanity closely resembling monomania, one focusing on 'moral' faculties such as emotion and will. Of great notoriety in contemporary philosophy of medicine and psychiatry is New Orleans alienist Samuel Cartwright's monomanic notion of 'draepetomania', the compulsion of slaves to flee their masters (Cartwright 1851), which has served as a cautionary tale about social value judgments in clinical diagnosis in contemporary philosophy of medicine.

## Why Nineteenth-Century Monomanias Fell into Disfavor

The American historian of modern France, Jan Goldstein, in her book *Console and Classify* (Goldstein 2001) examined the rise of French psychiatry in the nineteenth century for an English-speaking audience. In a related paper (Goldstein 1998) she described the 'fall' of monomania by the late nineteenth century.

Goldstein's account is much more substantive that I can summarize here. Her prevailing thesis about the role of Esquirol and monomania concerns the role that the concept had in establishing the value and importance of psychiatric expert testimony in French criminal courts, and derivatively, the growth of alienist (psychiatric) autonomy in managing and expanding French asylums for the insane. Goldstein summarizes her account below (1998, p. 389):

> The key to the rise and fall of monomania in 19th-century France, my research showed, was not the internal cogency of the diagnosis (I will have more to say about that later) but the extra-scientific uses to which the diagnosis was put by its advocates. Monomania was the linchpin in the effort made by the Esquirol circle from the 1820s to the 1840s to gain public recognition for psychiatry by carving a place for expert psychiatric testimony in the courts of law. Now the evaluation of the madness or sanity of a person accused of a crime was already an accepted part of the French criminal trial under the Old Regime, and it had been reaffirmed in Article 64 of the Napoleonic penal code. But, before the "discovery" of monomania, that evaluation could just as well be done by a lay person as by a physician.

Prior to Esquirol, the man-on-the-street juror had no difficulty recognizing madness in a defendant, because the defendant's rantings and ravings were evident to all. What need was there for psychiatric expert testimony? Anyone could identify a madman. What the monomania concept introduced was the requirement for the skills and knowledge of the alienist to elicit the core of psychopathology when the jury was confronted with puzzling, seemingly sane, offenders. Goldstein, however, is not making a simplistic claim that Esquirol was a scheming opportunist who simply wanted to invent expertise to advance his field:

> I would argue against drawing the cynical, reductionistic conclusion that professional knowledge is simply an ideological mask for professional self-interest. I would favor a more nuanced claim that the knowledge bases of many professional domains—among them, psychiatry—are not generated by insulated, purely scientific procedures of inquiry but are notably porous to the sociopolitical environment, especially insofar as that environment influences the requirements of their bid for professional status. (Goldstein 1998, p. 393)

The spread of the monomania concept was substantive, from Germany to England and the new United States (Bynum 2003; Fear et al. 1998; Hunter and Macalpine 1963; Ray 1853; Amdur and Messinger 1993; De Saussure 1946; Haack et al. 2010; Mora 1972; Parsons 1887). But as soon as the 1870s the backlash against the diagnosis had been established (Goldstein 2001). Why?

Goldstein identifies several factors that contributed, both internal to the field and external to it. One was the change over the 30 years or so in French psychiatry that established the field both clinically and in the criminal-forensic arena. Goldstein recounts the revealing story of the distinguished French psychiatrist Falret's early rejection of the monomania concept but maintaining a silence about it for decades because he recognized the delicate place the emerging profession was in (Goldstein 1998). Falret was worried an attack on monomania would provoke setbacks in psychiatry's accomplishments and influence. (Falret's critique follows shortly.) Once the profession of psychiatry had established itself, the field could be more open in its reflection on its concepts. A second factor for Goldstein was that with extensive use, psychiatrists increasingly recognized that the strict symptomatic boundaries of monomania would break down when the patient was closely and persistently examined – other areas of irrationality and mental disturbance could be discovered: "A number of psychiatrists now stated that their extended observation of so-called monomaniacs had long ago revealed that the delirium was not confined to a single subject but invariably spilled over to other subjects and introduced errors into the very processes of reasoning" (Goldstein 1998, p. 392).

The delayed critique of Falret was emblematic of these issues. Falret was a pioneer of the idea that simple signs and symptoms were of limited value in understanding and diagnosing madness; for a more complete clinical picture, a longitudinal developmental narrative was required. When the patient's symptoms were placed into a larger developmental, life-story context, either whole madness or insanity was demonstrated, or the lack of illness was uncovered (Lepoutre and Dening 2012).

A third factor in the decay of monomania was the emergence of alternative formulations of forensic psychopathology, most notably Morel's concept of hereditary degeneracy, which moved forensic assessments out of the domain of descriptive psychopathology and into the domain of etiological psychopathology, further enhancing the scientific credibility of the field. (The rise and fall of degeneracy theory is another story; see [Rafter 1997, 2008])

Huneman (2008) identifies a fourth factor, which involved the conceptual distinction between normal and pathological passions (emotions) in monomania, a distinction that many nineteenth-century alienists found hard to make as they applied the concept in clinical and forensic settings in Europe and the U.S. This latter concern is contemporary, in that psychiatry still struggles with the distinction between normal and pathological sadness (Horwitz and Wakefield 2007) or social anxiety (Lane 2006).

Perhaps most important was the extreme negative reaction from many judges, jurists, and psychiatric colleagues, who saw the use of monomanias in forensic testimony as an undoing of the concept of criminal responsibility and an abuse of criminal excuse and the emerging 'insanity defense' (Goldstein 1998; Goldstein 2001; De Saussure 1946). While use of the monomania diagnosis had progressively declined by the end of the nineteenth century, laudatory comments about Esquirol and his coinage persisted well into the twentieth century (Amdur and Messinger 1993; De Saussure 1946; Mora 1972; Parsons 1887) but faded in the later part of the century (Bynum 2003) to the residual categories we see in the DSMs.

## Twenty-First Century Monomanias: DSM-5

At this point in the chapter we can jump ahead to the present day and examine the legacy of monomanias in DSM-5. Some of the DSM-5 categories may suggest qualification as monomanias through their name alone: Kleptomania, Pyromania, Trichotillomania are all official DSM-5 disorder names (American Psychiatric Association 2013) and carry the '-mania' suffix as in the classical monomanias. To identify other contemporary 'monomanias' will require some conceptual analysis to link DSM-5 disorders to the monomania concept. My purpose in this paper is not to establish a historical continuity between Esquirol's monomanias and DSM categories, but rather to describe a set of conceptual problems in formulating categories of psychopathology that were historically shared by Esquirol's monomania concept, on the one hand, and a select group of current DSM-5 diagnostic categories, on the other. Having defined these conceptual problems, the nineteenth- and early twentieth-century critical reactions to monomanias are relevant, and I go on to explore the conceptual problems in DSM-5 monosymptomatic disorders. My overall goals here are philosophical, not historical.

One of the fundamental and most useful distinctions in philosophy is the difference between word and concept. A word is a denotation for an idea (concept) that makes up the word's semantic meaning. A single concept may have many words: for example, blue, blau, bleu mean the same color concept in three different languages. Alternatively, a single word may denote multiple concepts. For example, in English, my proper name (John) is also slang for (1) a client of a prostitute or (2) a toilet or bathroom. A lot of philosophy is done teasing out word from concept and vice versa, and a lot of poetry is written playing with word/concept relationships.

Importantly, the meanings of words – their associated concepts – can change over time and historical circumstances, a phenomenon described nicely for the word 'schizophrenia' by the philosopher Joseph Margolis (1994). What 'schizophrenia' meant to Breuer in nineteenth-century Vienna and what it means to NIMH in twenty-first-century Bethesda are two quite different things. For my goals for this paper, what I need is a technical definition of 'monomania' that both suits contemporary examples in the DSM-5, as well as the historical concept described by Esquirol and many, perhaps most, of his followers and elaborators.

The 'mono' in 'monomania' provides a crucial ingredient to such a definition. Monomanias involve a single symptom, or single cluster of closely related symptoms. As Esquirol originally described, the associated clinical features of monomanias are reasonable derivatives from the fixed idea or preoccupation. Regarding Esquirol's intellectual monomania, we are reminded of today's Delusional Disorder, a condition defined by a relatively systematized delusion which, once the delusional premise is accepted, the implications follow logically, or at least credibly. As another example with contemporary resonance, the woman with Erotomania pursues her would-be lover as if he is committed to her; his refusals are interpreted in light of external barriers to their happiness. The child with Pyromania justifies his behavior as fun and intrinsically rewarding, and the adverse consequences as unintended. Monomaniacal syndromes are contrasted with complex syndromes like DSM-5 Schizophrenia, where core symptoms like hallucinations or delusions bear no practical or semantic relationship to other core features like abulia or loosening of associations. Monomania is distinct from today's DSM concept of a pervasive disorder. Esquirolian monomanias may include delusions or hallucinations, but they are circumscribed by the domain of the *idée fixe* or preoccupation, and are not accompanied by other marks of psychotic thinking, again in today's language would be items like thought disorder or pervasive mood changes. So a required feature of a contemporary monomania would be a focus on a single symptom or single cluster of closely-related symptoms.

A second feature of Esquirolian monomania is the notion of the overvalued idea, *idée fixe*, or what we might call a pathological (morbid) preoccupation today. Monomaniacal individuals may be preoccupied by ideas like love by another, fire-setting, gambling, or shoplifting. However, in Esquirol's own classification, discussed earlier, for the category of 'instinctive monomania', these latter syndromes involved impulsive, uncontrollable behaviors or actions, which were poorly motivated even by the patient's own admission. These cases of instinctive monomania often manifested in impulsive aggression, leading to the nineteenth-century category of homicidal monomania or homicidal insanity. These categories were to be important to pioneering American forensic clinicians like Isaac Ray, who wrote extensively about homicidal insanity along with Esquirol (Ray 1853; De Saussure 1946; Colaizzi 2002; Haack et al. 2010). So if Esquirol's work is the example, monomanias may involve morbid preoccupations as well as impulsive behaviors. A refinement of the monomania concept encompasses these two, nonexclusive, possibilities.

A third feature of Esquirol's monomania concept is the requirement for extremity in either thought or behavior. The single symptom/complex becomes a dominating motivational force in the patient's life. A single idea, emotion, or behavior is indulged *in extremis* leading to impairments and/or adverse consequences. Inversely, monomanias are not deficiency states, in the sense that the person lacks conviction, emotion, intent, or will. Monomanias involve too much of something.

Other features of Esquirol's concept are unclear or contradictory. For instance, instinctive monomanias like homicidal monomania are noted to retain some degrees of insight, for example, the patient recognizes her murderous act as wrongful, or, for Pyromania, that fire-setting is illegal or wrongful. However, the presence of delusions in intellectual monomania would seem to contradict the idea of the preservation of rational insight. Consider, for example, Erotomania, which carries a delusional belief that the patient is beloved by a higher-status individual. However, the capacity for reasoning outside the delusional domain is maintained in monomanias as well, again raising the resemblance to contemporary notions like Delusional Disorder, where thought disorder is not involved, and hallucinations are not a prominent clinical feature [7, p. 90; 11]. These ambiguities and contradictions were important, as noted earlier, to the 'fall' of the monomania concept.

My analysis leaves the core features of Esquirolian monomania as (1) presence of a single *idée fixe*, overvalued idea, or morbid preoccupation, (2) manifested in focused excesses of derivative thoughts, emotions, or behaviors, or some combination and (3) with relative preservation of rational capacities outside the domain of the morbid preoccupation and impulsive behavior. These core features, it will be shown, are shared by numerous DSM-5 categories (as well as their predecessors in DSM-III and IV).

By these criteria, which DSM-5 categories might fit this description as latter-day monomanias? Table 1 describes the DSM-5 disorders that meet these criteria, along with their associated preoccupations and behavioral excesses. Note that

**Table 1** DSM-5 candidate monomanias

| DSM-5 category | Preoccupation | Behavioral excess |
|---|---|---|
| Delusional Disorder 297.1 | Delusional belief(s) | Consequent |
| Body Dysmorphic Disorder 300.7 | Defect in appearance | Consequent |
| Hoarding Disorder 300.3 | Retaining possessions | Saving objects |
| Trichotillomania 312.39 | Hair-related discomfort | Hair-pulling |
| Excoriation Disorder 698.4 | Skin-related discomfort | Skin-picking |
| Illness Anxiety Disorder 300.7 | Having/acquiring serious illness | Consequent |
| Binge-Eating Disorder 307.51 | Consequent | Excessive eating |
| Pyromania 312.33 | Fires | Fire-setting |
| Kleptomania 312.32 | Stealing | Thefts |
| Intermittent Explosive Disorder 312.34 | Consequent | Impulsive aggression |

'consequent' means the respective preoccupation/excess is associated with closely-related meaningful consequences – for delusions, avoidance of the objects of concern (e.g., the CIA); for overeating, purchases of food and the monitoring of eating habits; for appearances, monitoring cosmetic appearances and seeking cosmetic surgery; for illness concerns, the seeking of health care; and so forth.

Before discussing these categories explicitly, I should mention some potential DSM-5 monomanias that did not, I would argue, make the cut. For instance, the classical monomania of 'dipsomania' might imply that Alcohol Use Disorder and related disorders might qualify. However, in my reading of the criteria, these DSM-5 categories fail because they are not monosymptomatic categories. Tolerance and withdrawal phenomena, as well as the reference to pathological denial of problems (criterion 9 for Alcohol Use Disorder, [American Psychiatric Association 2013, p. 491]) fall outside the domain of semantically consequent, derivative behaviors, as is typical of monomanias. (For example, the nausea and vomiting, insomnia, tremor, and seizures of withdrawal from alcohol are not meaning-driven, intentional actions consequent to the belief one needs alcohol. Instead, they are simply physiological consequences of alcohol overuse. Similarly, I need not deny my need for alcohol in order to remain convinced I need it. Denial is not a semantically consequent, derivative behavior.) Criterion 9 is also important in that it directly contradicts the monomania requirement for rational assessments of the preoccupation/behavioral excess.

One might raise the question about DSM-5 paraphilias as examples of monomanias, and I believe the argument could be made that paraphilias like DSM-5 Pedophilic Disorder could qualify. However, empirical data suggest that many paraphilias are accompanied by normative sexual behavior and are not 'exclusive' sexual behaviors, and may also be accompanied by other sexually-excessive behaviors (Laws and O'Donohue 2008). These findings erode the claim that paraphilias are preoccupations or involve an *idée fixe*; rather, the paraphilic interest may be one of a range of sexual interests. This leads paraphilias arguably outside the realm of morbid preoccupations, even though the DSMs tend to define them monosymptomatically. As noted earlier, Esquirol had a tripartite concept of monomanias: intellectual, affective, and instinctive, and the resulting pliable ambiguity of the monomania concept makes paraphilias apropos to instinctive, and perhaps affective, but not intellectual forms. Nevertheless, in discussing the conceptual issues with contemporary monomania concepts, I'll consider the paraphilias regardless.

Some childhood-associated disorders, like Conduct Disorder, could be construed as monosymptomatic, especially because the criteria are written solely along the lines of describing wrongful or criminal conduct. However, the associated features recognize Conduct Disorder as having other substantive unrelated clinical features, such as neuropsychological impairments (Moffitt 1993; Moffitt et al. 2008) which bring them outside the monosymptomatic profile. Moreover, because Esquirol's original concept was addressed to adult psychopathology, I have for simplicity's sake omitted disorders primarily associated with childhood and adolescence from this discussion. (See Sadler (2014) for a discussion of the conceptual issues with DSM Conduct Disorder.)

When one examines the DSM-5 diagnostic criteria for the disorders described in Table 1, one finds a monosymptomatic profile or complex of closely related symptoms. Typically, these monosymptomatic complexes are then further specified as 'not being due to another disorder' or differentiated from other conditions or etiologies. Some categories (e.g., Somatic Symptom Disorder) have specifications for duration of symptoms, or clarifications about the nature of the symptoms, or re-descriptions of symptoms. Still others invoke variations on the 'clinical significance criterion' (Spitzer and Wakefield 1999) as in Binge Eating Disorder's requirement for distress, or Hoarding Disorder's requirement for "impairment in social, occupational, or other important areas of functioning" (American Psychiatric Association 2013, p. 247). Notably, Kleptomania and Pyromania criteria describe a release of tension or pleasure in indulging in the behavior; but I would not consider this a non-consequent feature of the behaviors.

Having described a cluster of candidate DSM-5 monomanias, I now turn to my own discussion about the problems with contemporary monomanias.

## Conceptual Problems with Monosymptomatic Disorders

We can understand the core conceptual problems with monosymptomatic disorders through considering simple 'monosymptoms' from general medicine. What are the potential mistakes of classifying diseases on the basis of single symptoms like fever or headache?

### *Description vs. Etiological Diagnosis*

Calling a fever a disease (as in ancient Greek medicine [Porter 1997]) seems fallacious to contemporary medical ears because today we prefer classification of disease based upon causes – 'etiological classification' – rather than symptoms. What is powerful about this approach to classification of disease is that the condition is placed into a context or network of multilevel causal interactions (Schaffner 1993) – ideally, from molecular event to environmental exposure or influence – that constrain the identity of the condition. When we understand the causal pathways or networks of a condition, we place the condition (disease) in a context that severely restricts its identity as a condition. Fever from bacterial infiltration of the lung alveoli constrains that fever-condition to a narrow range of associated clinical features, and makes verification of the condition possible, with sufficient technology to do so.

The historical ideal of etiological classification might be Koch's postulates (King 1952; Inglis 2007) where four criteria must be met to establish infectious-disease causality: (1) the microorganism must be found in abundance in all organisms suffering from the disease; (2) the microorganism must be isolated from a diseased organism and grown in pure culture; (3) the cultured microorganism should cause disease when introduced into a healthy organism; and (4) the microorganism must be reisolated from the inoculated, diseased experimental host and identified as identical to the original specific causative agent (King 1952). Koch's practical tests for infectious disease are not feasible for many of our diseases today, but the existential features of his account – placement into an interdependent framework of defining operations and observations – still hold.

Unfortunately, the rich causal frameworks for most DSM mental disorders are yet to be described, leaving DSM-5 and its predecessors from DSM-III forward with a formulation of disease closer to Thomas Sydenham's description of clinical syndromes – collections of co-occurring observable features that are described and elicited by an examining clinician (Jablonski 1992). Sydenham, however, recognized the value of multiple descriptions of independent features of a disease, and thereby advanced the ideal that diseases be placed into a complex description, later to be articulated as multiple inclusionary and exclusionary features (Porter 1997). Thus 'dropsy' or the congestive heart failure syndrome is not simply described by a single independent symptom like exertional dyspnea or dependent edema, but rather described with multiple, seemingly-unrelated symptoms: jugular-venous distention, orthopnea, a 'gallop' of the heart sounds, etc.

The requirement for syndromes to include multiple phenomenologically independent symptoms (subjective observations by, or concerns of, the patient) and signs (observations made or elicited by the clinician) then provided the early descriptive frameworks for physician-scientists to try to draw out pathophysiological interactions that would cohere and point to causal explanations and new therapeutic opportunities.

In this latter regard we can recognize that the more classical syndromes/mental disorders in the DSM, categories such as Schizophrenia, Bipolar Disorder, Delirium, Major Depression (for instance) adhere well to the requirement for multiple phenomenologically independent signs and symptoms. That is, a flattened affect, loosening of associations, hearing of voices, motivational indifference, delusional beliefs may not bear any intuitive causal or meaningful relationship to each other (they are independent clinical indicators), though collectively they describe a schizophrenia syndrome.

In contrast, the diagnostic criteria for the conditions identified in Table 1 are notable in the marked absence of these classical features of syndromes. Such syndromes have few to no independent clinical indicators, few inclusionary and exclusionary criteria, and point to little in the way of complex causal networks that may aid in distinguishing true cases from resembling syndromes.

## *The Algebra of Impoverished Diagnostic Criteria*

A second, albeit related, conceptual problem with monosymptomatic categories can be understood by considering a simple thought experiment. If we define a bachelor as a 'man', then we can quickly recognize that we will in fact identify virtually all men, including married ones, as bachelors, a finding that is empirically incorrect. In order to define a bachelor, we have to add additional stipulations – inclusion and exclusion criteria – in order to sort out men who are bachelors from men who are not. If we add the criterion "unmarried" to the definition, we then restrict the list and get closer to the bachelor concept as used in ordinary discourse. However, priests and widowers are also unmarried men, though we don't conventionally regard them as bachelors. We might add the criterion "never married," but things get even more complicated, as a divorced man may currently be living as a 'bachelor' without contradiction in ordinary use. Defining concepts in ordinary language usage turns out to be more difficult than we might expect – as it is with diagnostic concepts!

The principle, nonetheless, is that to get clarity about sorting out a population of bachelors descriptively, one needs multiple descriptors and stipulations to tease out a relatively uniform population. If our descriptive criteria are 'impoverished' (e.g., 'bachelors are unmarried men') then our selected population is confounded by multiple examples of non-bachelors as defined by conventional usage.

So it is with monosymptoms. Using "headache' or 'fever' as a monosymptomatic disease today will be of little assistance in characterizing the patient or the condition. We recognize that countless disease entities can manifest by headache or fever. Only by circumscribing our definitions can we identify relatively uniform populations of people with 'mental disorders', in the absence of etiological classification. This problem is the problem of diagnostic 'heterogeneity' lamented in past DSM work (Krueger and Bezdjian 2009). Monosymptomatic disorders, in conclusion, get confounded by other conditions.

## *Defaulting to 'Loud' Symptoms*

When relatively little is known about associated or even defining clinical features for a syndrome, often the DSM defaults to relying on what I call 'loud' symptoms (Sadler 2010). Loud symptoms are the opposite of so-called 'soft signs'. Loud symptoms are the obvious, problematic, attention-getting, impossible-to-miss symptoms that patients may present: angry outbursts, assaults, lying, molestations, suicide attempts, panic attacks, grandiose delusions, etc. Psychiatric training and skills are not required to identify loud symptoms. Like the lay members of the jury in Esquirol's courtrooms, anyone can recognize global insanity and loud symptoms. In contrast, soft signs are clinical phenomena that require training, experience, and nuance to detect – injury to cranial nerve I or the early signs of masked facies in Parkinsonism. A brief examination of Table 1 will indicate a list of disorders based upon, with the possible exception of Delusional Disorder, loud symptoms. The only thing subtle about Kleptomania is the patient's stealth in stealing – once discovered, the problem is loud.

Loud symptoms in themselves are not problematic – all areas of medicine have them. The multiplex syndromes of DSM Schizophrenia, Bipolar Disorder, Delirium, Major Depression, Obsessive Compulsive Disorder and others have them. The problem is building a diagnosis around a single loud symptom. The DSM -IV and 5 Paraphilias (Paraphilic Disorders) were originally defined by loud symptoms, and to the relevant DSM-5 work group's credit, they have attempted to broaden the clinical descriptors in order to develop more focused, homogeneous populations for DSM-5 (Zucker 2010). Defining disorders with loud symptoms tends to focus clinical attention on them and away from more clinically important and indicative symptoms/signs – most notably evident from DSM-5 Conduct Disorder (CD), which is defined entirely upon loud symptoms describing wrongful and criminal conduct (Sadler 2014). Treatment then dwells upon problem behaviors and tends to ignore the underlying (in this case) neuropsychological and emotion impairments in the CD child (Moffitt 1993; Moffitt et al. 2008; Dery et al. 1999; Fishbein et al. 2006; Hobson et al. 2011; Narhi et al. 2010).

Moreover, because loud symptoms are obvious, they are the low-hanging fruit in writing diagnostic criteria, and appear, over several decades of DSM tradition, to deflate the motivation for finding independent symptoms and signs to expand and elaborate criteria. For example, CD diagnostic criteria are based entirely on a listing of antisocial and wrongful behaviors. Paraphilia symptoms are defined in terms of their divergent sexual focus. Similarly, symptoms/signs in the other Table 1 disorders are defined solely on the basis of the loud, defining symptom. While associated clinical features may only be associated, they may also open doors for a more complex etiological network for the condition, as has been described for CD in the earlier-cited neuropsychological literature.

## *Misleading Hierarchies*

Related to all the above problems is the problem of monosymptomatic categories' placement into hierarchical taxonomy (nosology) of mental disorders. In the recent DSMs, all the mental disorders are placed into broad, superordinate groupings of presumably related disorders (e.g., Mood Disorders, Anxiety Disorders, Impulse Control Disorders). Sometimes these groupings are based upon research evidence of individual disorders' relatedness to each other (biological, genetic, etiological, etc.), other times they are grouped solely on one or more similar features. Monosymptomatic disorders lend themselves well to comorbidities with other disorders or conditions because only a single shared clinical feature is required before the monosymptomatic disorder can be co-diagnosed with another disorder. The etiological and treatment-response relationships between substance abuse, psychotic, and mood disorders for monosymptomatic conditions remain obscure as the co-occurrence of these conditions is screened out of research subject policies, and we tend to not know about how the monosymptomatic condition relates to a more common condition. Hierarchical relationships of monosymptomatic disorders as subtypes of other conditions may be lost. For instance, impulse control disorders like Kleptomania and Pyromania may well just be behavioral subtypes of a more encompassing pathology of impulse control.

## Conclusion and Suggestions

The failure of monomanias to maintain influence and diagnostic prominence in psychiatry is well deserved. Monosymptomatic disorders are prone to blend into other conditions, collect epidemiologically diverse populations of patients, provoke simplistic criteria sets with potentially high rates of comorbidity and false-positives, complicate treatment planning, and obfuscate etiological studies.

Until etiological alternatives to classifying psychopathology are available psychiatric nosologists should be wary of monosymptomatic disorders and search for pragmatically and semantically independent, corroborating clinical features, and incorporate these into diagnostic criteria (Sadler 2005, 2008). Nosologists should be willing to question their hierarchical arrangements of monosymptomatic disorders carefully, and consider monosymptomatic disorders as potential subtypes of more encompassing disorders. If no additional clinical features can be discerned which uniformly apply to monosymptomatic disorders, then research efforts should be directed toward finding them and incorporating them into diagnostic criteria. If no additional features can be found which statistically aggregate, then the disorder's validity should be questioned.

## References


Amdur, B., & Messinger, E. (1993). Jean- Etienne-Dominique Esquirol. *American Journal of Psychiatry, 96*, 129–135.

American Psychiatric Association. (2013). *Diagnostic and statistical manual of mental disorders, fifth edition (DSM-5)*. Arlington, VA: American Psychiatric Publishing.

Bynum, W. (2003). Discarded diagnoses: Monomania. *Lancet, 362*, 1425.

Cartwright, S. (1851). Report on the disease of physical peculiarities of the Negro. *The New Orleans Medical and Surgical Journal, 7*, 707–708.

Colaizzi, J. (2002). *Homicidal insanity 1800–1985*. Birmingham: University of Alabama Press.

De Saussure, R. (1946). The influence of the concept of monomania on French medico-legal psychiatry (from 1825 to 1840). *Journal of the History of Medicine and Allied Sciences, 1*(3), 365–397.

Dery, M., Toupin, J., Pauze, R., et al. (1999). Neuropsychological characteristics of adolescents with conduct disorder: Association with attention-deficit-hyperactivity and aggression. *Journal of Abnormal Child Psychology, 27*(3), 225–236.

Esquirol, J. E. D. (1845). *Mental maladies: Treatise on insanity*. (trans: Hunt, E. K.). Philadelphia: Lea and Blanchard.

Fear, C. F., McMonagle, T., & Healy, D. (1998). Delusional disorders: Boundaries of a concept. *European Psychiatry, 13*, 210–218.

Fishbein, D. H., Hyde, C., Eldreth, D., et al. (2006). Neurocognitive skills moderate urban male adolescents' responses to preventive intervention materials. *Drug and Alcohol Dependence, 82*(1), 47–60.

Goldstein, J. (1998). Professional knowledge and professional self-interest: The rise and fall of monomania in the 19th-century France. *International Journal of Law and Psychiatry, 21*(4), 385–396.

Goldstein, J. (2001). *Console and classify: The French psychiatric profession in the nineteenth century*. Chicago: University of Chicago Press.

Haack, K., Kumbier, E., & Herpertz, S. C. (2010). Illnesses of the will in 'pre-psychiatric' times. *History of Psychiatry, 21*, 261–277.
Hobson, C. W., Scott, S., & Rubia, K. (2011). Investigation of cool and hot executive function in ODD/CD independently of ADHD. *Journal of Child Psychology Psychiatry, 52*(10), 1035–1043.
Horwitz, A. V., & Wakefield, J. C. (2007). *The loss of sadness: How psychiatry transformed normal sorrow into depressive disorder*. Oxford: Oxford University Press.
Huneman, P. (2008). Montpellier vitalism and the emergence of alienism in France (1750–1800): The case of the passions. *Science in Context, 21*(04), 615–647.
Hunter, R. A., & Macalpine, I. (Eds.). (1963). *Three hundred years of psychiatry, 1535–1860: A history presented in selected English texts*. London/New York: Oxford University Press.
Inglis, T. J. (2007). Principia aetiologica: Taking causality beyond Koch's postulates. *Journal of Medical Microbiology, 56*(11), 1419–1422.
Jablonski, S. (1992). Syndrome–a changing concept. *Bulletin of the Medical Library Association, 80*(4), 323.
King, L. (1952). Dr. Koch's postulates. *Journal of the History of Medicine and Allied Sciences, VII*(4), 350–361.
Krueger, R. F., & Bezdjian, S. (2009). Enhancing research and treatment of mental disorders with dimensional concepts: Toward DSM-V and ICD-11. *World Psychiatry, 8*, 3–6.
Lane, C. (2006). How shyness became an illness: A brief history of social phobia. *Common Knowledge, 12*(3), 388–409.
Laws, D. R., & O'Donohue, W. T. (2008). *Sexual deviance: Theory, assessment, and treatment* (2nd ed.). New York: Guilford.
Lepoutre, T., & Dening, T. (2012). 'De la non-existence de la monomanie', by Jean-Pierre Falret (1854): Introduction and translation (part 2). *History of Psychiatry, 23*, 488–495.
Margolis, J. (1994). Taxonomic puzzles. In J. Z. Sadler, O. P. Wiggins, & M. A. Schwartz (Eds.), *Philosophical perspectives on psychiatric diagnostic classification* (pp. 104–128). Baltimore: The Johns Hopkins University Press.
Moffitt, T. E. (1993). The neuropsychology of conduct disorder. *Development and Psychopathology, 5*(1–2), 135–151.
Moffitt, T. E., Arseneault, L., Jaffee, S. R., et al. (2008). Research review: DSM-V conduct disorder: Research needs for an evidence base. *Journal of Child Psychology and Psychiatry, 49*(1), 3–33.
Mora, G. (1972). On the bicentenary of the birth of Esquirol (1772–1840), the first complete psychiatrist. *American Journal of Psychiatry, 129*(5), 562–567.
Narhi, V., Lehto-Salo, P., Ahonen, T., & Marttunen, M. (2010). Neuropsychological subgroups of adolescents with conduct disorder. *Scandinavian Journal of Psychology, 51*(3), 278–284.
Parsons, R. L. (1887). Nomenclature in psychiatry: Monomania or oligomania, which? Paranoia, which? *Journal of Nervous and Mental Disease, 14*(4), 247–255.
Porter, R. (1997). *The greatest benefit to mankind: A medical history of humanity*. New York: W. W. Norton.
Rafter, N. H. (1977). *Creating born criminals*. Urbana: University of Illinois Press.
Rafter, N. H. (2008). *The criminal brain: Understanding biological theories of crime*. New York: NYU Press.
Ray, I. (1853). *A treatise on the medical jurisprudence of insanity* (3rd Ed.). Boston: Little, Brown.
Sadler, J. Z. (2005). *Values and psychiatric diagnosis*. London: Oxford University Press.
Sadler, J. Z. (2008). Vice and the diagnostic classification of mental disorders: A philosophical case conference. *Philosophy Psychiatry Psychology, 15*, 1–17.
Sadler, J. Z. (2010, December 2). Watch out for 'loud' symptoms. Psychiatry Times [Internet] http://www.psychiatrictimes.com/blog/DSM-5/content/article/10168/1745539
Sadler, J. Z. (2014). Conduct disorder as a vice-laden diagnostic concept. In C. Perring & L. Wells (Eds.), *Diagnostic dilemmas in child and adolescent psychiatry: Philosophical perspectives* (pp. 166–181). Oxford: Oxford University Press.

Schaffner, K. F. (1993). *Discovery and explanation in biology and medicine*. Chicago: University of Chicago Press.

Spitzer, R. L., & Wakefield, J. C. (1999). DSM-IV diagnostic criterion for clinical significance: Does it help solve the false positives problem? *American Journal of Psychiatry, 156*(12), 1856–1864.

Zucker, K. J. (2010). Reports from the DSM-V work group on sexual and gender identity disorders. *Archives of Sexual Behavior, 39*, 217–220.

# The Loss of Grief: Science and Pseudoscience in the Debate over DSM-5's Elimination of the Bereavement Exclusion

**Jerome C. Wakefield**

**Abstract** Was the DSM-5 revision process based on careful evaluation of scientific evidence, as the DSM-5 Task Force repeatedly claimed? The counterfeit of science is pseudoscience, the systematic motivated deviation from basic canons of rational scientific evaluation of evidence to create the false appearance of scientific support for a favored hypothesis. In this chapter, I consider the arguments that were used to support the DSM-5's controversial decision to eliminate the bereavement exclusion (BE) to major depressive disorder (MDD). I consider three central arguments: that the BE had to be eliminated for reasons of consistency; that the BE excluded cases from MDD that would respond to treatment; and that the BE leads to missing suicidal cases. The analysis reveals forms of rhetoric by which the question at issue was obfuscated or misconstrued, and the scientific evidence sidelined, rendered impotent, or outmaneuvered to make it seem to support elimination, despite strong evidence to the contrary. I conclude that the arguments for elimination of the BE were largely pseudoscientific and the BE's elimination unwarranted by the evidence.

Was the DSM-5 (American Psychiatric Association 2013) revision process scientifically based? The DSM-5 Task Force repeatedly asserted that the changes in the manual would be made on the basis of scientific evidence. In this chapter, focusing on DSM-5's decision to eliminate the bereavement exclusion (BE) to major depressive disorder (MDD), I explore the scientific quality of some of the arguments used in the debate over a proposed DSM-5 change. I argue that key assertions by those favoring the BE's elimination ("eliminationists") were in crucial respects pseudoscientific.

Like many others, I believe that the BE (explained below) was a sensible way to protect against overpathologizing the grief process. In fact, as the debate over the BE proceeded, the scientific support for the BE's validity and thus for its retention in DSM-5 became increasingly persuasive, anchored in multiple epidemiological studies replicated across several major data sets (Gilman et al. 2012, 2013; Mojtabai

J.C. Wakefield (✉)
School of Social Work, Department of Psychiatry, and Center for Bioethics, New York University, New York, NY, USA
e-mail: jerome.wakefield@nyu.edu

 
S. Demazeux, P. Singy (eds.), *The DSM-5 in Perspective*, History, Philosophy and Theory of the Life Sciences 10, DOI 10.1007/978-94-017-9765-8_10

2011; Wakefield and Schmitz 2012a, b, 2013a, b, c, d, 2014; Wakefield et al. 2007, 2011a, b). The proposed elimination was criticized in editorials in leading medical journals, such as *Lancet* (Editorial 2012) and *New England Journal of Medicine* (Friedman 2012). Claims were made and tested, and a clear, scientifically satisfying outcome emerged supporting the validity of the BE (Wakefield 2013a, b, c).

Yet the BE was eliminated. For those bewildered by the BE's elimination, the post-mortem question arises of how this came to pass. Was it a matter of scientific debate and attention to the nature of the scientific evidence, as the DSM-5 Task Force insists? If not, by what manner of argument or forms of rhetoric was the question at issue obfuscated, diverted, or misconstrued, so that the scientific evidence was sidelined, rendered impotent, or outmaneuvered? I explore these questions by considering three central arguments from the DSM-5 BE debate: that the BE had to be eliminated for reasons of consistency; that the BE excluded cases from MDD that would respond to treatment; and that the BE leads us to miss suicidal cases.

Terminology is challenging in a discussion like this. Because the issue is whether certain periods of sadness are psychiatric disorders or normal reactions, and these two domains tend to involve quite different terminology, the discussion can be become quite tortured as one flips from one vocabulary to another, and one's choice of vocabulary can appear to beg the very question being disputed. In this paper, for convenience I adopt the standard terminological convention in discussions of the DSM-5, which is that I often use the medical vocabulary to describe the conditions whose status is being debated, but these terms are neutral in such contexts in that they do not imply disorder. Thus, terms generally associated with disorder such as "symptom," "depression," and "diagnosis" are used descriptively, such that normal grief has "symptoms" such as sadness and insomnia, "depression" is sometimes part of normal grief, and one can "diagnose" a normal condition. Obviously, the use of these terms does tend to medicalize the discussion, but properly understood, it need not bias the outcome. I also use the phrase "depressive episode" neutrally to denote any condition that satisfies the DSM's symptom and duration criteria for major depressive disorder (i.e., at least 5 symptoms for at least two weeks), but again it remains open to dispute whether such depressive episodes are sometimes part of normal grief or always instances of depressive disorder. In fact, "depression" has long been common as a description of both pathological depression and normal experiences of depression due to life's vicissitudes (e.g., Clayton et al. 1974).

## The DSM-III Through DSM-IV Bereavement Exclusion

Sadness is a biologically designed emotional reaction to loss and stress seen in other species as well, a view put forward by Darwin (1872) and further explored and supported in a vast theoretical and empirical literature on the nature and evolution of sadness (e.g., Bowlby 1980; Ekman and Friesen 1971; Freud 1917; Horwitz and Wakefield 2007; McGuire et al. 1997; Nesse 2000, 2009; Nettle 2004; Price 1967;

Price et al. 1994; Sloman et al. 1994; Watson and Andrews 2002; Welling 2003). The intensity of sadness generally tends to be roughly proportional to the magnitude of a loss and to ameliorate over time as the individual reconstructs his or her meaning system and adapts to the changed situation. However, the intensity and duration of the reaction and the nature of the specific events that trigger the reaction vary to some extent across cultures and among individuals within a culture. The precise functions of sadness remain a matter of scientific investigation, but negative emotions like sadness may be analogous to physical pain in focusing our attention on addressing a challenge, and evidence suggests that the withdrawal and rumination that occurs during grief may help us to readapt to a changed environment (Andrews and Thompson 2009; Horwitz and Wakefield 2007).

Throughout medical history, physicians have observed that intense normal sadness in response to life events can include many of the same symptoms of distress as occur in depressive disorder (Horwitz and Wakefield 2007). They have also observed that loss can trigger pathological depressive reactions that go beyond the range of normal response to loss and that continue in a morbid trajectory without adaptation and with severe symptoms (Parkes 1964). Thus, when diagnosing intense sadness with its associated symptoms, physicians traditionally have asked: all considered, is there a sufficient cause in the individual's circumstances to explain the individual's condition as likely a normal reaction to loss or stress, or is the condition so severe or enduring or independent of context to be better explained as a pathological failure of normal mood regulation?

Starting with the third edition of the *Diagnostic and Statistical Manual of Mental Disorders* (DSM-III) (American Psychiatric Association 1980), diagnosis of major depressive disorder (MDD) was based on operationalized descriptive diagnostic criteria that largely ignored the context of symptoms, essentially requiring that someone must experience 5 or more out of 9 specified symptoms for at least 2 weeks to be diagnosed with MDD. There was one exception, however, to the DSM's context-free approach to MDD diagnosis. Paula Clayton et al.'s (1968) classic longitudinal investigation of normal grief, using a nonclinical sample of relatives of individuals who had died, empirically established for the specific context of bereavement what the physicians had observed since antiquity, that normal grief after the death of a loved one routinely includes many of the symptoms that the DSM was using to diagnose MDD. Among the DSM's symptoms of depression, Clayton et al. distinguished those that are also common manifestations of normal general distress reactions, such as sadness, insomnia, fatigue, difficulty concentrating, decreased appetite, and loss of interest in usual activities, from the symptoms that are more patho-suggestive and tend to be distinctive of depressive disorder. For example, in Clayton's normal sample, in the first weeks post-loss, 87 % of bereaved subjects reported depressed mood, 85 % sleep disturbance, 79 % crying, and about half reported each of diminished interest in usual activities, difficulty concentrating, and lessened appetite, all of which appear among the DSM symptoms used to diagnose depression. Such bereavement-related episodes containing only general-distress-type depressive symptoms remitted on their own over time and did not cause the kind of marked impairment that frequently leads to psychiatric consultation and care. In contrast, such symptoms as psychotic ideation, self-condemnation, suicidal ideation, and psychomotor retardation were rare in these normal cases but more common in a comparison group of pathological cases of depression unrelated to bereavement (Clayton et al. 1974). Some degree of withdrawal and thus role impairment was common among the normally bereaved, whereas severe impairment or very prolonged course were rare.

Clayton's research yielded an empirically supported approach to distinguishing between normal and disordered reactive depressions triggered by bereavement. In effect, the symptoms used to identify depression fell into two categories noted above, namely, those that occur in depressive disorder but are not distinctive of disorder and are present in normal distress responses as well, and those that are more distinctive of pathological episodes. The critical point was that with major depression having a 5-symptom threshold for diagnosis, one could have enough of the general distress symptoms during normal bereavement that one could be misdiagnosed as having a depressive disorder. Indeed, many individuals among nonclinical grievers – 42 % in Clayton's original studies (Hensley and Clayton 2013) – do reach the DSM's 5-symptom threshold for MDD at some point in the first weeks after loss, and almost all of the rest have some subthreshold depressive symptoms as part of their normal grieving.

Thus, to prevent a massive number of false positive diagnoses among normally grieving individuals, an exclusion clause was added to DSM-III criteria for MDD. This clause became known as the "bereavement exclusion" (BE) and, although changing over time in its details, it was retained in similar form through to DSM-IV. It specifies which depressive episodes during bereavement should be presumed to be normal and which are likely pathological. In its most recent incarnation, the BE specifies that, when a depressive episode satisfies the DSM's symptom and duration criteria for MDD but follows the death of a loved one, it can be considered a depressive disorder only if:

> E. The symptoms are not better accounted for by Bereavement, i.e., after the loss of a loved one, the symptoms persist for longer than 2 months or are characterized by marked functional impairment, morbid preoccupation with worthlessness, suicidal ideation, psychotic symptoms, or psychomotor retardation. (American Psychiatric Association 2000, p. 356)

The BE states that a depressive reaction should not be diagnosed as MDD if it can be better explained as part of an intense normal reaction, and it offers guidelines as to which sorts of symptoms generally suggest that it cannot be so explained. Thus, in effect it says that to qualify for exclusion from MDD diagnosis, a bereavement-related DSM-defined depressive episode must meet six tests: (1) no psychotic ideation (2) duration of no more than 2 months, by which point it must remit; (3) does not cause severe impairment in role functioning; (4) no suicidal ideation; (5) no psychomotor retardation (i.e., no general and observable slowing down of thought and movement); and (6) the bereaved individual must not suffer from a morbid preoccupation with his or her worthlessness as a human being. Episodes during bereavement that meet these six requirements are considered "uncomplicated" and classified as normal. All other depressive episodes during bereavement that fail one or more of these six tests are classified as "complicated" and are diagnosed as MDD despite the recent loss.

## The Definition of the BE and the "Eliminate or Extend" Argument

You cannot have a rational scientific debate about whether a claim is true if there is lack of agreement or lack of clarity about the nature of the claim. Of course, as sociologists of science would no doubt observe, illogical arguments can sometimes lead to scientific progress. However, that is selective attention; mostly, confusion about what is being disputed leads to nonsense. Scientific discourse generally starts with consensually agreed and precise statements of the propositions for which evidence is being marshaled so that the evidence can be rationally brought to bear on a fixed target claim over time. Illogic can be fertile at times, but generally not when the question is specifically whether the evidence logically supports a publicly stated and reasonably precise claim.

This commonsense principle, that a target claim under dispute should be accurately and nontendentiously stated to advance scientific understanding, was repeatedly violated by the BE eliminationists, making rational discussion of the BE's validity virtually impossible. In particular, the eliminationists frequently misstated the meaning of the BE itself in ways that made it appear less valid or more open to counterargument than it was. Misstating one's opponent's position for argumentational advantage is called the "straw-man fallacy," and it is one of the most common forms of pseudoscientific argument. I focus my discussion on only the most egregious examples of misstatements by those who had unusual authority in the discussion of the BE and thus potentially did the most damage to the scientific integrity of the dispute.

Professor Kenneth Kendler was without doubt the most scientifically eminent of the staunch eliminationists. He is a world-renowned research psychiatrist specializing in the genetic underpinnings of psychiatric disorder, an expert on depression, and a sophisticated analyst of the conceptual foundations of psychiatry. He was also one of the few individuals to have actually done empirical research on the BE. Consequently, when in 2010 the proposal to eliminate the BE was criticized in an opinion piece in the New York Times by Allen Frances (2010), the DSM-5 mood disorders work group, which was responsible for the proposal, turned to Kendler, who was at that time a member of the work group, to write an official rebuttal and explanation for why the BE should be eliminated. Kendler's (2010) official statement was placed on the DSM-5 website, and remains the most detailed and serious attempt by the work group to officially state its reasoning. (Kendler later left the work group and ascended to Chair of the Scientific Review Committee reviewing DSM-5 proposals from all the work groups for their scientific merit, a committee whose proceedings and findings remain secret to this day [see Demazeux, this volume])

Kendler begins his statement by explaining that misconceptions about the proposal to eliminate the grief exclusion from DSM-5 have been presented in the media, so he aims to provide "some insight into thinking behind the proposal." He begins his argument proper with the point that the BE did not exist in earlier diagnostic systems, in the course of which he provides a definition of the BE:

> First, the grief exclusion criterion – which states that someone who has experienced a recent bereavement is not eligible for a diagnosis of major depression – was not present in the two major psychiatric diagnostic systems that formed the basis for the DSM-III – the diagnostic manual that is the immediate precursor of our current DSM-IV. (Kendler 2010)

Kendler (2010) asserts that no BE existed before the DSM-III, presumably suggesting that the DSM's exclusion was a novel and potentially arbitrary deviation from the baseline of medical thought. However, Kendler's assertion that such exclusions were not present in earlier systems of diagnosis is simply untrue (Wakefield 2011). The two pre-DSM-III diagnostic systems to which Professor Kendler refers are the Feighner (1972) criteria (named after the first author of the paper reporting them) and the subsequently refined Research Diagnostic Criteria (RDC) (Spitzer et al. 1975a, b, 1978). With respect to the Feighner criteria, the BE was not stated in the formal diagnostic criteria but was included in the instruction to raters. We know this because when Eli Robins, a coauthor of the Feighner paper, worked with Robert Spitzer and Jean Endicott on the RDC, he explained to them that Feighner-criteria raters had been routinely cautioned not to diagnose an individual as having a depressive disorder if the individual had recently suffered the loss of a loved one, even if the individual met full Feighner criteria for depression (Jean Endicott, personal communication, October 15 2009).

Consequently, Spitzer, Endicott, and Robins took a similar approach in the RDC criteria, and placed the BE in manuals of instructions to raters (Robert Spitzer, personal communication, October 29, 2010; New York State Department of Mental Hygiene 1980). Kendler's account is contradicted by the fact that the BE is cited as a requirement of the RDC and applied in major pre-DSM-III epidemiological studies of depression. Weissman and Myers (1978) state, for example, that "Because of the overlap in presenting symptoms, there is an effort in the RDC to separate clinical depression from normal grief reactions secondary to the death of a 'significant other'- termed grief....Symptoms that lasted more than a year were considered symptomatic of major depression" (p. 1306).

Kendler's claims are not just inaccurate regarding the Feighner and RDC criteria but misleading about the earlier history of psychiatry as well. To take one salient example, Emil Kraepelin, the preeminent nineteenth-century diagnostic theoretician who inspired the DSM approach to depression and who Kendler (1990) elsewhere cites as the seminal thinker about depression, believed that intense normal sadness in response to grief and other reverses in life circumstances could symptomatically look just like major depressive disorder but have a different prognosis. The clinician's diagnosis, according to Kraepelin, depends on an examination of the context of the symptoms:

> Several times patients have been brought to me, whose deep dejection, poverty of expression, and anxious tension tempt to the assumption of a circular [pathological] depression, while it came out afterwards, that they were cases of moodiness, which had for their cause serious delinquencies and threatened legal proceedings... [T]he slighter depressions of manic-depressive insanity, as far as we are able to make a survey, may wholly resemble the well-founded moodiness of health. (Kraepelin 1917)

However, historical inaccuracies aside, what is most stunning about the passage in Kendler's statement quoted above is his characterization of the BE. He asserts that "the grief exclusion criterion ... states that someone who has experienced a recent bereavement is not eligible for a diagnosis of major depression." That is, Kendler starts his discussion by defining the BE as excluding *all* bereaved people from MDD diagnosis.

This, as we saw, is a mischaracterization of the BE, which excludes a limited subgroup of those experiencing bereavement-related depression who must satisfy six demanding criteria that indicate a normal-range depressive response to loss. Kendler's incorrect characterization of the BE is not random error but rather highly tendentious. It makes the BE look like a much broader exclusion than it is, thus potentially making it look less reasonable than it is and setting up a straw-man position as a target for criticism. As far as I know, no one in the debate over the BE held that all bereaved individuals (or even all bereaved individuals with shorter than 2-month episodes) should be excluded from MDD diagnosis. As noted, it has been well-known since antiquity and confirmed repeatedly in modern studies that some people do develop a depressive disorder during an extreme grief reaction to the loss of a loved one, sometimes even to the point of sinking into chronic psychotic melancholia. So, if the BE claimed that "someone who has experienced a recent bereavement is not eligible for a diagnosis of major depression," it would be manifestly invalid. Having misstated the BE at the outset and set up a weak straw-man position that nobody holds as the target for his critique, Kendler's official statement becomes potentially irrelevant to the question of whether the real BE is valid.

Is it possible that Kendler's initial characterization of the BE is just an abbreviated introductory explanation that is elaborated later in the statement? No, nowhere later in his statement does Kendler accurately restate the BE. To the contrary, he reasserts the same interpretation in his pivotal argument. I (Wakefield et al. 2007) had earlier proposed that the BE should be expanded to apply to uncomplicated depressive reactions to losses and stressors other than bereavement (e.g., marital dissolution, job loss). Kendler argues against this proposal by stating that if the BE were extended to other stressors, then "no depression that arises in the setting of adversity would be diagnosable." This statement presupposes (like his initial statement) that the BE excludes all bereavement-related depressive episodes, so if extended to other stressors it would similarly exclude all stressor-related episodes. But, of course, that is not how the BE works; it only excludes episodes that meet certain stringent conditions. So, the distortion in the initial statement was not a transient approximation but rather the target of Kendler's entire analysis. (I return below to the argument about extending the BE.)

Is it possible that Kendler, an expert on depression diagnosis, never noticed or grasped the details of the BE's diagnostic criteria that he is attacking? No, lack of a clear understanding of the BE cannot be the explanation because in a scientific paper he senior-authored in the prestigious *American Journal of Psychiatry* just 2 years before issuing his official statement (Kendler et al. 2008), Kendler had quite correctly defined the BE, including, and even specifying, the features of the BE that he could not apply in the analysis:

> [W]e operationalized the DSM-IV "normal grief criterion" as a duration of ≤2 months and an absence of psychomotor retardation, suicidal ideation, and severe work impairment. We could thus determine whether each depressive episode met these proposed criteria for a normal grief response. Our interviews did not inquire about psychotic symptoms related to bereavement or "morbid preoccupation with worthlessness," the other DSM-IV criteria for normal grief. (Kendler et al. 2008, p. 1450)

Clearly, Kendler knew that the BE says nothing remotely like "someone who has experienced a recent bereavement is not eligible for a diagnosis of major depression." But perhaps the statement of the correct criteria at the beginning of the 2008 research article was just paying lip service to the correct criteria for scholars, and the research article then went on to use the broader notion that appears in Kendler's statement in the analyses? No, in the research article, the correct definition of the BE is systematically and consistently used throughout. In the article's Table 1, in a column listing analyzed variables, there is a sublist of variables titled "DSM-IV exclusion criteria" that lists the features relevant to evaluating the BE's applicability: "Duration of > 2 months"; "Psychomotor retardation"; "suicidal ideation"; and "Severe work impairment" (Kendler et al. 2008, p. 1451). At the end of the list, there is a summary variable that is in part, 'Meets criteria for 'uncomplicated bereavement-related disorder . . . '." The interaction analysis performed in the study specifically uses these correct "normal grief" criteria.

Even if Kendler knew and applied the correct BE exclusion, perhaps he was under the impression that virtually all bereavement-related depressive episodes in fact do satisfy the BE's six requirements for exclusion? If so, then he would have believed that in effect, as a practical matter, the BE eliminates virtually anyone whose depressive episode followed loss of a loved one, and his statement's definition would be only technically incorrect.

However, Kendler knew to the contrary that the BE eliminates only a minority of bereavement-related depression, and that his mischaracterization was not even approximately correct. We know this because one of the major results of Kendler's et al. (2008) study was that only a minority of bereavement-related depressive episodes is excluded by the BE. In Table 1, the row labeled partly "Meets criteria for 'uncomplicated bereavement-related disorder" indicates that only about a quarter (28 %) of all bereavement-related depressive episodes actually qualified for exclusion (p. 1451). Could this one row in a large table have gone unnoticed by Kendler, lost among the many analyses he performed? No, that is impossible too, because Kendler specifically commented on this result in his Discussion section: "A low percentage of individuals with bereavement-related depression met criteria for symptoms and a course of illness consistent with 'normal grief'" (p. 1453). The finding of Kendler's own study showed that the BE excludes a modest "low percentage" of bereavement-related cases, a finding radically different from Kendler's claim in his Statement that all such cases are excluded.

One might still ask: did the misstatement of the BE really matter to Kendler's argument? Some of his points, such as the claim that no BE appeared in earlier diagnostic manuals, are relatively independent of the precise content of the BE and would not have been affected by his mischaracterization.

The answer is that Kendler's misstatement of the BE is not a side issue but is directly implicated in his most important argument for eliminating the BE. Kendler's central argument in his statement, and the argument that ended up being most influential in the BE debate, was based on findings that if you apply the BE's six criteria to depressive reactions to other stressors, you get a similarly benign-looking group. Wakefield et al. (2007) were the first systematically to apply the BE's criteria for normal depressive episodes to reactions to losses other than death of a loved one, such as relationship problems, marital dissolution, job loss, financial ruin, negative medical diagnosis in oneself or a loved one, loss of possessions in a disaster, and other such stresses and losses that are known to trigger depressive feelings. Using a series of 11 validators that indicate the degree to which a condition is likely pathological versus normal, ranging from measures of service use to recurrence and duration, they found that the excluded bereavement-related cases so closely resembled the other-stressor cases that satisfied the same BE criteria that they could not be statistically differentiated. Moreover, both these groups scored much lower on level of pathology than the episodes – both bereavement and other-stressor triggered – that did not satisfy the BE criteria. These results were replicated both in Kendler et al.'s (2008) subsequent study as well as in later follow-up studies that were methodologically more rigorous (Wakefield and Schmitz 2012a, 2013a, b, c, d, 2014). Thus, it is inconsistent with the evidence to exclude uncomplicated bereavement-related episodes but not to exclude uncomplicated other-stressor related episodes. How this inconsistency is resolved – by eliminating the BE or by extending its rules to the exclusion of depressive reactions to other stressors – was framed as a central issue by Kendler.

Kendler consistently dismissed out of hand the alternative of extending the BE, but demanded consistency with the evidence showing that the BE must be eliminated or extended, thus reduced to absurdity the notion of retaining the BE. This central argument emerged in his statement as follows:

> [A] broad range of evidence agreed to by both sides of this debate shows that there are little to no systematic differences between individuals who develop a major depression in response to bereavement and in response to other severe stressors... So the DSM-IV position is not logically defensible. Either the grief exclusion criterion needs to be eliminated or extended so that no depression that arises in the setting of adversity would be diagnosable. This latter approach would represent as major shift, unsupported by a range of scientific evidence, in the nature of our concept of depression as epidemiologic studies show that the majority of individuals develop major depression in the setting of psychosocial adversity. (Kendler 2010)

Certainly no one wants to eliminate all adversity-triggered episodes from MDD. Also, it is correct that by far most episodes of MDD in community studies are adversity-triggered. However, when he penned this passage, Kendler knew that the argument's dramatic pivotal claim, that extending the BE to other stressors would eliminate *all* adversity-triggered episodes from MDD so that "no depression that arises in the setting of adversity would be diagnosable," and that therefore depression as we know it would mostly disappear from psychiatric diagnosis except for untriggered engoneous depression, was untrue. Table 1 of his own 2008 study of the BE and its extension shows that, of all adversity-triggered MDD episodes (including both grief and other triggers), only about a quarter (25.7 %) are excluded by an extended BE (28.1 % of bereavement-related episodes and 24.6 % of other-stressor triggered episodes) (Kendler et al. 2008, p. 1451). Moreover, Kendler highlighted the low percentage of other-stressor cases that would be excluded by an extended BE in his paper's Discussion section. There, immediately after noting that "a low percentage" of bereavement-related episodes satisfied the BE's exclusion requirements, he observes that "the same percentage" of other-stress triggered episodes satisfied the extended BE (p. 1453).

However, Kendler's *reductio ad absurdum* argument is no longer valid if, as his and other research establishes, only a minority of other-stressor-triggered MDD is eliminated by extending the BE. The *reductio* step is no longer compelling because, to many observers of the extraordinarily high community prevalence rates of DSM-defined MDD, it is not at all absurd on its face that some percentage of DSM-defined MDD cases are in fact intense normal sadness reactions to various losses other than death of a loved one (Clayton et al. 1968; Maj 2011a,b; Regier et al. 1998). If this issue is taken seriously, then rather than a simple armchair reduction to absurdity, the question of the BE's status becomes the scientific challenge of understanding the nature of other-stressor-triggered uncomplicated depression cases, and of looking at the proper status of uncomplicated triggered depressions overall. This is an issue that has never been addressed by any DSM review group, and was never even broached by the DSM-5 work group. It was left to others to address the question empirically.

So, there is no longer an automatic reduction to absurdity of retaining and extending the BE once the implications are accurately stated. Yet, in all likelihood at least partly due to Kendler's influence, the notion that extending the BE to other stressors would lead to a diagnostic disaster in which virtually all cases of depression as currently diagnosed would no longer be diagnosable became firmly presupposed by the eliminationists and the DSM-5 work group, so that the work group offered no serious analysis at all of the evidence for and against this option. It instead took the pseudoscientific path of insisting that its own position was supported even as it refused to address the major alternative hypothesis. Even when directly challenged (Wakefield et al. 2009), Kendler refused to provide analyses from his 2008 article's data that would address the validity of extending the BE by comparing all uncomplicated to all complicated triggered cases (Kendler and Zisook 2009). The "extend" option was ruled out of bounds based on the made-up idea that it would depathologize most MDD, then the "consistency" argument that demanded either to extend or to eliminate was used to perform the *coup de grâce* on the BE.

Kendler wrote one way, with scientific accuracy, when addressing his colleagues in research journals, then wrote in an entirely different and contradictory way when arguing for the elimination of the BE in the context of the DSM-5 revision. His research correctly states the BE criteria, the low percentage of bereavement-related cases that the BE excludes, and the comparably low percentage of other-stressor reactions that an extended BE would exclude, but his official DSM-5 website statement supporting the elimination of the BE offers radically different and incorrect versions of all these facts and definitions. That gap reveals the failure of the DSM-5 revision process, and specifically the elimination of the BE, to be scientifically defensible.

In fact, the evidence now overwhelmingly supports the validity of the BE itself and of the "extend" option. Study after study has shown that both BE-excluded cases and the cases excluded under an extended BE do not have the core features of MDD, such as recurrence and suicide attempts (Mojtabai 2011; Paksarian and Mojtabai 2013; Gilman et al. 2012, 2013; Wakefield 2013a, b; Wakefield and Schmitz 2012a, b, 2013a, b, c, d, 2014; Wakefield et al. 2007, 2011a, b). Indeed, given Kendler's great concern about consistency, he might consider the following inconsistency. In a classic article on validation of disorder categories, Kendler (1990) pointed out that "for a number of disorders, one or two [validators] are implicitly more important than the others, because they reflect the key defining features or 'construct' of the disorder" (p. 970). He points out that this is particularly the case when it comes to the recurrent course of mood disorders: "For Kraepelin, the 'construct' of … manic-depressive insanity assumed a relapsing disorder without deterioration" and thus "course and outcome would be the most important validators" (p. 970). The notion that course is a critical validator was also expressed by other eliminationists who, on the basis of no relevant data at all (Wakefield and First 2012), generally claimed that BE-excluded uncomplicated bereavement-related depressive episodes had the same course and outcome as other major depression (Pies 2009; Zisook and Kendler 2007; Zisook et al. 2007). However, once appropriate studies were done, it consistently emerged that adversity-triggered uncomplicated depression cases, whether bereavement- or other-stressor related, do not have elevated rates of recurrence over background population levels of depression incidence among those who have never had a depressive disorder, whereas other depressive disorder has highly elevated recurrence rates, as the Kraepelinian formulation would predict. Thus, the extended BE excludes cases that, judged by Kendler's own analysis performed outside the politics of the DSM-5, do not satisfy the crucial validating criterion for MDD.

Kendler is by no means the only one to engage in brazen straw-man tactics. Such tactics were routinely used by those defending the BE's elimination to exaggerate the impact of the BE and make it seem unreasonable. For example, in a *Scientific American* blog published just a few months before DSM-5 was published, Professor Sidney Zisook, the primary consultant to the DSM-5 mood disorders work group on the BE to whom the committee outsourced much of its work on the BE (and who had been arguing for the elimination of the BE since 1991), explained:

> The "exclusion" essentially detailed a two-month period of "normal grief" that people would experience after the loss of a loved one. During this period, it was all but forbidden to diagnose a patient with major depression—even if the individual had all the symptoms (which are, in important and sometimes life-threatening ways, different from grief). (Zisook 2013)

Zisook's claim that, according to the BE, diagnosis of depression during the first months of grief was "all but forbidden," is of course incorrect. Indeed, he was a coauthor of the study senior-authored by Kendler that showed that the BE actually excluded from MDD only a "low percentage" of bereavement-related depressive episodes overall. It is instructive to consider what the 2008 study of which Zisook was a coauthor actually showed about the first 2 months in particular. According to that study's Table 1, about 67 % of bereavement-related depressive episodes lasted 2 months or less, but only 28 % of bereavement-related episodes satisfied criteria for BE exclusion. So, of bereavement-related episodes lasting 2 months or less, 42 % qualified for exclusion, whereas 58 % were diagnosable as MDD despite the recent loss. That is quite different from diagnosis being "all but forbidden" by the BE in the first 2 months. One might argue – and I believe the research shows quite convincingly – that in fact the extent of misdiagnosis of normal reactions as major depression was underestimated by the BE and that the threshold for diagnosis should be even higher. For example, the research suggests that a much more generous durational limit for normal depressive episodes after loss, closer to between 6 months and 1 year, is more valid than the DSM-IV 2-month limit (Wakefield et al. 2011a, b). However, that is another matter. The point here is that even the BE in its overly constrained DSM-IV form was misrepresented.

Particularly egregious instances of misleading statements of the BE are those issued by the DSM-5 Task Force and asserted by its leaders, thus giving an official imprimatur to such distortions. An example is the authoritative explanation offered by the Chair of the DSM-5 Task Force himself, Professor David Kupfer, in a short video featured on the DSM informational website aimed at explaining the exclusion's elimination. Kupfer (2013) states: "[A]fter reviewing the literature and having a number of our advisors go over all the material that was available, we decided to remove the fact that clinicians should not make a diagnosis of clinical depression in anybody who has suffered a loss before two or three months." But, of course, the BE states no such thing. Similarly, in the DSM-5 itself, in the appendix chapter presenting the highlights of the changes to the manual, it is explained that "In DSM-IV, there was an exclusion criterion for a major depressive episode that was applied to depressive symptoms lasting less than 2 months following the death of a loved one (i.e., the bereavement exclusion)" (DSM-5, p. 811).

A further official "fact sheet" on the bereavement exclusion from the DSM-5 Task Force that is posted on the DSM-5 website explains:

> Using DSM-IV, clinicians were advised to refrain from diagnosing major depression in individuals within the first two months following the death of a loved one in what has been referred to as the 'bereavement exclusion.' By advising clinicians not to diagnose

> depression in recently bereaved individuals, the DSM-IV bereavement exclusion suggested that grief somehow protected someone from major depression. (American Psychiatric Association 2013b)

However, the BE, we have seen, classified most depression during bereavement, and even most depression during the first 2 months after loss, as MDD. The Fact Sheet's misstatement that the BE prohibited diagnosis of MDD during the first 2 months yields a bogus *reductio* argument because it is implausible that there is no pathology among depressive episodes early in grief. The BE is constructed so as to identify those early episodes based on five symptom and impairment criteria.

## The Medication Responsiveness Argument

One argument repeatedly put forward for eliminating the BE is that there is evidence that medication works with excluded cases, therefore the cases should be considered depressive disorders. When Jan Fawcett (2010), Chair of the DSM-5 mood disorders work group, initially presented the work group's proposed changes, he credited treatment responsiveness as the sole reason for eliminating the BE: "The Mood Disorders Workgroup has decided to remove the bereavement exclusion from the major depressive episode diagnosis based on data indicating that when a patient meets the criteria for a major depressive episode, the response to treatment is identical to that for any major stressor preceding a major depression" (p. 536). Fawcett (2012) repeated this claim in a later commentary on why the BE was eliminated: "People who develop major depression from or after bereavement respond the same way to treatment, even to medications, as people who develop depression that comes out of nowhere."

The medication response claim is dubious from the start as an argument for considering a condition to be pathological, since many psychotropic medications have an effect across a large swath of normal and disordered conditions. So, it would not be particularly surprising for there to be a treatment response even if BE-excluded depressive feelings are perfectly normal. For example, stimulants make everyone, not just those with ADHD, more focused and alert, including, say, those getting tired while staying up all night studying for college exams, but that does not support the claim that the inability to stay up all night studying for college exams without getting tired is a disorder. It is true that sometimes the fact that two similar conditions do not respond to the same medication may suggest that they are different disorders, an approach to validation known as "pharmacological dissection." However, there is no similar basis for a "pharmacological assimilation" thesis that when two conditions do respond to the same medication, they are likely the same disorder, because a given drug can influence an array of normal and disordered conditions.

Has the claim that medication is helpful in BE-excluded depression at least been scientifically demonstrated by standard empirical methods, as Fawcett's claims imply? Despite the logical flaw pointed out above in the argument against the BE based on medication effectiveness, the scientific status of such claims about treatment effectiveness remain important. Aside from the fact that people want to know the answer to this question due to the extremely painful nature of grief, in the BE debate there are many who would be swayed by a pragmatic argument that we know that medication helps, therefore we might as well provide a way to diagnose it. The frequency with which this argument was put forward by the eliminationists testifies to its perceived persuasive power.

Fawcett's support for his claim can be traced back to Zisook and Kendler's (2007) mention of one earlier paper by Zisook et al.: "The only treatment study of individuals who the DSM would diagnose with 'bereavement' rather than MDD based on time since death also found [bereavement-related depression] to respond to antidepressant medication similar to other studies of SMD (Zisook et al. 2001)" (Zisook and Kendler 2007, p. 789). There have been no relevant studies since, and the same single reference is cited by several other authors to support eliminating the BE (e.g., Gilman et al. 2012; Shear 2011).

However, the Zisook et al. study is so weak as to be scientifically meaningless, and it would not be taken seriously as scientific evidence in any other medical specialty. Zisook et al. (2001) treated 22 bereaved individuals who satisfied DSM-IV MDD criteria at about 6–8 weeks post-loss on average. Only the BE criteria requiring no psychotic or suicidal ideation were applied. Subjects were treated for 8 weeks, and 13 out of the 22 subjects (59 %) experienced a reduction of ≥50 % in symptom scores on a standard inventory of depressive symptoms. The study's fatal flaw is that it contains no control group, so the sample's modest "response rate" is impossible to interpret. This is because the medication was administered during a period – roughly between 6 and 14 weeks post-loss – in which, without treatment, normal bereavement-related depressions have precipitous drops in symptoms anyway, at roughly the same rate as was observed in the study. Zisook et al. themselves acknowledge that "because of the open, uncontrolled design, it is impossible to be sure that the observed changes were due to the effects of bupropion SR . . . " (p. 229).

For comparison purposes, in Clayton et al.'s (1968) prospective study of normal bereavement, in the period from the first month to on average about 3 months after loss, roughly comparable to Zisook et al.'s period of treatment, the percentages of Clayton's sample having six high-prevalence depressive symptoms decreased as follows: depressed mood, 87–12 %; sleep disturbance, 85–27 %; crying, 79–12 %; difficulty concentrating, 47–27 %; loss of interest in TV, news, friends, 42–19 %; anorexia and/or weight loss, 49–27 %. Virtually none of these subjects (3 %) were taking medication. In Zisook's own earlier study of depression in bereavement (Zisook and Shuchter 1991, 1993), out of about 75 non-recurrent depressives who had early depressions triggered by a loss, 30 (40 %) were no longer depressed by 2 months, almost all without medication, and Zisook himself observed the need for placebo controls in medication studies. Zisook's bupropion data are thus quite consistent with trajectories of resolution for depressive symptoms during early grief without medication.

## The Suicide Risk Argument

There is a great emphasis in contemporary discussion of depression on this disorder's potential for suicide. Nonetheless, one would think that making statements about heightened suicide risk without appropriate evidence, apart from being antithetical to reasoned scientific discourse, is inappropriate as a way to score points in a dispute over a DSM-5 proposal. Yet, this is precisely what happened in the BE debate. Without supportive evidence to back up their assertions, proponents of eliminating the BE repeatedly raised the specter of suicide in individuals excluded from MDD diagnosis by the BE. In doing so, they carefully avoided addressing or even mentioning the fact that the BE is constructed to prevent any such error, since the BE exclusion requires absence of suicidal ideation. The suicidal ideation criterion directly challenges the plausibility of the eliminationists' claims, so it was effectively banished from their discussions. This was pseudoscience, in which reliance on emotions of fear and selective attention to facts replaced careful assessment of evidence.

For example, in a review paper, Zisook, Shear, and Kendler (2007) cited the risk of suicide as a reason for not waiting for BE-excluded depressive feelings to subside on their own without early treatment. To support their point, they noted that "a recent study demonstrated that both lack of a partner and time in depression were significant predictors of suicidality among people meeting criteria for MDE" (p. 104). The study that Zisook et al. cited (Sokero et al. 2005) concerned largely severely pathological inpatient subjects, many of whom had prior suicide attempts. For example, "preceding the follow-up phase, 15 % of the cohort had attempted suicide during the index episode and 24 % before that" (pp. 316–317). This is far cry from a sample with only general-distress symptoms and no suicidal ideation, and it is a sample wholly irrelevant to predicting behavior by individuals with uncomplicated bereavement-related depressions that satisfy the six demanding conditions required for exclusion. Zisook, Shear, and Kendler's logic seemed to be that depressed people without partners are generally more likely to commit suicide, and bereaved individuals (if the loss was of a spouse) do not have partners, thus depressed bereaved individuals are at elevated risk for suicide, thus BE-excluded cases – which are bereaved and depressed – are at elevated risk for suicide. This reasoning commits the elementary fallacy of ignoring a crucial fact that influences the outcome of interest, namely, that the BE-excluded individuals are specifically screened to avoid suicide risk by prohibiting suicidal ideation and other risky symptoms such as a sense of worthlessness.

Zisook (2010), in arguing for elimination of the BE, declared on public radio that "I'd rather make the mistake of calling someone depressed who may not be depressed, than missing the diagnosis of depression, not treating it, and having that person kill themselves." He failed to mention to the public that cases are screened for suicidal ideation as part of the assessment for exclusion.

Shear et al. (2011), in explaining why they agree with Zisook (who was a coauthor of the paper) that the BE should be eliminated, argue that early intervention with depressive feelings even during acute grief is warranted because "bereavement may increase the risk of suicide" (p. 111). Shear et al. fail to mention that BE exclusion requires that there be no suicidal ideation. They cite two references to support their contention that the BE poses a suicide risk (Ajdacic-Gross et al. 2008; Stroebe et al. 2005). Both concern increased suicide risk in general during bereavement, and neither address uncomplicated cases or cases in which suicidal ideation is absent, rendering them irrelevant to assessing suicidal risk in BE-excluded, uncomplicated cases.

In these citations of supposed support for eliminationist claims, there is a general lack of thoughtfulness about whether the findings of the study are pertinent and revealing in the way claimed about the question at issue. For example, Stroebe et al.'s entire study uses suicidal ideation as a proxy for suicide, concluding that: "Bereaved persons are at excess risk of suicidal ideation compared to nonbereaved people. Heightened suicidal ideation in bereavement is associated with extreme emotional loneliness and severe depressive symptoms" (p. 2178), and Stroebe et al. remark that "ideation would seem a precursor to suicidal acts" (p. 2178). Yet, suicidal ideation is by definition not present in excluded cases. Moreover, the heightened suicidal ideation Stroebe et al. found in bereavement was associated specifically with severe depressive symptoms but not with more moderate depressive symptoms, whereas excluded cases are generally mild or moderate, because severe symptoms tend to be the pathosuggestive symptoms that block exclusion according to the BE. The findings of the cited study do not fit the context of their citation by Shear et al.

In a further article, Shear (2011) defends the elimination of the BE by asserting that the standard DSM depression criteria have been developed by experts and stood the test of time and thus should be applied during bereavement (thus refusing to consider the hypothesis that uncomplicated cases are distinct), and that "among the things we have learned are that MDD is associated with a high mortality from suicide (as many as 15 % of people with severe MDD die by suicide)." No citation is provided for these startling claims about suicide potential. However, the source of the 15 % rate is well-known (Guze and Robins 1970) and has long been repudiated (Bostwick and Pankratz 2000). It was the rate for a very severe population of hospitalized depressed individuals, many of whom had bipolar disorder, and the statistics were biased in several other ways as well. Shear simply assumes that whatever is known about MDD in general applies to BE-excluded cases as well, yet the homogeneity of uncomplicated cases with standard MDD is precisely what is denied by the BE. In the guise of a scientific paper, she is simply asserting her authority on the question based on no specific evidence pertaining to BE-excluded cases.

There is a basic point of scientific logic involved here. The claim regarding the BE is that excluded, uncomplicated episodes are a different and less pathological kind of condition from standard MDD. Consequently, citing general points about the seriousness of standard MDD to argue for the elimination of the BE, as eliminationists tended to do, begs the question at issue of whether BE-excluded cases are different from standard MDD. This sort of argument is in effect simply a way of refusing to seriously consider an alternative hypothesis, and that is the mark of pseudoscience. Ignoring the fact that the BE includes a requirement for no suicidal ideation, which is directly relevant to suicide risk and thus renders irrelevant most studies of standard MDD suicide risk, is a further way of avoiding a scientific evaluation of an opposing view.

What is the truth about suicide risk in uncomplicated depression? The only published evidence regarding suicide risk in uncomplicated depression comes from my own studies with Mark Schmitz of uncomplicated depression. The evidence is overwhelmingly reassuring and supportive of the BE's validity (Wakefield and Schmitz 2014). We used four major epidemiological data sets to explore the relationship between uncomplicated depression and suicide attempt risk among those who experienced a period of sadness. Our results decisively falsified the claim by the eliminationists that exclusion of uncomplicated depression in an extended BE would risk missing suicidal MDD cases. In terms of both concurrent risk during the uncomplicated episode and predictive risk over 1- and 3-year follow-up periods, our results indicated that those depressive episodes that satisfy the demanding requirements for being uncomplicated depression predict rates of suicide attempt that are no more than, and often less than, those of the general population.

In our study, the "no suicidal ideation" component of the uncomplicated criteria by itself reduced suicide attempt rates to background population levels. However, the additional uncomplicated criteria reduced the rate further. It turned out to be simply false that excluding uncomplicated depression from MDD diagnosis poses a risk that the elevated rates of suicide among MDD patients will lead to missed suicidal cases, because uncomplicated depression does not have the elevated suicide attempt rates that are a major feature of standard MDD. In terms of concern about suicide risk, one might as well be concerned about the average person on the street as BE-excluded cases, because the BE-excluded rates are the same as or lower than background population levels in those who have never had MDD. Proponents of BE elimination in effect engaged in fear mongering in raising the specter of suicide without any relevant evidence.

## Concluding Remarks

My analysis has led me to the conclusion that the arguments for elimination of the BE were largely pseudoscientific and the BE's elimination unwarranted by the evidence. In considering the line of argument leading me to this conclusion, three caveats are important to keep in mind. First, in arguing that pseudoscientific thinking was decisive in eliminating the BE, I do not deny that there may have been considerable nonsense and pseudoscience on the part of those defending the BE's retention as well. However, as is indicated in a sketchy way in the course of the discussion, the difference is that the scientific evidence supported retention. Second, although I have perhaps been rather harsh in criticizing the assertions of several colleagues for their unscientific performance in the BE debate, these same colleagues have provided many admirable and scientifically invaluable accomplishments as well, and the concerns expressed herein on the specific issue of the BE in no way reflect a blanket judgment about any individual investigators or their work. Finally, the debate over the fate of the BE was long and complex and multifaceted, and unraveling fallacious reasoning is often much more demanding than putting forward the fallacious reasoning in the first place. Consequently, the debate could not be fully excavated in this one chapter, and many points and arguments made by the eliminationists and those defending the BE could not be addressed. Instead, I settled for three pivotal illustrative examples that offer a précis for a longer analysis of the debate hopefully to come.

The usual indicator that pseudoscience has replaced science is the systematic motivated deviation from rational assessment of evidence by a community in arguing for a claim that the community insists is scientifically supported but which in reality is not supported by the evidence. In pseudoscience, the basic canons of scientific discourse are systematically violated in such as a way as to attempt to create the false appearance of scientific discourse. Moreover, those putting forward the bogus arguments must be in a position to more appropriately evaluate the evidence but fail to do so. It is this sort of suspension of the usual scientific rules that I argue occurred on the part of the eliminationists in the debate over DSM-5's elimination of the BE.

When I say that support for the elimination of the BE was pseudoscientific, I mean this quite literally, and not as hyperbole for a vaguer assertion that the arguments for elimination were weak or incorrect. There was systematic abandonment of basic scientific and medical canons of reasoning. In my view, the arguments by the eliminationists were intellectually no sounder than claims that we routinely dismiss as pseudoscientific, such as the claims of astrology.

It is of course a potentially questionable and even desperate strategy to label one's opponents as pseudoscientists and rhetoricians, and to suggest that a process claimed to be scientific was merely a masquerade of science. It is generally better to get on with the difficult scientific work necessary to advance understanding, and this work has been undertaken by investigators cited above. Nevertheless, I believe "pseudoscience" is the accurate description of what happened in the BE debate, and it deserves to be labeled as such. This was after all a debate about whether millions of people experiencing grief should or should not be considered by psychiatry to have a mental disorder. Pretending that there was a scientific dispute that went one way rather than another as opposed to a gross deviation from scientific standards of reasoning would be to distort reality and to further betray the individuals newly subject to diagnosis.

A hidden cost of the BE debacle is that, in spuriously disputing the non-disorder classification of the cases identified by the BE and placing those cases within major depression, the DSM-5 work group effectively prevented serious exploration of whether other, less immediately obvious candidates might also be normal sadness or grief rather than depressive disorder. Surely not everyone who experiences suicidal ideation or feels worthless is mentally disordered, yet a wider-ranging discussion of the boundary between major depression and normal sadness never occurred once the most scientifically supportable cases of normal sadness were spuriously reclassified.

Recognizing the pseudoscientific status of the BE decision potentially opens up further discussion of other boundary questions.

The quality of the BE debate potentially has broader implications. The proposal to eliminate the BE was one of the most controversial in the DSMs history. This issue was "in the radar" of the Task Force Chair and Vice-Chair, who were confronted with a tidal wave of public challenges regarding this proposal. Thus, the quality of the BE discussion tells us something about the quality of the overall regulation of the DSM-5 revision process and the sorts of arguments that were taken as adequate for making decisions on proposals, although the quality of the scientific reasoning underlying DSM-5 revision proposals no doubt varied with the work groups and the individuals involved.

Science is an achievement that requires discipline of the mind's unruly elements, and pseudoscience is the use of the frame and vocabulary of science without the developed discipline of science. It is precisely the unruly antiscientific elements that are tempted to emerge in the DSM political process. Given emotionally charged issues such as grief and suicide risk for which many potential patients seek solace from mental health professionals, lack of quality control from a strong Chair may allow the revision process to spin out of control towards pseudoscience rather than the scientific evaluation of alternative hypotheses. The DSM-5 process in the case of the BE debate is a prototype and a cautionary tale of how a manual revision can thus quickly move psychiatry from science to pseudoscience, undermining rather than enhancing its status in the long run.

## References

Ajdacic-Gross, V., Ring, M., Gadola, E., Lauber, C., Bopp, M., Gutzwiller, F., & Rossler, W. (2008). Suicide after bereavement: An overlooked problem. *Psychological Medicine, 38*, 673–676.

American Psychiatric Association. (1980). *Diagnostic and statistical manual of mental disorders: DSM-III* (3rd ed.). Washington, DC: American Psychiatric Association.

American Psychiatric Association. (2000). *Diagnostic and statistical manual of mental disorders: DSM-IV-TR* (4th ed.). Washington, DC: American Psychiatric Association.

American Psychiatric Association. (2013a). *Diagnostic and statistical manual of mental disorders: DSM-5* (5th ed., p. 2013). Arlington, VA: APA.

American Psychiatric Association. (2013b). DSM-5 fact sheet: Major depressive disorder and the bereavement exclusion. Available at http://www.psychiatry.org/dsm5. Accessed 1 Jan 2014.

Andrews Paul, W., & Anderson Thomson, J., Jr. (2009). The bright side of being blue: Depression as an adaptation for analyzing complex problems. *Psychological Review, 116*, 620–664.

Bostwick, J. M., & Pankratz, V. S. (2000). Affective disorders and suicide risk: A reexamination. *American Journal of Psychiatry, 157*(12), 1925–1932.

Bowlby, J. (1980). *Sadness and depression* (Attachment and loss, Vol. 3). New York: Basic Books.

Clayton, P. J., Desmarais, L., & Winokur, G. (1968). A study of normal bereavement. *American Journal of Psychiatry, 125*, 168–178.

Clayton, P. J., Marijan, H., Murphy, G. E., & Woodruff, R. A., Jr. (1974). Mourning and depression, their similarities and differences. *Canadian Psychiatric Association Journal, 19*(3), 309–312.

Darwin, C. (1872). *The expression of the emotions in man and animals*. London: John Murray.

Editorial. (2012). Living with grief. *Lancet, 379*, 589.

Ekman, P., & Friesen, W. V. (1971). Constants across cultures in the face and emotion. *Journal of Personality and Social Psychology, 11*, 1124–1129.

Emil, K. (1917). *Lectures on clinical psychology* (3rd ed.). New York, NY: Wood.

Fawcett, J. (2010). An overview of mood disorders in the DSM-5. *Current Psychiatry Reports, 12*, 531–538.

Fawcett. (January 2012). DSM-5: A 'living document' that may impact practice, patients health. *Psychiatric Annals* December 21, 2012. Accessed http://www.healio.com/psychiatry/practice-management/news/online/%7B48A37EECE458-48A7-8A02-A932D35FD4EA%7D/DSM-5-a-living-document-that-may-impactpractice-patients-health. 17 Jan 2013.

Feighner, J. P., Robins, E., Guze, S. B., Woodruff, R. A., Winokur, G., Jr., & Munoz, R. (1972). Diagnostic criteria for use in psychiatric research. *Archives of General Psychiatry, 26*, 57–63.

Frances, A. (2010). Good grief. *New York Times*, 15 Aug.

Freud, S. (1917). Mourning and melancholia. In J. Strachey (Ed. & Trans.). *The standard edition of the complete psychological works of Sigmund Freud* (Vol. 15, pp. 243–258). London: Hogarth. (1957).

Friedman, R. A. (2012). Grief, depression, and the DSM-5. *New England Journal of Medicine, 366*, 1855–1857.

Gilman, S. E., Breslau, J., Trinh, N.-H., Fava, M., Murphy, J. M., & Smoller, J. M. (2012). Bereavement and the diagnosis of major depressive episode in the National Epidemiologic Survey on Alcohol and Related Conditions. *Journal of Clinical Psychiatry, 73*, 208–215.

Gilman, S. E., Breslau, J., Trinh, N.-H., Fava, M., Murphy, J. M., & Smoller, J. M. (2013). Dr. Gilman and colleagues reply (letter). *Journal of Clinical Psychiatry, 74*(7), 741–743.

Guze, S. B., & Robins, E. (1970). Suicide and primary affective disorders. *British Journal of Psychiatry, 117*, 437–438.

Hensley, P. L., & Clayton, P. J. (2013). Why the bereavement exclusion was introduced in DSM-III. *Psychiatric Annals, 43*(6), 256–260.

Horwitz, A. V., & Wakefield, J. C. (2007). *The loss of sadness: How psychiatry transformed normal sorrow into depressive disorder*. New York: Oxford University Press.

Kendler, K. S. (1990). Towards a scientific psychiatric nosology: Strengths and limitations. *Archives of General Psychiatry, 47*, 969–973.

Kendler, K. S. (2010). A statement from Kenneth S. Kendler, M.D., on the proposal to eliminate the grief exclusion criterion from major depression. http://www.dsm5.org/Pages/Default.aspx. Accessed 29 Apr 2013.

Kendler, K. S., & Zisook, S. (2009). Drs. Kendler and Zisook reply (letter). *American Journal of Psychiatry, 166*, 491–492.

Kendler, K. S., Myers, J., & Zisook, S. (2008). Does bereavement-related major depression differ from major depression associated with other stressful life events? *American Journal of Psychiatry, 165*, 1449–1455.

Kupfer, D. J. (2013). Why was the bereavement exclusion removed from DSM-5? http://www.psychiatry.org/practice/dsm/dsm5/dsm-5-video-series-bereavement-exclusion. Accessed 1 Jan 2014.

Maj, M. (2011a). When does depression become a mental disorder? *British Journal of Psychiatry, 199*, 85–86.

Maj, M. (2011b). Refining the diagnostic criteria for major depression on the basis of empirical evidence. *Acta Psychiatrica Scandinavica, 123*, 317.

McGuire, M. T., Troisi, A., & Raleigh, M. M. (1997). Depression in evolutionary context. In S. Baron-Cohen (Ed.), *The maladapted mind* (pp. 255–282). Hillsdale, NJ: Lawrence A. Erlbaum Associates.

Mojtabai, R. (2011). Bereavement-related depressive episodes: Characteristics, 3-year course, and implications for DSM-5. *Archives of General Psychiatry, 68*, 920–928.

Nesse, R. M. (2000). Is depression an adaptation? *Archives of General Psychiatry, 57*, 14–20.

Nesse, R. M. (2009). Explaining depression: Neuroscience is not enough, evolution is essential. In C. M. Pariante, R. Nesse, R. Nutt, & D. Wolpert (Eds.), *Understanding depression: A translational approach* (pp. 17–35). Oxford: Oxford University Press.

Nettle, D. (2004). Evolutionary origins of depression: A review and reformulation. *Journal of Affective Disorders, 81*, 91–102.

New York State Department of Mental Hygiene. (1980). *SADS, SADS-L, and RDC clarifications*. Albany, NY: New York State Department of Mental Hygiene.

Paksarian, Diana, and Ramin Mojtabai. (2013). *Psychiatric Annals* 43(6), 276–282

Parkes, C. M. (1964). Recent bereavement as a cause of mental illness. *British Journal of Psychiatry, 110*, 198–204.

Pies, R. (2009). Depression and the pitfalls of causality: Implications for DSM-V. *Journal of Affective Disorders, 116*, 1–3.

Price, J. S. (1967). The dominance hierarchy and the evolution of mental illness. *Lancet, 290*, 243–246.

Price, J. S., Leon, S., Gardner, R., Jr., Gilbert, P., & Rohde, P. (1994). The social competition hypothesis of depression. *British Journal of Psychiatry, 164*, 309–315.

Regier, D. A., Kaelber, C. T., Rae, D. S., Farmer, M. E., Knauper, B., Kessler, R. C., & Norquist, G. S. (1998). Limitations of diagnostic criteria and assessment instruments for mental disorders: Implications for research and policy. *Archives of General Psychiatry, 55*, 109–115.

Shear, M. K. (2011). Bereavement and the DSM-5. *Omega, 64*(2), 101–118.

Shear, M. K., Simon, N., Wall, M., Zisook, S., Neimeyer, R., Duan, N., Reynolds, C., Lebowitz, B., Sung, S., Ghesquiere, A., Gorscak, B., Clayton, P., Ito, M., Nakajima, S., Konishi, T., Melhem, N., Meert, K., Schiff, M., O'Connor, M.-F., First, M., Sareen, J., Bolton, J., Skritskaya, N., Mancini, A. D., & Keshaviah, A. (2011). Complicated grief and related bereavement issues for DSM-5. *Depression and Anxiety, 28*, 103–117.

Sloman, L., Price, J. S., Gilbert, P., & Gardner, R. (1994). Adaptive function of depression: Psychotherapeutic implications. *American Journal of Psychotherapy, 48*, 401–416.

Sokero, T. P., Melartin, T. K., Rytsala, H. J., Leskela, U. S., Lestela-Mielonen, P. S., & Isometsa, E. T. (2005). Prospective study of risk factors for attempted suicide among patients with DSM-IV major depressive disorder. *British Journal of Psychiatry, 186*, 314–318.

Spitzer, R. L., Endicott, J., & Robins, E. (1975a). Clinical criteria for psychiatric diagnosis and the DSM-III. *American Journal of Psychiatry, 132*, 1187–1192.

Spitzer, R. L., Endicott, J., & Robins, E. (1975b). *Research diagnostic criteria*. New York: New York State Department of Mental Hygiene, New York State Psychiatric Institute, Biometrics Research.

Spitzer, R. L., Endicott, J., & Robins, E. (1978). Research diagnostic criteria: Rationale and reliability. *Archives of General Psychiatry, 35*(6), 773–782.

Stroebe, M., Stroebe, W., & Abakoumkin, G. (2005). The broken heart: Suicidal ideation in bereavement. *American Journal of Psychiatry, 162*, 2178–2180.

Wakefield, J. C. (2011). Should uncomplicated bereavement-related depression be reclassified as a disorder in DSM-5?: Response to Kenneth S. Kendler's statement defending the proposal to eliminate the bereavement exclusion. *Journal of Nervous and Mental Disease, 199*, 203–208.

Wakefield, J. C. (2013a). Uncomplicated depression: New evidence for the validity of extending the bereavement exclusion to other stressors. *Acta Psychiatrica Scandinavica, 128*, 92–93.

Wakefield, J. C. (2013b). The DSM-5 debate over the bereavement exclusion: Psychiatric diagnosis and the future of empirically supported practice. *Clinical Psychology Review, 33*, 825–845.

Wakefield, J. C. (2013c). DSM-5 grief scorecard: Assessment and outcomes of proposals to pathologize grief. *World Psychiatry, 12*(2), 171–173.

Wakefield, J. C., & First, M. B. (2012). Validity of the bereavement exclusion to major depression: Does the empirical evidence support the proposal to eliminate the exclusion in DSM-5? *World Psychiatry, 11*, 3–10.

Wakefield, J. C., & Schmitz, M. F. (2012a). Recurrence of depression after bereavement-related depression: Evidence for the validity of the DSM-IV bereavement exclusion from the epidemiologic catchment area study. *Journal of Nervous and Mental Disease, 200*, 480–485.

Wakefield, J. C., & Schmitz, M. F. (2012b). Beyond reactive versus endogenous: Should uncomplicated stress-triggered depression be excluded from major depression diagnosis?: A review of the evidence. *Minerva Psichiatrica, 53*, 251–276.

Wakefield, J. C., & Schmitz, M. F. (2013a). Normal vs. disordered bereavement-related depression: Are the differences real or tautological? *Acta Psychiatrica Scandinavica, 127*, 159–168.

Wakefield, J. C., & Schmitz, M. F. (2013b). Can the DSM's major depression bereavement exclusion be validly extended to other stressors?: Evidence from the NCS. *Acta Psychiatrica Scandinavica, 128*, 294–305.

Wakefield, J. C., & Schmitz, M. F. (2013c). When does depression become a disorder? Using recurrence rates to evaluate the validity of proposed changes in major depression diagnostic thresholds. *World Psychiatry, 12*, 44–52.

Wakefield, J. C., & Schmitz, M. F. (2013d). Predictive validation of single-episode uncomplicated depression as a benign subtype of unipolar major depression. *Acta Psychiatrica Scandinavica*. Published online, 17 Aug 2013. doi: 10.1111/acps.12184.

Wakefield, J. C., & Schmitz, M. F. (2014). Uncomplicated depression, suicide attempt, and the DSM-5 bereavement-exclusion debate: An empirical evaluation. *Research on Social Work Practice, 24*(1), 37–49.

Wakefield, J. C., Schmitz, M. F., First, M. B., & Horwitz, A. V. (2007). Should the bereavement exclusion for major depression be extended to other losses? Evidence from the national comorbidity survey. *Archives of General Psychiatry, 64*, 433–440.

Wakefield, J. C., Schmitz, M. F., First, M. B., & Horwitz, A. V. (2009). The importance of the main effect, even within an interaction model: On expanding the bereavement exclusion for major depression. *The American Journal of Psychiatry, 166*, 491–492.

Wakefield, J. C., Schmitz, M. F., & Baer, J. C. (2011a). Relation between duration and severity in bereavement-related depression. *Acta Psychiatrica Scandinavica, 124*, 487–494.

Wakefield, J. C., Schmitz, M. F., & Baer, J. C. (2011b). Did narrowing the major depression bereavement exclusion from DSM-III-R to DSM-IV increase validity? Evidence from the national comorbidity survey. *Journal of Nervous and Mental Disease, 199*, 66–73.

Watson, P. J., & Andrews, P. W. (2002). Toward a revised evolutionary adaptationist analysis of depression: The social navigation hypothesis. *Journal of Affective Disorders, 72*, 1–14.

Weissman, M. M., & Myers, J. K. (1978). Affective disorders in a U.S. urban community: The use of research diagnostic criteria in an epidemiological survey. *Archives of General Psychiatry, 35*, 1304–1311.

Welling, H. (2003). An evolutionary function of the depressive reaction: The cognitive map hypothesis. *New Ideas in Psychology, 21*, 147–156.

Zisook, S. (2010). Today in your health: Bereavement. *National public radio, Morning edition*, 2 Aug 2010.

Zisook, S. (2013). Getting past the grief over grief|Guest blog, scientific american blog network, scientific american, 25 Feb 2013. http://blogs.scientificamerican.com/guest-blog/2013/02/25/getting-past-the-grief-over-grief/. Accessed 26 Feb 2013.

Zisook, S., & Kendler, K. S. (2007). Is bereavement-related depression different than nonbereavement-related depression? *Psychological Medicine, 37*, 779–794.

Zisook, S., & Schuchter, S. R. (1993). Uncomplicated bereavement. *Journal of Clinical Psychiatry, 54*, 365–372.

Zisook, S., & Shuchter, S. R. (1991). Depression through the first year after the death of a spouse. *American Journal of Psychiatry, 148*, 1346–1352.

Zisook, S., Shuchter, S. R., Pedrelli, P., Sable, J., & Deaciuc, S. C. (2001). Bupropion sustained release for bereavement: Results of an open trial. *Journal of Clinical Psychiatry, 62*, 227–230.

Zisook, S., Katherine Shear, M., & Kendler, K. S. (2007). Validity of the bereavement exclusion criterion for the diagnosis of major depressive disorder. *World Psychiatry, 6*, 102–107.

# Against Hyponarrating Grief: Incompatible Research and Treatment Interests in the DSM-5

**Şerife Tekin**

> *"There is nothing more terrible, I learned, than having to face the objects of a dead man. Things are inert: they have meaning only in function of the life that makes use of them. When that life ends, the things change, even though they remain the same. They are there and yet not there: tangible ghosts, condemned to survive in a world they no longer belong to. What is one to think, for example, of a closetful of clothes waiting silently to be worn again by a man who will not be coming back to open the door? . . . Or a dozen empty tubes of hair coloring hidden away in a leather traveling case? – suddenly revealing things one has no desire to see, no desire to know. There is poignancy to it, and also a kind of horror. In themselves the things mean nothing, like the cooking utensils of some vanished civilization. And yet, they say something to us, standing there not as objects but as remnants of thought, of consciousness, emblems of solitude in which a man comes to make decisions about himself: whether to color his hair, whether to wear this or that shirt, whether to live, whether to die. And the futility of it all once there is death."*
>
> (Paul Auster 1982, pp. 10–11)

**Abstract** The controversial debate on whether to remove the bereavement exclusion from the DSM's depression criteria has mostly focused on whether depression and grief related distress are in fact distinct. Those who argue for the removal provided scientific evidence for the truth of this claim, while those argue against it suggested that the cited evidence base is slim. Despite heated controversy, the change took place. In this article, I use a different argument to address the problems with this change in the DSM-5. Even if we assume that there is no meaningful difference between the properties of grief-related distress and depression symptoms, diagnosing the grieving individual with depression is not the best therapeutic approach to address their needs.

Ş. Tekin (✉)
Department of Philosophy and Religious Studies, Daemen College, 4380 Main Street, Amherst, NY, USA
e-mail: serife.tekin@gmail.com; http://www.serifetekin.com

 
S. Demazeux, P. Singy (eds.), *The DSM-5 in Perspective*, History, Philosophy and Theory of the Life Sciences 10, DOI 10.1007/978-94-017-9765-8_11

## Introduction

Suppose your colleague Steve has not been quite himself lately. He has not been stopping by your office as he usually does, to see how your weekend went. He keeps conversations brief. More than a few times, you have run into his students in the hallway waiting for him. They complain that they cannot find Steve during his office hours, and he does not respond to e-mails. They also tell you that he forgot to give the scheduled test last week. "He is lecturing like a robot," one adds. You are puzzled, as the person they are talking about does not sound like Steve at all. You meet him one day on your way to lunch and ask him to join you. Steve follows but seems disengaged, making you think the students are on to something. He is quiet, his eyes are red, and he has lost weight. During lunch he is spaced out and restless. When you ask if he'd like to go to the play being put on by the theatre students, he says "No," without explanation. When you ask him if he is okay, he says he did not sleep at all last night. You are worried and want to help. You send him an e-mail to follow up on your lunch, but he does not respond. A few days later you learn from your department chair that Steve lost his partner of 30 years a few weeks ago in a car accident, and since then, he has not been well. "Now it all makes sense," you think; you know how much Steve loved his partner.

Suppose Steve's situation gets worse and he is no longer able to meet the demands of his job. He even fails to fulfil the smallest tasks in his daily life. You encourage him to seek clinical support, and he goes to a family doctor. Because Steve's complaints, such as not being able to sleep or eat, losing interest in normally enjoyed activities, and being trapped in a depressed mood resemble the symptoms of Major Depressive Disorder (depression for short) in the *Diagnostic and Statistical Manual of Mental Disorders* (DSM), Steve's clinician may diagnose him with depression and address his complaints with antidepressants or other medications such as sleeping pills. Despite the obvious connections between Steve's loss and his experiences, he can be diagnosed this way because of the removal of the bereavement exclusion criterion from the depression diagnostic criteria in the most recent edition of the DSM, the DSM-5.

The rationale for removing the bereavement exclusion is that the observable and distressing experiences of grieving individuals and those in depression are similar. The argument is that there is no scientific evidence for characterizing bereavement related distress and depression as distinct conditions; hence, whatever treatment helps the latter will also help the former (Zisook and Kendler 2007; Zisook et al. 2001; APA 2011). I call the first premise of this argument grief erosion (GE) because it assumes that depression and grief related distress are not distinct.[1] GE is a conceptual assumption that equates grief and depression and thereby obscures, or even erases, the distinction between the two. It promotes a certain type of

[1]I use the word "erosion" because the DSM's removal of the bereavement exclusion criterion from the depression category has been an extended process, starting with the publication of the DSM-III (1980) and ending with the APA (2013). In other words, the "erosion problem" existed long before

therapeutic approach indicated in the conclusion of the argument above, which I call grief erosion therapy (GET), the therapy that treats grief in the same manner as depression.

Proponents of this change in the DSM-5 further argue that if those struggling with grief are not diagnosed with depression, they will not receive clinical support for coping with their loss even if they need it. For instance, the chair of the DSM-5 Task Force, psychiatrist David Kupfer, argues that without the change, a person who is suffering from severe depression symptoms 1 or 2 months after a loss can't be diagnosed as depressed and "may then not get the treatment they need" (Carey 2012). Some have even proposed there is an increased risk of suicide among those who suffer from bereavement-related distress and highlight the importance of early intervention (Shear et al. 2011).[2]

The controversial debate on whether to remove the bereavement exclusion from the depression criteria has mostly focused on GE, i.e., the assumption that depression and grief related distress are not distinct or the grief erosion assumption. Those who argue for the removal provide scientific evidence for the truth of GE (Zisook and Kendler 2007; Zisook et al. 2001; Carey 2012), but those arguing against it suggest the cited evidence base is slim (Horwitz and Wakefield 2007; First and Wakefield 2012; Kleinman 2012; Frances 2013; see also Wakefield's chapter in this volume). Despite heated controversy, the change took place.

In this article, I use a different argument to challenge the change in the DSM-5. In my view, even if we consider that there is no meaningful difference between grief and depression, the GET is not the best therapeutic approach to address the needs of grieving individuals. We do not need to show the falsity of the assumption that grief and depression are similar in order to argue that grieving individuals may not necessarily benefit from a depression diagnosis.[3] What it takes for the bereaved to effectively cope with grief is more complex than envisioned by the proponents of the removal of bereavement exclusion criterion, and it is wrong to assume that diagnosing grief-related distress as depression will help the bereaved to cope. Receiving a depression diagnosis during a struggle with grief may help *some* individuals to develop effective coping strategies, but there are strong reasons to claim that a depression diagnosis may shrink grieving individuals' psychological

---

the DSM-5 was published. Section "DSM Hyponarrativity and Grief Erosion" summarizes this process. Inspiration for this concept comes from Alan Horwitz and Jerome Wakefield's *Loss of Sadness* (2007).

[2]In a National Public Radio interview, Sidney Zisook said: "I'd rather make the mistake of calling someone depressed who may not be depressed, than missing the diagnosis of depression, not treating it, and having that person kill themselves" (Zisook 2010).

[3]I do not claim that all those in grief must be clinically treated. Rather, my focus is on those who have difficulty coping with their loss on their own, and who need clinical assistance. This is known as complicated grief in the medical literature, including the DSM-5. Throughout the essay, whenever I indicate "grief related distress" I refer to complicated grief.

resources for coping. More specifically, the DSM-5's symptom-based approach makes its mental disorder categories poor constructs for clinical purposes, even though they have arguable advantages in research contexts.

In the first part of the paper, I identify the source of GE as the symptom-based approach to mental disorders in the DSMs, also known as hyponarrativity (Sadler 2005). Next, I discuss how GE, because of the DSM hyponarrativity, might negatively influence individuals who struggle with grief because the DSM hyponarrativity has negative implications for both the clinical treatment of the grieving individual and her self-narratives about her loss. Finally, I offer a strategy to overcome these negative implications. I conclude with the more general point that research and treatment goals are incompatible in the DSM-5 and that the DSM must abandon the ambitious task of developing representations of mental disorders that are amenable to being used in both context. While it may be fruitful for research to explore whether grief related distress and depression are in fact similar, it is false to assume that they can be treated by receiving the same medical attention. Methodologically, the paper features first-person accounts of grieving individuals to balance the available literature on the removal of the bereavement exclusion criterion, which has downplayed or ignored the actual experiences of grieving individuals even though they were the target population for the DSM revisions.

## DSM Hyponarrativity and Grief Erosion

Despite its ubiquity, there are many varieties of grief.[4] Some rely on their own cognitive and emotional resources, others welcome support from their friends, relatives, and pastors, and still others need clinical assistance. Addressing the needs of the latter group who struggle with coping has historically challenged medical practitioners. There are significant cultural, social, and individual differences in the way grief is experienced, and such variations constitute an obstacle for the development of a standard therapeutic approach. Starting in the late 1970s, the creators of the DSM have consistently entertained the possibility that grief may turn into a mental disorder, based on its frequently observed emotional, cognitive, and physiological features, including sadness, sleep disruptions, changes in appetite, fatigue, diminished interest or pleasure in previously enjoyed activities, and difficulties in concentration. The similarity between the features of grief-related distress and the symptoms and signs of depression has been highlighted to make clinicians aware of the risk of grief turning into a pathological condition, or complicated grief (Horwitz and Wakefield 2007; Jackson 1986; Zisook et al. 2007). The challenge has been to distinguish the cases in which the experiences of grieving individuals are within the range of appropriate response to loss from those in which they are manifestations of depression.

[4]Grief generating loss includes the death of a loved one, loss of home, property, employment, as well as losses due to a natural disaster. In this article, I limit my discussion to bereavement.

To avoid a possible false positive problem, where the distress associated with grief could yield a misdiagnosis of the grieving individual's condition as depression (First and Wakefield 2012; Frances 2013), the psychiatric community introduced the bereavement exclusion criterion into the depression diagnostic criteria. More specifically, the DSM-III (APA 1980) stated that if the depression-like symptoms of the individual in grief exceeded 1 year, a depression diagnosis could be made and the individual could receive appropriate treatment. In the DSM-IV (APA 1994), the length of appropriate grieving was reduced to 2 months. In the fifth and latest edition, as noted above, the bereavement exclusion criterion has been removed, based on the assumption that grief related distress and depression are not distinct. Clinicians can now diagnose those individuals who struggle with grief related distress, like Steve, with depression, if their grief is manifested in experiences resembling the symptoms of depression, as early as 2 weeks after the loss – the required time for those with depression symptoms to receive a clinical diagnosis.

The push to remove the bereavement exclusion criterion must be understood in the historical context of psychiatry's attempts to establish itself as a scientific discipline. In the DSM this manifests itself in a shift from the etiological approach to classification in the DSM-I and DSM-II, to a symptom-based approach in the DSM-III (APA 1980) onwards.[5]

The DSM-I and the DSM-II were grounded in the psychoanalytic approach to understanding human psyche. Mental disorders, also called "reactions," were represented in relation to the causal factors thought to underlie them. These causal factors were taken either as a dysfunction in the brain or a general adaptational difficulty to environmental stressors due to unresolved sexual conflicts of childhood. For example, psychotic disorders were defined as "disorders of psychogenic origin or without clearly defined physical cause or structural change in the brain" (APA 1952; Gruenberg et al. 2005). As such, theoretical presuppositions about the human mind underlay the descriptions of mental disorders, including for example, the unresolved sexual tensions of childhood causing adult psychosis. As these presuppositions were not grounded on the observable features of psychopathology, they were criticized as lacking scientific validity and reliability (Beck 1962; Katz et al. 1968; Schwartz and Wiggins 1987a, b). A scientifically valid category of mental disorder requires external validators, such as symptoms (experienced by the patient), signs (observed by others), and neurobiological markers (Robins and Guze 1970; First et al. 2004), not simply theories. In addition, research showed that psychiatric diagnoses differed markedly between Europe and the USA (Cooper et al. 1969). Because the diagnosis of mental disorder relied on theoretical presuppositions about the mind, as opposed to observable and measurable evidence, it was difficult to individuate a set of behaviors expressing the same mental disorder across settings, throwing the validity of the DSM's categories into question.[6]

[5]The symptom-based approach is also known as "atheoretical" or "purely descriptive."

[6]See also Tsou's chapter in this volume for a detailed discussion about the theoretical and descriptive views of psychiatry.

For many, focusing on the observable aspects of mental disorders, i.e., symptoms and signs, could overcome these problems of scientific validity and reliability. Thus, scientific psychiatrists sought to ground descriptions of mental disorders in scientific data derived from studies in the natural sciences,[7] and starting with the DSM-III, emphasis shifted from the theoretical explanations of the causes of mental disorders, to the observable and measurable features of psychopathology such as signs and symptoms. Under the symptom-based schema, categories of mental disorders, such as depression, became individuated through a list of criteria identifying the symptoms (observed by the patient) and signs (observed by others).

As a result, mental disorders are now represented through a list of observable properties, i.e., the physical or psychological experiences of the patient.[8] Polythetic criteria sets, as opposed to monothetic criteria sets, are used to determine the boundaries of a disorder category. Monothetic classifications are based on the characteristics that are both necessary and sufficient for the identification of members of a class; each member of a class must have at least one property shared with all members of the class. In polythetic classifications, each individual member of a class shares a large proportion of its properties with other members but all members do not necessarily share any one property (Sokal 1974; Guze 1978). For instance, according to the criteria for depression in the DSM-5, at least five symptoms have to be present during a 2-week period and must represent a change from the patient's previous functioning: either "depressed mood" or "diminished interest in and pleasure from daily activities" must be among the five. Additional symptoms include: significant weight loss or gain; insomnia or hypersomnia; psychomotor agitation; fatigue or loss of energy; feelings of worthlessness or inappropriate guilt; and diminished ability to think or concentrate, indecisiveness, recurrent thoughts of death, and suicidal ideation (APA 2013, 125).

There are also clauses that attempt to prevent misdiagnosis. If the symptoms are accounted for by the direct physiological effects of a substance or by a general medical condition, a depression diagnosis cannot be made. Until the publication of the DSM-5, bereavement was one of these conditions (e.g., APA 1994, 327).

Symptom-based descriptions of mental disorders are thought to enhance both the scientific and the clinical utility of the DSMs because they constitute both valid and reliable scientific categories (APA 2013, xli; APA 1994, xv). Consider first the scientific utility. A diagnostic category based on symptoms is believed to have validity because the diagnosed individual exhibits behaviors or feelings that are typical of depression (insomnia, loss of interest in previously enjoyable activities, etc.). It also has reliability because the same set of observable behaviors

[7]This move was influenced by the rise of logical positivism and the development of computer assisted diagnosis, i.e., DIAGNO by Robert Spitzer and Jean Endicott (1968).

[8]Among the examples for psychological symptoms comes depression as mood related; obsessive thoughts as thought related; memory problems as related to cognitive functioning. Symptoms of anxiety (e.g., palpitation) and disturbances of vegetative functions (e.g. appetite loss, weight gain) are included among physical symptoms (APA 1994; Fulford et al. 2006;).

will be individuated as the same mental disorder regardless of the context. Thus, the symptom-based approach, by simplifying a complex target such as mental disorder, helps validate the mental disorder constructs by identifying common signs and symptoms, and creates reliable categories across different settings. In addition, a focus on the observable features of mental disorders helps further research into the genetic and neural underpinnings of mental disorders, by, say, performing imaging studies on the brains of depressed individuals who suffer from insomnia. All these are believed to ensure the scientific utility of the DSM.

The validity and reliability of diagnostic categories afforded by symptom-based approach are also considered useful for clinical treatment by easing the diagnostic process, allowing a clinician to evaluate the patient's complaints against a list of symptoms. In addition, the clinician may aid the management of some symptoms, say, by recommending certain types of psychotherapy or drugs. Finally, the patient would be given the same diagnosis and treatment plan regardless of the context in which they are diagnosed.

In short, the DSMs are intended to serve as a "practical, functional, and flexible" guide for both research and treatment (APA 2013, xli). Starting with the DSM-III, the DSM creators committed themselves to establishing diagnostic classificatory systems that emphasize descriptions (as opposed to causal explanations), remain atheoretical with respect to etiology, and operationalize terminology so that communication is facilitated across various contexts that require a diagnostic label (Poland 2001; Sadler 2005; Tekin 2010, in press).[9]

GE is a manifestation of the DSM's symptom-based approach to mental disorders. The proponents of the removal of bereavement exclusion have argued for the similarity of grief-related distress and depression based on the similarity of the observable experiences of individuals in the two groups: "evidence does not support separation of loss of loved one from other stressors" (APA 2011).[10] If the DSM categories relied on a psychoanalytically oriented causal etiology, as in the DSM-I and the DSM-II, these two conditions would not be considered the same. What

[9]Another argument for a symptom-based approach is to model mental disorders following physical diseases. For instance, from a nosological point of view it does not matter if someone has lung cancer because he smokes too much, or because he lives in a polluted city, or because he is unlucky. Whatever the cause, his disease is lung cancer. A similar reasoning is at play with depression and the DSM-5: the cause of the symptoms does not matter; a disease is not defined by its cause, but by its symptoms and signs. Context does not dissolve disorder, even if it seems to explain it. In fact, to further research into the genetic and neural underpinnings of mental disorders, it might be necessary to isolate the content from the context, and focus exclusively on the observable properties of mental disorders. I thank Steeves Demazeux and Patrick Singy for their helpful input on this point.

[10]The "evidence" is *one* review paper (Zisook and Kendler 2007) which purports to show that bereavement-related experiences are generally similar to the symptoms of standard depression; including extreme sadness, disturbed sleep, disturbed appetite and energy, agitation, difficulty concentrating, etc. Some have taken issue with the vagueness of the notion of "similarity" and the slim evidence provided, but their concerns have not affected the DSM outcome (First and Wakefield 2012).

follows GE is the grief erosion therapy (GET): since depression and grief related distress are similar, grief related distress can be better addressed therapeutically if labeled as depression.

I argue that even if there may be grounds to believe the truth of GE, the effectiveness of GET can be challenged: while the symptom-based approach has advantages for scientific research, its utility in clinical contexts is arguable, especially in the context of addressing the needs of the bereaved (e.g., Sadler 2005; Tekin, in press). In his evaluation of how the symptom-based approach is used in treatment related contexts, Sadler coins the term hyponarrativity (Sadler 2005). A DSM category of a mental disorder is a hyponarrative, when we consider it in the larger schema of an individual's unique and rich life narrative. Even though mental disorders are sensitive to an individual's life story, including developmental history, biological and environmental risk factors, interpersonal relationships, race, gender, and other important aspects of personal identity, these elements of the story are not included in mental disorder categories. In other words, the DSM categories abstract (or bracket) the self-related and context-specific aspects of the encounter with mental disorders. By saying little about how the illness experience is integrated to the patient's life as a whole, they become nothing more than a "repertoire of behavior" (Cohen 2003). Hence, the symptom-based approach makes the DSMs hyponarrative.[11]

Though it is important for scientific purposes to be wary of descriptions that are too subjective to be generalizable, stripping away these self- or context-related aspects of the encounter with mental disorder may prevent an effective addressing of a patient's concerns in treatment contexts (Sadler 2005; Tekin in press). As I show in the rest of the chapter, even if we were to agree that grief-related distress and depression are similar or even identical, there are grounds to argue that diagnosing grief-related distress as depression may have negative implications to therapeutically addressing grieving individuals' distress, due to the DSM hyponarrativity.

## Negative Implications of Grief Erosion on the Bereaved

Diagnosing grief-related distress as depression has implications for grieving individuals not just in the clinical context, where they seek help from health practitioners, but also in the personal context, where they try to adjust to a new life without the loved one by creating self-narratives about their loss. It is an empirical matter whether and how the new depression description in the DSM-5 will affect grieving individuals who need help. There will be individual and cultural variations in the way they may respond to receiving a depression diagnosis, and some may benefit

[11] It is beyond the scope of this paper to discuss whether re-appraising the self-related aspects of encounter with mental disorder is even possible within the framework of the DSM project.

from it. My intention is not to make an overarching claim about how depression diagnosis may cause harm, but rather, to identify potentially problematic implications of diagnosing a grieving individual with depression, and more generally to explore the problematic implications of the DSM's symptom-based strategy.

## *Grief Erosion in the Clinic*

DSM hyponarrativity has disadvantages when addressing the therapeutic needs of patients in a clinical context. For one, this schema, insofar as it defines what to look for in a patient, turns the clinician's attention away from the "self" as a more or less integrated multidimensional unity (Tekin 2014), to her symptoms (Sadler 2007). This type of therapeutic intervention is geared towards symptom management; it is not an integrated approach seeking to address the individual's concerns in the larger context of her life. Although the DSM does not advise the clinician to "ignore the patient's story" (Sadler 2005), it guides clinical practice by directing the clinician's interest towards the symptoms, rather than encouraging the development of an integrated approach to improve the individual's condition by relating her complaints to the larger historical, cultural, social, and personal context of her life. In this sense, the DSM exhibits "a redirection, a marginalizing, a setting aside in lieu of something else" (Sadler 2005, 177). What it sets aside are the patients' stories that play an important role in their encounter with mental disorder (Sadler 2005; Tekin, in press; 2011; 2010).

Consider how Steve, who presents with experiences such as depressed mood, sleep disruptions, etc., may be affected by the DSM hyponarrativity. We can imagine scenarios where he is diagnosed with depression by a clinician who does not know about or inquire into his significant loss, as he meets the diagnostic criteria for depression. Unfortunately, this diagnosis may not provide him with the help he may need because the DSM-5's individuation of grief as a cluster of symptoms fails to align with the true complexity of grief. Focusing on complaints that can possibly be manifestations of a mental disorder may fail to guide the clinician's therapeutic attention to the right place in understanding and helping Steve. A more integrated approach is crucial.

Grief is an emotional, physical, cognitive, behavioral, social response to the loss of someone to whom the individual has a strong attachment. It is an adjustment process that involves lamenting for the deceased and for the self. The individual is distressed about the loss of an inherently and independently valuable person who can no longer participate in life and engage in the completion of certain projects. The age of the deceased, the way he died (e.g., sickness or accident), and his character, all influence how the individual grieves. She must recognize and negotiate with the fact that the deceased will not be returning back to life. For example, Paul Auster notes the "kind of horror" involved in making decisions about what to do with the "closetful of clothes waiting silently to be worn again by a man who will not be coming back to open the door" (Auster 1982, 10).

William James suggests that a person's self includes "not only his body and his psychic powers, but . . . his wife and children, his ancestors and friends . . . and if they dwindle and die away, he feels cast down" (James 1890/1983, 279–280). A key dimension of grief is to salvage one's own identity. However, this is not an automatic process. As Sigmund Freud writes, grieving individuals have to reclaim the energy invested in the deceased (Freud 1917). The more their identity was wrapped up with the deceased, the more difficult it will be to adjust to the loss and redesign their lives. Consider how Michel de Montaigne illustrates his feelings following the loss of his dear friend Étienne de la Boétie by citing Gaius Valerius Catullus:

> What shame or limit should there be to grief for one so dear? . . . How wretched I am, having lost such a brother! With you died all your joys, which your sweet love fostered when you were alive. You, brother, have destroyed my happiness by your death: all my soul is buried with you. Because of your loss I have chased all thoughts from my mind and all pleasures from my soul. (Catullus as quoted in Montaigne 2003/1580, 217–218)

Grief process requires decoupling one's self from the lost person, and recreating a life with new joys without that person in it. Arthur Kleinman, a psychiatrist and an anthropologist, explains that with his wife's death a part of him "was gone forever":

> In March, 2011, my wife died and I experienced the physiology of grief. I felt greatly sad and yearned for her. I didn't sleep well. When I returned to a now empty house, I became agitated. I also felt fatigued and had difficulty concentrating on my academic work. My weight declined owing to a newly indifferent appetite. This dark experience lightened over the months, so that the feelings became much less acute by around 6 months. But after 46 years of marriage, it will come as no surprise to most people that as I approach the first anniversary of my loss, I still feel sadness at times and harbor the sense that a part of me is gone forever. I'm not even sure my caregiving for my wife, who died of Alzheimer's disease, ended with her death. I am still caring for our memories. Is there anything wrong (or pathological) with that? (Kleinman 2012)

The potential failure of clinicians to engage with the deceased-related and self-related dimensions of grief which are missing from the DSM-5 individuation of grief related distress may present an obstacle to the development of effective therapeutic strategies. In these kinds of situations, if a grieving individual were to seek clinical help, a clinician must recognize the details of the individual's loss, and what that relationship meant to her, to best address her needs. It may be important for the clinician to know that Steve's partner died in a car accident to help him process his loss. Or that Steve used to co-author articles with his partner; even his academic identity was enmeshed with hers. As George Graham has recently discussed, "solidarity with the beloved does not evaporate just because or when the beloved dies and the love can no longer be shared" (Graham 2013, 168). One must actively work on recreating the self. In fact, as philosopher and psychiatrist Hanna Pickard writes, a narrative self-understanding achieved in the psychotherapeutic context enables self-creation and enhances self-autonomy which are both central for the recovery process (Pickard, in press).

One might oppose this argument by suggesting that the DSM-5 offers guidelines to the clinician about the appropriate way to approach the individual. After all, the definition of depression includes the following caveat:

> Note: Responses to a significant loss (e.g., bereavement, financial ruin, losses from a natural disaster, a serious medical illness or disability) may include the feelings of intense sadness, rumination about the loss, insomnia, poor appetite, and weight loss... which may resemble a depressive episode. *Although such symptoms may be understandable or considered appropriate to the loss, the presence of a major depressive episode in addition to the normal response to a significant loss should also be carefully considered.* The decision inevitably requires the exercise of clinical judgment based on the individual's history and the cultural norms for the expression of distress in the context of loss. (APA 2013, 161, my emphasis)

Even though the note mentions the likelihood of encountering certain kinds of distress during grief, it fails to offer guidelines on how to discriminate these from symptoms of depression. Rather, it prioritizes a definition of grief as a symptom-cluster involving "feelings of intense sadness, rumination about the loss, insomnia, poor appetite, and weight loss," encouraging the clinician *to look for* a possible *mental disorder*, as opposed to developing an integrated approach to understand the patient's distress.

In effect, by prioritizing the definition of grief as a symptom-cluster, the DSM-5 fails to embrace the individual-context-sensitive nature of grief. One can easily imagine scenarios in which the overworked family doctor, within the 15 min he is allotted for each patient, may diagnose a grieving individual with depression, basing his diagnosis on the similarity of the complaints to the symptoms of depression and failing to recognize that the underlying stressor of those complaints is the loss of a loved one. Thus, by slotting grief under depression, the DSM-5 erases the therapeutic import of distinguishing the two. It does not actively guide the clinician to engage with the personal details of the individual's particular loss. In addition, given that the DSMs are used for teaching purposes in medical schools, the more hyponarrative the mental disorder descriptions are, the less likely it will be for psychiatry residents to receive special training in psychotherapy and other narrative treatment techniques, as the symptom-management approach lends itself naturally to a medication based treatment strategy.

These problems can be compounded when grieving people demand medications to manage their symptoms. My suggestion is not that medications will not help these individuals, but rather that prioritizing medical treatment in lieu of psychotherapy may delay the adjustment process following the loss by masking the bereaved person's authentic feelings (Tekin 2013b).

Hence, even if we were to agree that there is no distinction between grief-related distress and depression, the DSM hyponarrativity makes the DSM-5 be a poor guide for clinically addressing the needs of grieving individuals.

## *Grief Erosion in Self-Narratives*

DSM hyponarrativity also has negative implications for an individual's understanding of, and response to, her mental illness (Tekin 2010, in press).[12] The DSM's neglect of the "self" leaves mental disorder descriptions irreconcilable with subjective experiences and directs an individual's attention away from her own understanding of her mental disorder to rely on what the DSM says. It becomes a challenge for those with mental disorders to understand their condition and its significance,[13] severely limiting their ability to develop coping strategies. In this respect, grief erosion may negatively affect a grieving individual's self-narratives.

I define self-narratives as selective representations of the states of affairs in an individual's life, which are organized in a more or less sequential, coherent and meaningful manner (Tekin 2013a, 2011). Important work on the relationship between the self and self-narratives in philosophy of cognitive science, as well as feminist philosophy, attests to the ways self-narratives are used by individuals to make sense of the states of affairs in their lives, including experiences of mental disorders, the loss of a loved one, or experiences of violence (Tekin 2010, 2011, 2013a; Flanagan 1996; Lindemann Nelson 2001; Baylis 2011; Lumsden 2013; Jopling 2000). These self-narratives are not mere expressions of an individual's conception of an experience, but tools for developing self-understanding, forming self-concepts, and shaping appropriate responses to the encountered states of affairs. In this sense, self-narratives shape the self.

Self-narratives, as cognitive representations of the states of affairs in an individual's life, regulate that person's responses to these states of affairs. For instance, following a break-up, a college student may reiterate the series of events that led to the break up, re-examine the role played by each partner, identify potential reasons for the break-up, etc. These self-narratives, incomplete and selective by nature, have varying degrees of connection with the states of affairs that have actually occurred: they may be veridical or far removed from the reality. In this respect, they are plastic;

[12]The limitations of the symptom-based approach are acknowledged in the introduction to DSM-5, which emphasizes that a categorical approach to classifications works best when all members of a class are homogeneous, when there are clear boundaries, and when the different classes are mutually exclusive. It acknowledges that this is not the case in mental disorders. Patients, even if they share a diagnosis, form heterogeneous classes, as each has a unique encounter with mental disorder, due to the contingencies of her own life. Further, using the categorical system does not assume clear boundaries between different mental disorders (APA 2013, 19–20). Despite these shortcomings, the DSM is considered able "to assist trained clinicians in the diagnosis of their patients' mental disorders as part of a case formulation assessment that leads to a fully informed treatment plan for each individual" (APA 2013, 20).

[13]I am not suggesting that a symptom-based approach to mental disorder is completely wrong, I am suggesting that it is incomplete, i.e., it only captures some aspects of a mental disorder, and leaves out the subjective aspects. This may have some advantages, but when it comes to individual's reflection on her condition, a symptom-oriented approach, as opposed to a more integrated one, has negative implications.

they are influenced not only by the actual states of affairs, but also by the social, cultural, and scientific norms that evaluate these experiences. Upon reading a book about male-female power dynamics in relationships, the college student may create another version of her self-narrative. Creating one that makes sense of the break-up through an understanding of power relations may help her adjust better to the break-up, and overcome negative experiences, such as abuse, in a more resourceful way.

Others' narratives of the states of affairs in an individual's life can be made part of that person's own self-narratives (Tekin 2011, 2013a). Depending on their quality, they may help or hinder her adjustment, shaping her available cognitive and social resources (Jopling 2000). There are salient examples throughout human history of dominant groups shaping the narratives of certain underrepresented groups through socially shared narratives, "damaging" their identity so that they cannot exercise their moral agency freely (Lindemann Nelson 2001). Others' narratives shape moral agency in individual contexts as well. For example, if the college student's self-narrative highlights her partner's insistence on her negative personality traits and posits these as the sole reason for the break-up, she may develop feelings of insecurity that hinder adjustment and deprive her of self-understanding.

Sciences of mental disorder, in addition to offering methods of treatment, also serve as epistemic sources for individuals to create self-narratives of their illness (Tekin 2010, 2011). In other words, the individual may be inclined to use the causal framework offered by psychiatry to shape her narratives of her own disorder, affecting her experience of and response to that disorder. A DSM diagnosis of a mental disorder informs an individual's self-narrative via the clinician's description of the mental disorder to her patient and the DSM culture, i.e., the cultural context in which information and speculations about mental disorders are disseminated (Tekin 2010, 2011, in press; Hacking 1995). Used as cognitive tool, a DSM-based self-narrative helps the individual make sense of her otherwise unfathomable mental distress. Yet because of an absence of self in the DSM's depression category, she may be less inclined to rely on the subjective life context in which she encounters mental disorder and more inclined to use available DSM-based explanations and typical treatment strategies, e.g. drug therapy.[14]

The following is a good example of a DSM-based self-narrative. Elyn Saks writes about browsing through the DSM and reading the definition of her diagnosis of schizophrenia:

> I had discovered the DSM .... I read it cover to cover. Knowledge had always been my salvation, but with my immersion into the DSM, I began to understand that there were some truths that were too difficult and frightening to know .... And now, here it was, in writing: The Diagnosis. What did it mean? Schizophrenia is a brain disease which entails a profound loss of connection to reality. It is often accompanied with delusions, which are fixed yet false beliefs—such as you have killed thousands of people—and hallucinations, which are false sensory perceptions-such as you have just seen a man with a knife. Often

[14]The "self" is introduced in the DSM for the first time via the general definition of personality disorders in the DSM-5 (APA 2013). My focus here is on mood related disorders.

> speech and reason can become disorganized to the point of incoherence. The prognosis: I would largely lose the capacity to take care of myself. I wasn't expected to have a career, or even a job that might bring in a pay check. I wouldn't be able to form attachments, or keep friendships, or find someone to love me, or have a family of my own—in short I'd never have a *life* .... I'd always been optimistic that when and if the mystery of me was solved, it could be fixed; now I was being told that whatever had gone wrong inside my head was permanent, and from all indications, unfixable. Repeatedly, I ran up against words like "debilitating," "baffling," "chronic," "catastrophic," "devastating" and "loss." For the rest of my life. *The rest of my life*. It felt more like a death sentence than a medical diagnosis. (Saks 2007, 167–168; italics in original)

This self-narrative reflects the formal definition of schizophrenia in the DSM and the speculative and unverified knowledge about schizophrenia disseminated in the DSM culture. For instance, the DSM does *not* say that schizophrenia is a brain disease, but the substantive literature on schizophrenia, including research articles, pharmaceutical studies, neuroscientific literature, not to mention popular literature, refers to it as such. Some terms, including "permanent," "debilitating," "catastrophic" are *not* found in the DSM.

This particular self-narrative is constraining, as it accounts for Saks's experiences with hallucinations and delusions superficially, without engaging with how these states are lived out in different contexts of her life. For instance, it is relevant in the development of Saks's understanding and response to her illness that she know the times she is more disposed to certain hallucinations, for example, the highly stressful situation created by examinations at Oxford, so that she can develop ways to deal with them; however, a symptom-oriented narrative does not engage with those factors.

The kind of narrative that disengages Saks' symptoms from her larger life context might shut down other, more integrated accounts of her experience with schizophrenia. In fact, Saks has a fulfilling life despite her illness, demonstrating her ability to create alternative, more integrated self-narratives in which she is not a collection of symptom-clusters, but an individual person with a meaningful life. Saks writes in her memoir that she has flourished by replacing the above narratives with more resourceful ones where she engages with the meaning of her illness and its significance in her life (Saks 2007). Not relying on the DSM-based narratives has helped her.

Thus, given the self-narrative shaping power of a DSM diagnosis, receiving a depression diagnosis for grief-associated distress may influence the individual's emotional and cognitive responses to her loss. For instance, Steve's self-narratives about his distress in the wake of his loss might present his experience as a psychopathological condition, as opposed to an appropriate response to loss. Making sense of his grief as an abnormal condition could alienate Steve from his own feelings. More specifically, he may be inclined to think of his experiences as a cluster of medical symptoms rather than engaging with the deceased-related and self-related aspects of his grief. However, full engagement with his personal feelings and experiences may help Steve create a meaningful self-narrative that will allow him to develop cognitive and emotional coping strategies.

Some research supports the intuition that it matters whether the grieving person engages with the subjective features of her grief response by using her own language, rather than the psychiatric language of depression.[15] James Pennebaker's work in cognitive science supports the claim that if the subject is able to connect her symptoms to her specific life experiences, she will develop adaptive cognitive responses. Pennebaker demonstrates that when individuals write about traumatic experiences by deeply engaging with the associated emotional difficulties in their lives, significant physical and mental health follows. This is referred to as the disclosure phenomenon. Further, talking and writing about emotional experiences yield comparably higher biological, mood and cognitive effects than talking or writing about superficial subjects (Pennebaker 1997).[16] While some of the positive outcomes of the disclosure phenomenon are due to a reduction in inhibition, researchers argue that basic cognitive and linguistic processes during writing are more significant factors. There is a connection between language used and health outcomes (Pennebaker 1993). More thoughtful writing, it is suggested, leads to better health outcomes, as does the use of positive emotion words.[17]

Similar forces might be at play in grief-related distress. Perhaps the individual will better cope with the loss of a loved one if she is encouraged to explore her feelings connected to his presence in her life, the sadness that comes with her loss, how she feels about living in the world without him, and so on. If she establishes causal and insightful relationships between her symptoms and the deceased-related and the self-related dimensions of her grief, she may develop rich insights into her grief. Active engagement with the components of her grief may give her an enhanced ability to cope with it.

---

[15]Research in psychology indicates the positive influence of a detailed language of emotion on subjects' responses to traumatic life experiences. For instance, in a study conducted among families with preadolescent children, Robyn Fivush et al. found that family narratives in which *specific* emotions were expressed and explained in a collaborative fashion, while acknowledging especially the negative emotions, were positively related to preadolescents' reported competencies and self-esteem. However, family narratives expressing very *general* positive emotions and neglecting negative emotions were negatively related to preadolescents' competencies (Marin et al. 2008). These studies support the idea that a detailed understanding of subjective experience of mood disorders has a significant influence on an individual's ability to cope with her mood disorder.

[16]Pennebaker et al. write: "When individuals write or talk about personally upsetting experiences in the laboratory, consistent and significant health improvements are found. The effects are found in both subjective and objective markers of health and well-being. The disclosure phenomenon appears to generalize across settings, most individual differences, and many western cultures, and is independent of social feedback" (Pennebaker et al. 1997, 164).

[17]Pennebaker et al.'s analysis of data from six writing studies found three linguistic factors reliably improved physical health. First, the more that individuals use positive emotion words, the better their subsequent health. Second, a moderate number of negative emotion words predicted health. Both very high and very low levels of negative emotion words correlated with poor health. Third, an increase in both causal and insight words over the course of writing was strongly associated with improved health (Pennebaker et al. 1997, 165).

## Looking Forward

The DSM seeks to represent mental disorders in a way that will serve both research and treatment. But as seen in the example of grief, research and treatment interests are not automatically compatible. A particular feature of a DSM category – hyponarrativity – can yield positive outcomes in some research contexts and undesirable outcomes in clinical and personal contexts. Isolating depression like symptoms from the individual as a more or less integrated unit may lead to the DSM's failure to adequately address the needs of individuals in grief. Even if grief erosion is a viable assumption, it does not follow that diagnosing grieving individuals with depression will help improve their condition.

One way to look forward is to challenge the fundamental ambition of the DSM to serve disparate purposes and to use different constructs of mental disorder (with more or less self-related features) depending on our interests in using these constructs. As seen in the grief case, the details that need to be abstracted from a mental disorder description to guide scientific research may also be required to address the therapeutic needs of the individual. The DSM's schema for grief related distress is not useful in the clinical context, as it neglects the important properties of grief required in an effective therapeutic approach. The following statement is misguided:

Although this edition of DSM was designed first and foremost to be a useful guide to clinical practice, as an official nomenclature it must be applicable in a wide variety of contexts. DSM has been used by clinicians and researchers from different orientations . . . all of whom strive for a common language to communicate the essential characteristics of mental disorders presented by their patients (APA 2013, xli).

It is a mistake to assume one representation of mental disorders can be useful for a wide variety of contexts, including research, treatment, and various administration related affairs. Similar incompatibilities may be found in the other contexts where the DSMs are used (e.g., insurance related interests and treatment interests).[18] For example, what looks like a *prima facie* useful feature of the DSM for research purposes, i.e., hyponarrativity, may have negative implications in treatment contexts. The best way for psychiatry to move forward might be to determine the primary purpose of the DSM categories and develop models of mental disorders that will fit that purpose. If it is solely used as a clinical treatment guide, it must abandon hyponarrativity. If it is to be used for research purposes, some hyponarrativity can remain, depending on the purposes of a particular research area.

**Acknowledgments** I would like to thank Patrick Singy, Steeves Demazeux, George Graham, Hanna Pickard, Owen Flanagan, and Jennifer Radden for their helpful comments on this article. The paper was presented at the University of Western Michigan Medical Humanities conference in September 2013; I also thank the audience for their comments.

[18]Similar points in favour of pluralism in assigning roles for a psychiatric taxonomy can be found in the chapters by (Tsou 2015; Faucher and Goyer 2015) in this volume.

## References

American Psychiatric Association. (1952). *Diagnostic and statistical manual: Mental disorders* (1st ed.). Washington, DC: American Psychiatric Association.

American Psychiatric Association. (1968). *Diagnostic and statistical manual of mental disorders* (2nd ed.). Washington, DC: American Psychiatric Association.

American Psychiatric Association. (1980). *Diagnostic and statistical manual: Mental disorders* (3rd ed.). Washington, DC: American Psychiatric Association.

American Psychiatric Association. (1994). *Diagnostic and statistical manual of mental disorders* (4th ed.). Washington, DC: American Psychiatric Association.

American Psychiatric Association. (2012). Major depressive episode. Rationale. www.dsm5.org. March 2012.

American Psychiatric Association. (2011). http://www.dsm5.org/Pages/Default.aspx. March 2011.

American Psychiatric Association. (2013). *Diagnostic and statistical manual: Mental disorders* (5th ed.). Washington, DC: American Psychiatric Association.

Auster, P. (1982). *The invention of solitude*. New York: Penguin Books.

Baylis, F. (2011). The self in situ: A relational account of personal identity. In J. Downie & J. J. Llewellyn (Eds.), *Being relational: Reflections on relational theory and health law*. Vancouver: UBC Press.

Beck, A. T. (1962). Reliability of psychiatric diagnoses: A critique of systematic studies. *American Journal of Psychiatry, 119*, 210–216.

Carey, B. (2012). Grief could join list of disorders. *The New York Times*. January 24. http://www.nytimes.com/2012/01/25/health/depressions-criteria-may-be-changed-to-include-grieving.html?pagewanted=all&_r=0

Cohen, B. (2003). *Theory and practice of psychiatry*. New York: Oxford University Press.

Cooper, J. E., Kendell, R. E., Gurland, B. J., Sartorius, N., & Farkas, T. (1969). Cross-national study of diagnosis of the mental disorders: Some results from the first comparative investigation. *The American Journal of Psychiatry, 10*(Suppl), 21–29.

Faucher, L., & Goyer, S. (2015). RDoC: Thinking outside the DSM box without falling into a reductionist trap. In S. Demazeux, & P. Singy (Eds.), *The psychiatric babel: Assessing the DSM-5, history, philosophy and the theory of the life sciences series* (pp. 199–224). Dordrecht: Springer.

First, M. B., Pincus, H. A., Levine, J. B., Williams, J. B. W., Ustun, B., & Peele, R. (2004). Clinical utility as a criterion for revising psychiatric diagnoses. *American Journal of Psychiatry, 161*, 946–954.

First, M. B., & Wakefield, J. C. (2012). Validity of the bereavement exclusion to major depression: Does the evidence support the proposed elimination of the exclusion in DSM-5? *World Psychiatry, 11*, 3–11.

Flanagan, O. (1996). *Self-expressions: Mind, morals, and the meaning of life*. New York: Oxford University Press.

Frances, A. (2013, January 7). Last plea to DSM-5: Save grief from the drug companies. *Huffington Post*. http://www.huffingtonpost.com/allen-frances/saving-grief-from-dsm-5-a_b_2325108.html

Freud, S. (1917). Mourning and melancholia. In *The complete works of Sigmund Freud* (pp. 243–258). London: The Hogard Press.

Fulford, K. W. M., Thornton, T., & Graham, G. (2006). *Oxford textbook of philosophy and psychiatry*. Oxford: Oxford University Press.

Graham, G. (2013). *The disordered mind: An introduction to philosophy of mind and mental illness* (2nd ed.). New York: Routledge.

Gruenberg, A. M., Goldstein, R. D., & Pincus, H. A. (2005). Classification of depression: Research and diagnostic criteria: DSM-IV and ICD-10. In J. Licinio & M.-L. Wong (Eds.), *Biology of depression: From novel insights to therapeutic strategies*. Weinham: Wiley-VCH Press.

Guze, S. B. (1978). Validating criteria for psychiatric diagnosis: The Washington University approach. In H. S. Akiskal & W. L. Webb (Eds.), *Psychiatric diagnosis: Exploration of biological predictors* (pp. 47–59). New York: Spectrum Publications.

Hacking, I. (1995). The looping effects of human kinds. In D. Sperber & A. J. Premack (Eds.), *Causal cognition* (pp. 351–383). Oxford: Oxford University Press.

Horwitz, A. V., & Wakefield, J. C. (2007). *The loss of sadness: How psychiatry transformed normal sadness into depressive disorder*. New York: Oxford University Press.

Jackson, S. W. (1986). *Melancholia and depression: From Hippocratic times to modern times*. New Haven: Yale University Press.

James, W. (1890/1983). *Principles of psychology*. Cambridge: Cambridge University Press.

Jopling, D. A. (2000). *Self-knowledge and the self.* New York: Routledge.

Katz, M. M., Cole, J. O., & Barton, W. E. (1968). *The role and methodology of classification in psychiatry and psychopathology* (Public Health Service publication, Vol. #1584). Washington, DC: Department of Health, Education and Welfare.

Kleinman, A. (2012). The art of medicine: Culture, bereavement, psychiatry. *The Lancet, 379*(9816), 608–609.

Lindemann Nelson, H. (2001). *Damaged identities: Narrative repair*. New York: Cornell University Press.

Lumsden, D. (2013). Whole life narratives and the self. *Philosophy, Psychiatry, and Psychology, 20*(1), 1–10.

Marin, K. A., Bohanek, J. G., & Fivush, R. (2008). Positive effects of talking about the negative: Family narratives of negative experiences and preadolescents' perceived competence. *Journal of Research on Adolescence, 18*(3), 573–593.

Montaigne, M. (2003/1580). *The complete essays*. London: Penguin Books.

Pennebaker, J. W. (1997). Writing about emotional experiences as a therapeutic process. *Psychological Science, 8*(3), 162–166.

Pennebaker, J. W. (1993). Putting stress into words: Health, linguistic, and therapeutic implications. *Behavior Research & Therapy, 31*(6), 539–548.

Pickard, H. (in press). Stories of recovery: The role of narrative and hope in overcoming PTSD and PD. In J. Sadler (Ed.), *The Oxford handbook of psychiatric ethics*. Oxford University Press.

Poland, J. (2001). Review of DSM-IV sourcebook, volume 1. *Metapsychology Online Reviews, 5*(14).

Robins, E., & Guze, S. B. (1970). Establishment of diagnostic validity in psychiatric illness: Its application to schizophrenia. *American Journal of Psychiatry, 126*(7), 983–987.

Sadler, J. (2005). *Values and psychiatric diagnosis*. Oxford: Oxford University Press.

Sadler, J. Z. (2007). The psychiatric significance of the personal self. *Psychiatry, 70*(2), 113–129.

Saks, E. (2007). *The centre cannot hold: My journey through madness*. New York: Hyperion.

Schwartz, M. A., & Wiggins, O. P. (1987a). Typifications the first step for diagnosis in psychiatry. *The Journal of Nervous and Mental Disease, 175*, 65–77.

Schwartz, M. A., & Wiggins, O. P. (1987b). Diagnosis and ideal type: A contribution to psychiatric classification. *Comprehensive Psychiatry, 28*(4), 227–291.

Shear, M. K., Simon, N., Wall, M., et al. (2011). Complicated grief and related bereavement issues for DSM-5. *Depression and Anxiety, 28*, 103–117.

Sokal, R. R. (1974). Classification: Purposes, principles, progress. Prospects. *Science, 185*(671), 1115–1123.

Spitzer, R., & Endicott, J. (1968). DIAGNO: A computer program for psychiatric diagnosis utilizing differential diagnostic procedure. *Archives of General Psychiatry, 18*, 746–756.

Tekin, Ş. (2011). Self-concept through the diagnostic looking glass: Narratives and mental disorder. *Philosophical Psychology, 24*(3), 357–380.

Tekin, Ş. (2013a). How does the self adjudicate narratives? *Philosophy, Psychiatry, and Psychology, 20*(1), 25–28.

Tekin, Ş. (2013b). "Will I be Pretty, Will I be Rich?" The missing self in antidepressant commercials. *American Journal of Bioethics, 13*(05), 19–21.

Tekin, Ş. (2014). The missing self in Hacking's looping effects. In H. Kincaid & J. Sullivan (Eds.), *Classifying psychopathology: Mental kinds and natural kinds* (pp. 227–256). Cambridge, MA: MIT University Press.

Tekin, Ş. (2010). *Mad narratives: Self-constitutions through the diagnostic looking glass*. Ph.D. dissertation, York University, Print.

Tekin, Ş. (in press). Self-insight in the time of mood disorders: After the diagnosis, before the treatment. *Philosophy, Psychiatry, Psychology, 21*(2), 139–155.

Tsou, J. (2015). DSM-5 and psychiatry's second revolution: Descriptive vs. theoretical approaches to psychiatric classification. In S. Demazeux, & P. Singy (Eds.), *The psychiatric babel: Assessing the DSM-5. History, philosophy and the theory of the life sciences series* (pp. 43–62). Dordrecht: Springer.

Wakefield, J. C. (2015). The loss of grief: Science and pseudoscience in the debate over DSM-5's elimination of the bereavement exclusion. In S. Demazeux, & P. Singy (Eds.), *The psychiatric babel: Assessing the DSM-5, history, philosophy and the theory of the life sciences series* (pp. 157–178). Dordrecht: Springer.

Zisook, S. (2010, August 2). Today in your health: Bereavement. *National Public Radio, Morning Edition*.

Zisook, S., & Kendler, K. S. (2007). Is bereavement-related depression different than non-bereavement-related depression? *Psychological Medicine, 37*, 779–94.

Zisook, S., Shuchter, S. R., Pedrelli, P., et al. (2001). Bupropion sustained release for bereavement: Results of an open trial. *Journal of Clinical Psychiatry, 62*, 227–230.

Zisook, S., Shear, K., & Kendler, K. S. (2007). Validity of the bereavement exclusion criterion for the diagnosis of major depressive episode. *World Psychiatry, 6*, 102–107.

# RDoC: Thinking Outside the DSM Box Without Falling into a Reductionist Trap

**Luc Faucher and Simon Goyer**

**Abstract** Just as the DSM-5 was about to be finalized, the National Institute for Mental Health (NIMH) launched its "Research Domain Criteria" (RDoC) initiative, a project that has been seen by many as a disavowal of the type of nosological enterprise incarnated by the DSM itself, from DSM-III to DSM-5. In our paper, we first want to describe the context in which RDoC appeared and demonstrate that, if it is not a disavowal of the DSM-5's work, it certainly signals the abandonment of a method of trying to establish a valid nosology; a paradigm shift in nosology so to speak. We will then question if RDoC is a reductionist enterprise. We will explain why RDoC is not reductionist in a strong and naïve sense, but why it could be understood as reductionist in a weaker sense. If this weaker form of reductionism does not possess the problems the stronger forms of reductionism do, it might nonetheless generate problems of its own that researchers should be aware of. We will try to delineate some of these problems.

Just as the DSM-5 was about to be finalized, the National Institute for Mental Health (NIMH) launched its "Research Domain Criteria" (RDoC) initiative,[1] a project that has been seen by many as a disavowal of the type of nosological enterprise

[1] As stated on the website of the NIMH, the RDoC project is an initiative that aims to "define basic dimensions of functioning (such as fear circuitry or working memory) to be studied across multiple units of analysis, from genes to neural circuits to behaviors, cutting across disorders as traditionally defined. The intent is to translate rapid progress in basic neurobiological and behavioral research to an improved integrative understanding of psychopathology and the development of new and/or optimally matched treatments for mental disorders" (NIMH, "Research Domain Criteria" Web. April 5th, 2014, http://www.nimh.nih.gov/research-priorities/rdoc/index.shtml).

L. Faucher (✉)
Université du Québec à Montréal, Montréal, Canada
e-mail: lucfaucher@hotmail.com

S. Goyer
Université de Montréal (CHUM), Montréal, Canada



S. Demazeux, P. Singy (eds.), *The DSM-5 in Perspective*, History, Philosophy and Theory of the Life Sciences 10, DOI 10.1007/978-94-017-9765-8_12

incarnated by the DSM itself, from DSM-III to DSM-5.[2] The latter interpretation of the initiative has been fuelled by statements from individuals either involved in the RDoC initiative or working close to it. For instance, Thomas Insel, on his blog on the NIMH website, celebrated the future arrival of the DSM-5 by saying that "[p]atients with mental disorders deserve better" (April 2013). In the same spirit, Hyman (a former NIMH Director) has been writing that "it now appears that the accreting failures of the current diagnostic system cannot be addressed simply by revising individual criterion sets and certainly not by adding more disorders to DSM-5 [ . . . ] DSM-III was a brilliant advance; *it is now time to move on*" (2011, 3 and 14; our emphasis).

Indeed, RDoC calls for a paradigm shift in research that would align psychiatry with the rest of medicine. Current research in medicine has unearthed etiological mechanisms, discovered biomarkers, identified risk factors—all of which, separately or in combination—allow for the development of new treatments. While the rest of medicine is making incredible, progressive strides in the treatment of cancer or Alzheimer's disease, research for the treatment of major mental disorders has stalled almost completely (Akil et al. 2010; Hyman and Fenton 2003). Insel et al. (2012) describe the situation thusly: "the world is experiencing a crisis in drug development for mental illness; drug companies are withdrawing from the field or redirecting the investments" (269). It is believed that the discovery of new treatments necessitates a more sophisticated understanding of mental disorders. RDoC advocates[3] argue that this understanding will come only if we abandon the DSM's current constructs for new constructs developed in a bottom-up fashion to better mirror the "reality" of mental illness. Since RDoC conceptualizes mental illness as brain disorders (Insel et al. 2010, p. 749), it looks for "reality" at the level of neural circuits. In this respect, RDoC's initiative is brain-centered and might appear reductionist in spirit.

In our paper, we first want to describe the context in which RDoC appeared and demonstrate that, if it is not a disavowal of the DSM-5's work, it certainly signals the abandonment of a method of trying to establish a valid nosology; a

---

[2]At least, this is how it has been interpreted in the popular press; see for instance, Belluck and Carey 2013; Campbell 2013; Horgan 2013; Koven 2013.

[3]The NIMH lists a number of publications on its website which describes the RDoC project (http://www.nimh.nih.gov/research-priorities/rdoc/nimh-rdoc-publications.shtml). Among the authors of these papers, we find Thomas Insel (actual director of the NIMH), Bruce Cuthbert (director of NIMH Division of Translation Research), Sarah E. Morris (chief of project of the NIMH Schizophrenia Spectrum Disorders Research Program), Charles A. Sanislow (who now works at Yale Psychiatric Institute, but used to be the program chief of NIMH Extramural Research in the Mood and Sleep Disorders Research), Daniel S. Pine (chief of the NIMH Section of Development and Affective Neuroscience). These researchers, as well a few other co-authors, are responsible for presenting, articulating, and defending (if necessary) the RDoC project. It is to them that we are referring when we are talking about the "RDoC advocates." We are aware that other members of the project—for instance, participants to the workshops that were convened to define the constructs that figure in the RDoC framework—might not share the views expressed by the leading advocates of the project.

paradigm shift in nosology so to speak (section "DSM vs. RDoC: The Problem and Its Solution"). We will then question if RDoC is a reductionist enterprise (section "Is RDoC Reductionistic?"). We will explain why RDoC is not reductionist in a strong and naïve sense, but why it could be understood as reductionist in a weaker sense (section "If Not Reductionism, Then What?"). If this weaker form of reductionism does not possess the problems the stronger forms of reductionism do, it might nonetheless generate problems of its own that researchers should be aware of. We will try to delineate some of these problems (section "Problems with the RDoC").

## DSM vs. RDoC: The Problem and Its Solution

In this section, we will review some of the DSM's problems that are claimed to provide reasons for arguing for the necessity of a new approach in the domain of mental illness. We will then argue that these issues are not only well known outside the RDoC circle, but that the need for a paradigm shift in research is also commonly recognized. Despite these well-acknowledged shortcomings, we will see that RDoC advocates are not, in fact, arguing for the dismissal of the DSM (at least not in the near future). We will conclude by presenting the RDoC's solution to the issues raised within this section.

### *If RDoC Is the Solution, What Is the Problem?*

To understand the DSM's current problems, one should return to the source of the recent DSM paradigm, i.e., to the DSM-III. DSM-III can be seen as a response to a series of problems that were plaguing psychiatry in the 1970s. These problems were both epistemological (for instance: providing a classification that identified natural kinds, furthering research, countering anti-psychiatry arguments concerning the reality of [certain] mental disorders, reflecting complexity and interrelationships between biopsychosocial aspects of the individual under evaluation) and practical (assisting patient treatment, reducing inter-rater variability, reducing insurance company concerns related to who genuinely deserves care, etc.). The solutions adopted were numerous; among them were the introduction of a multi-axial system, a definition of mental illness, and so forth. Most importantly for the preoccupations of this paper is the constitution of a non-etiological and atheoretical system of classification using diagnostic criteria based on easily observable features of subjects' behavior. This system is categorical (instead of dimensional) and polythetic (instead of monothetic)—that is, it considers mental illnesses as discrete, non-overlapping entities (with zones of rarity between them and between disordered and healthy conditions) that can be identified by different combinations of operationalized criteria. As Poland and his colleagues remarked—and this is fundamental to what follows—the DSM is "supposed to be a 'purely descriptive' classification system that mirrors a theoretical taxonomy based on (as yet unknown) etiologies and pathologies" (Poland et al. 1994, p. 241).

The main contention with the DSM for RDoC advocates (see also Meehl 1995; Poland et al. 1994) is not that psychiatry has adopted the medical model of disease—as anti-psychiatrists have accused psychiatry to do (i.e., that it considers mental illness as a medical condition such as diabetes or cancer)—but that psychiatry is lagging behind the rest of medicine. The rest of the medical field has turned from defining diseases in terms of reported symptoms and externally observed clinical symptoms, to defining them in terms of their etiological mechanisms, but psychiatry has not yet taken this turn. And there are reasons to think that the current categorization of mental disorders proposed by the DSM might be an impediment to the process of alignment of psychiatry with the rest of medicine.

When the Feighner's Criteria (i.e., clinical descriptions, laboratory studies, delimitation from other disorders, follow-up studies, family studies; Feighner et al. 1972) were introduced, it was thought that these criteria would be efficacious in identifying psychiatry's natural kinds (establishing DSM's mental disorder as "real entities"). However, it was quickly realized that "the surface phenomena of psychiatric illness (i.e. the clustering of symptoms, signs, course and outcome) provide no secure basis for deciding whether a diagnostic class or rubric is valid in the sense of delineating a specific, necessary and sufficient biological mechanism" (Kendell and Jablensky 2003, p. 7). For this reason, in the 1990s, Nancy Andreasen (1995) argued that Feighner's criteria were insufficient and that genetics and neuroimagery would offer diagnostic categories their "etiopathogenic validity" (Schaffner 2002). These disciplines would provide "[n]ew models of validation [which] probe beneath such surface features and seek to identify actual neural and genetic mechanisms [and which] ... can give mental illnesses a powerful credibility" (Andreasen 1995, p. 162). In fact, these new models of validation never did validate diagnostic categories; instead, they seem to have shown that the DSM is "a poor mirror of nature" (Miller 2010, p. 1437). More specifically, there were three major complications that plagued the DSM:

1. The heterogeneity of the diagnostic categories: As we have stated, it was hoped that genetic and neuroscientific research would lead to the validation of current diagnostic categories by showing that people thought to belong to the same diagnostic categories would share the same basic etiologic basis, but that hope never panned out. This state of affairs (the genetic heterogeneity) should come as no surprise. In a way, this heterogeneity might just reflect that which is present at the clinical level of description (otherwise called "phenomenological heterogeneity"). Indeed, as it has been repeatedly observed (Wardenaar and de Jonge 2013; Hickie et al. 2013), current diagnostic categories group a heteroclite collection of individuals that belong to mechanistically different subtypes together, with potentially different course trajectories and different

reactions to treatment. Again, this should come as no surprise, because with the adoption of polythetic diagnostic criteria for disorders, the DSM was already opening the door to such a possibility.

2. Comorbidity and NOS: Comorbidity refers to the state of individuals who have two or more disorders with distinct etiologies, which are concomitantly present. The problem is that comorbidity is so extensive in the DSM that it has led some to question the DSM's underlying structure and categories. For instance, one commentator remarked, "more than one third of all cases in the population have at least two comorbid diagnoses and only 15 % of patients in long-term in-patient treatment were reported to have only a single personality disorder" (Aragona 2009, p. 6). The extensive character of comorbidity has led some to think that comorbidity might not be the result of the simultaneous presence of multiple disorders, but rather different expressions of a single, underlying disorder. If such is the case, this would argue against what have been a driving assumption behind the DSM (Robins and Guze 1970; Feighner et al. 1972; Spitzer et al. 1975), i.e., that mental disorders can be represented as non-overlapping categories.

   A related problem is that DSM categories are understood as rigid and narrow (Hyman compares them to "silos"): you either fit in or you don't. If you don't fit, you can be assigned to residual "not otherwise specified" (NOS) diagnoses. Hyman remarks that "the need to use NOS diagnoses varies according to disorder clusters and clinical communities but appears to represent a large fraction of diagnoses made in certain domains, including developmental and other childhood mental disorders, eating disorders, and personality disorders" (2011, p. 7). This means that in certain categories, a majority of patients do not fit into specified diagnostic criteria.

   Both comorbidity and NOS categories indicate an excessive tendency to split on the part of DSM creators, and the necessity to reject the sharp boundaries inherent to actual DSM categories.

3. The Reification of the Categories: Hyman (2010) and Cuthbert and Insel (2012) identify another problem with the DSM: the "reification of categories." In the mind of the architects of the DSM-III, the categories that figured in the manual were supposed to be research heuristics. The problem is that, rather than being hypotheses, researchers (and funding agencies) have come to assume their validity wholesale and organized research accordingly. Largely due to the two preceding points, it has been claimed that a different strategy should be adopted in research, "[s]pecifically, it may be necessary to deconstruct currently defined higher order clusters of complex behaviors (or subsets of these clusters) into intermediate functions that are not themselves clinical symptoms in order to understand the relationship of higher order 'criterion' symptoms to lower order causal networks that include cognition, emotion, hormones, neural circuits, and their molecular pathways and structures" (Sanislow et al. 2010, p. 632). We will explain in section "Should We Get Rid of the DSM?" how RDoC articulates this very strategy.

## *Should We Get Rid of the DSM?*

Given the problems with the DSM reviewed in the previous sub-section, what should we do with the DSM? What does RDoC plan to do with it—does it plan to just ditch it?

Not surprisingly, advocates of the RDoC claim that a "shift in thinking about the classification of psychiatric disorders" (idem; see also Cuthbert and Insel 2013, p. 129) is called for. Despite this invitation for a change of classification paradigm, RDoC affirms that it wants to remain "largely agnostic with respect to contemporary diagnostic classifications (i.e., the *DSM* and the ICD)" (Sanislow et al. 2010, p. 636; see also Cuthbert and Insel 2012, p. 67). RDoC does not seek to replace the DSM and ICD's classifications, and instead hopes to provide a research framework that serves to integrate existing data and generate new ones through different research designs. That is what, according to Morris and Cuthbert, the RDoC is explicitly aiming for: "If the project [the RDoC] is successful, future versions of the DSM and ICD—perhaps not even DSM-6, but DSM-7—will be informed by the findings that emerge from RDoC-guided research" (2012, p. 35). Moreover, RDoC explicitly focuses on targeted disorders, not intending to "'cover the waterfront' of symptoms and illnesses" (Morris and Cuthbert 2012, p. 33), so it will not (at least in the next decade or so)[4] try to produce an account of each and every mental disorder listed in the DSM. For the time being, the RDoC's objective is modest as "it seems unlikely that we will replace the 300-disorder taxonomy of the DSM-5 with an alternative biologically based classification system anytime soon. Therefore the real opportunity for psychiatry is to use the emerging advances in genetics, molecular biology, imaging and cognitive science to supplement, rather than replace, the symptom-driven diagnosis" (Kapur et al. 2012, p. 1176).

In conclusion, there are no suggestions, on the part of RDoC advocates, to ditch the DSM-5. On the other hand, assertions related to the need for a paradigm shift suggest that research should free itself from constraints imposed by the DSM. In the short term, the DSM-5 is not in grave danger, as there is at the present insufficient data to force a major revision of its conceptual organization, or of its categories. In the long term however, it is difficult to imagine how the DSM-5 (or something with a similar conceptual structure) could resist the impulsion for research provided by the RDoC (if it bears fruits as promised and delivers concepts that are both scientifically sound and clinically useful): the nosology of the future should be radically different from today's DSM.

---

[4]RDoC initial framework was completed in 2012. As Morris and Cuthbert are saying, "the iterative process of evaluating and refining the constructs will likely occur over a 5 to 10-year timeframe, followed by ongoing modification based on new scientific discoveries" (2012, p. 33). We are not aware of any set timeframe for the production of a complete nosology of the type of the DSM.

## *Description of the RDoC*

The *National Institute of Mental Health* (NIMH) aims to "transform the understanding and treatment of mental illnesses through basic and clinical research, [...] [and to pave] the way for prevention, recovery and cure" (NIMH 2008). In 2008, the NIMH unveiled a strategic plan with four goals:

1. Promote discovery in brain and behavioral sciences to fuel research on the causes of mental disorders;
2. Chart mental illness trajectories to determine when, where, and how to intervene;
3. Develop new and better interventions that incorporate the diverse needs and circumstances of people with mental illnesses;
4. Strengthen the public health impact of NIMH-supported research.

The *Research Domain Criteria*'s (RDoC) project is a part of the first goal mentioned above. It aims to "[d]evelop, for research purposes, new ways of classifying mental disorders based on dimensions of observable behavior and neurobiological measures." The goal of NIHM researchers is, indeed, to construct a nosology informed by genetics, neuroscience, and behavioral science in order to (better) classify patients in mental health research. Moreover, because it is expected that relevant findings will be produced rather slowly in the years to come, this new nosology will spend a sizeable amount of time under construction (Cuthbert and Insel 2013, pp. 131–32).[5] Eventually, it will either replace the DSM, or as Insel and Liberman claim, it will inform it substantially, and consequently will impact clinical practice (Insel and Lieberman 2013).

What is the driving logic behind the RDoC? The main idea is that normal behavior depends on normally functioning systems (some responsible of cognitive capacities like memory, control, decision making; others of affective ones, like emotions, pleasure, pain; others of arousal or social relationships) that in turn depend on intact neural circuits.[6] These different neural circuits have been studied (and continue to be studied) extensively by genetics, neuroscience, and behavioral sciences. So if one were to open a book overviewing the field of neuroscience (for instance, Gazzaniga (2009) *The Cognitive Neurosciences* or Kandel et al. (2000) *Principles of Neural Science*), one would find chapters on perception, movement, memory, attention, language, emotion, and social cognition, as well as higher cognition, arousal, and consciousness (but also development and plasticity). In these chapters, one would find description of the different neural circuits underlying these functions. In certain of these chapters, one would also find a description of the molecular events and genetic material deemed necessary for the construction and the

---

[5]For that reason, RDoC constructs are not planned to be ready to be used for clinical diagnostics for at least a decade or two (NIMH 2011a).

[6]For instance, normal response to threat (i.e. fear response and learning) depends crucially on a neural circuit involving different nuclei inside the amygdala (Ledoux 2012).

good functioning of these neural circuits. These functionally defined neural circuits are the axes around which current neuroscience organizes research.

By stipulating that what we call "mental disorders" are "are biological disorders involving brain circuits that implicate specific domains of cognition, emotion and behavior" (Insel 2013), RDoC attempts to develop a psychiatry that uses the currency of neuroscience. So the idea is to create general constructs, called "functioning domains," which are in harmony with the work coming from neuroscience and to study mental disorders using these constructs.

Because RDoC is a work in progress, an NIMH workgroup has first identified a number of promising "functioning domains" representing functional dimensions of behaviour that cut across traditional disorders categories. The domains identified are (a) negative valence systems, (b) positive valence systems, (c) cognitive systems, (d) systems of social processes, and (e) arousal/modulatory systems. Then workshops with external experts were convened on each domain to define and evaluate the constructs to be included within the domain. For instance, after discussions, the NIHM RDoC working group proposed four constructs inside the "negative valence systems" that reflect what the field believes about that domain: the constructs were (a) acute threat ('fear'), (b) potential threat ('anxiety'), (c) sustained threat, loss and (d) frustrative non-reward (Cuthbert and Insel 2013, p. 130; see also Morris et al. 2014).

RDoC also proposes that each of these constructs can be described at the following level of analysis: genetic, molecular, cellular, neural circuits, physiological, behavioral, and self-reported. The functioning domains, constructs, and levels of analysis are all represented in a research matrix. The functioning domains and their respective constructs form the rows while the levels of analysis form columns. The points where rows and columns meet are called "cells." In a cell, one finds information from empirical research that is relevant to the cell in question. For instance, the cell that joins (a) the construct "Sustained Threat" in the functioning domain "Negative Valence Systems" and (b) the row matched with the level of analysis "Physiology" might contain, among other things, results from research concerning the effects of a dysfunctional hypothalamic-pituitary-adrenal axis on the expression of certain genes in the brain that result from the modification of certain brain structures (such as the hippocampus). Information concerning the dysregulation of the HPA axis can assist in the better understanding of many mental disorders such as major depression and anxiety (for instance, see Bogdan and Hariri 2012 for an explanation as to how exposure to stress in childhood can cause deleterious effects on the HPA axis [especially in women] which can in turn cause mood and anxiety disorders).

It is important to note that the level of analysis that is deemed to be central in the explanation of mental disorders is the level of "circuits." Indeed, as one can read on the NIMH website,

> Circuits represent the core aspect of these classes of variables – both because they are central to the various biological and behavioral levels of analysis, and because they are used to constrain the number of constructs that are defined. (NIMH 2011a)

It is around "circuits" that the research will be organized, with experiments looking at the functioning of these circuits in a wide range of subjects from normal to impaired (either meeting or not meeting the actual clinical threshold or across different conditions). To put it in a different way, it is now through the lenses of these circuits that researchers will be asked to look at mental disorders.

Finally, one should note that in the matrix, there is a column entitled "paradigm." This column does not describe a level of analysis as such, but rather indicates various research paradigms used to generate data found in the cells of the matrix. For instance, for the construct "acute threat ('fear')," paradigms can be: "fear conditioning," "viewing aversive pictures or films," or "emotional imagery" (see NIMH 2011b).

Advocates of RDoC claim that there is much to be gain by adopting a strategy of research using this framework. Among the expected gains, one can mention the fact that:

(a) It could enable researchers to identify biological mechanisms underlying mental disorders or aspects of them and to develop therapies and drugs that target their biological causes (see Morris et al. 2014).
(b) It could help to shed light on comorbidity between disorders. For instance, researchers could explain the comorbidity between major depression and anxiety by showing that they are both the results of a hyperactive amygdala (as it is suggested by previous research Ressler and Mayberg 2007, p. 1117). The fact that these two conditions are closely related is also compatible with genetic findings according to which the same genes are implicated in both conditions (as it is suggested by Kendler et al. 2007a).
(c) It could allow researchers to explain why there exists such heterogeneity inside a single diagnostic category. For instance, the fact that there seems to be different types of major depression could be explained by the fact that different biological processes are involved in patients diagnosed with it. For instance, certain depressed individuals display high levels of cortisol, while others display low levels of the same substance (as suggested by Krishnan and Nestler 2008 pp. 898–899). Others who carry polymorphism C1019G on a gene coding for a serotonin receptor 5-HT$_{1A}$ and who also carry polymorphism G196A on a gene coding for protein Brain-Derived Neurotrophic Factor (BDNF), are more susceptible to be resistant to anti-depressant treatments if they develop a major depression (as suggested by Anttila et al. 2007). Ultimately, by taking into account this kind of data, it would be possible to refine the way major depression is currently conceived in the same way that the conception of cancer has been refined in the last few years.
(d) It could allow researchers to integrate biological neurodevelopmental considerations in their explanations (and in treatments that follow from them) of mental disorders. For instance, with the help of longitudinal studies, they could study the course of adolescent brain maturation and synaptic pruning to identify genes and circuit development factors associated with departures from normal developmental functioning, and points in prodromal stages where intervention might particularly be targeted. (NIMH 2011a)

(e) It could finally allow researchers to retranslate in biological terms the impact that environmental factors have on molecular activity inside the brain. For instance, it has been established that when a mother rat does not take care of its pups, this produces strong methylation of a gene coding for the production of glucocorticoids receptor in the hippocampus. It follows then, that when a mother rat does not lick its pups, they develop fewer of these receptors (as suggested by Krishnan and Nestler 2008, p. 899). For this reason, those rats are more sensitive to stress. The same kind of processes could explain susceptibility to depression or anxiety in humans (Bogdan and Hariri 2012).

As we have seen, the logic offered by RDoC advocates leads them to a certain conception of mental disorders. In the following, we will examine that very conception. More specifically, we will ask the following question: because RDoC identifies mental disorders with neural circuits operating in the brains of individuals, is it legitimate to maintain that RDoC is a reductionist enterprise (as some have claimed, for instance Parnas 2014, p. 46)? This is what will occupy us in the next two sections.

## Is RDoC Reductionistic?

In this Section, we will first attempt to be a bit more specific about what 'being reductionistic' means in the context of our discussion (section "What Is Reductionism?"). We will then explain why RDoC has sometimes been perceived as a reductionist enterprise (section "Is RDoC Epistemologically Reductionist?"). As we will explain, RDoC advocates themselves have at times contributed to establishing such a perception.

### *What Is Reductionism?*

There are three types of reductionism: *ontological*, *epistemological,* and *methodological. Ontological* reductionism concerns the fundamental structure of reality. Advocates of this form of reductionism maintain that higher-level phenomena (flowers, rocks, humans, etc.) are constituted by lower-level ones, and in a strong version of that thesis, they claim that higher-level phenomena are nothing but lower-level phenomena. For instance, an advocate of that position might want to argue that all the entities posited by high-level sciences like psychology or biology are fictions that are in reality nothing other than fermions and bosons (Rosenberg 2011, p. 21). In psychiatry, some see this type of reductionism as being implicit in the thesis that mental disorders are brain disorders (Gold 2009).

For its part, *epistemological* reductionism concerns the level of explanation that one should favour when explaining a phenomenon. According to this form of reductionism, one should always prefer lower level explanations to higher-level ones. Historically, epistemological reductionism has been defended on the basis that higher-level laws could be derived from lower level laws. The idea here was that one would gain in generality and deepness by adopting lower level explanations. The derivation of optics from the electromagnetic theory is an example of this type of reduction (Bunge 1977, p. R79). The reduction of explanatory models in cognitive psychology to explanatory models in molecular neurobiology is another example of such reductionism (Bickle 2003). In psychiatry, this form of reductionism is implied in the idea that explanatory models of mental disorders are reducible to explanatory models in neurobiology (Akil et al. 2010, p. 1580).

Finally, *methodological* reductionism concerns the choice of methods that we should use to understand natural phenomena. Advocates of methodological reductionism claim that the best methods or strategies to investigate reality are those that proceed by decomposing the various systems we study in lower-level components and operations (Griffiths and Stotz 2013, p. 57). To study mental disorders and their causes, advocates of methodological reductionism favour methods adopted by neuroscience (like the use of fMRI), or by molecular biology in animal studies (like the use of knockout; for an example of this method for the study of the genetic basis of ADHD in mice, see Lee and Silva 2011).

Note that though the three types of reductionism are sometimes adopted in one fell swoop, it does not have to be that way. For instance, an ontological reductionist can decide that interesting generalizations or laws about the world are to be found at higher levels (these laws would have "practical significance," so to speak, even if they are about metaphysically fictional objects).[7] In that case, one's ontological reductionism would not lead him to endorse an epistemological reductionism. Another example of this, is a non-(epistemological) reductionist who would consider that (at least initially) in a field inquiry, the best strategy is to adopt a reductionist method (for instance, by identifying and localizing components of the system; see for instance Bechtel 2009). What these examples demonstrate is that ontological reductionism does not as such constitute a sufficient basis for the adoption of either an explanatory or methodological reductionism.

Given what we have stated, we think that the interesting question in the context of our investigation is to try to determine if RDoC (as presented by the NIMH and as defended by its advocates) is reductionist in the *epistemological* sense. It is obvious to us that RDoC is an ontological reductionist project (after all, RDoC states explicitly that mental disorders *are* brain disorders). As we see it, the question as to whether it is methodologically reductionist depends on its epistemological position (the way it articulates its reductionism or non-reductionism). And since the question of epistemologically reductionism cannot be settled by looking at the ontological position, we think that we should focus on the question of the purported

[7]For instance, a neural reductionist might consider that psychological laws are useful because they enable him to formulate generalizations in a more compact and economical format than neural laws.

epistemological reductionism of RDoC in the next section. Does RDoC posit that lower level explanations (genetic, molecular, brain circuits) are more true, adequate, or relevant than those at higher levels (psychological, sociological, cultural)?

## *Is RDoC Epistemologically Reductionist?*

One might think that RDoC and its advocates are adopting a form of epistemic reductionism, since they seem to want to reduce explanatory models of mental illness to strictly biological models (see section "Description of the RDoC"). At times, Thomas R. Insel seems to adopt such an epistemic position. It might look like this position follows from the ontological position he adopts concerning mental illness. As we mentioned earlier, Insel claims that "[m]ental disorders are biological disorders involving brain circuits that implicate specific domains of cognition, emotion, or behavior" (Insel 2013). Commenting on a previous and similar claim made by Insel, philosopher Ian Gold (2009) proposes that the only charitable way to understand this kind of claim is to postulate that Insel means that psychological sciences are to be "relegated to the status of mere placeholder sciences awaiting replacement by neuroscience and molecular biology" (507).

Such an interpretation might also be inferred from papers such as "Faulty Circuits" (2010), which Insel wrote for a general audience (it was published in *Scientific American*) and that illustrates what could be a strictly epistemological reductionist interpretation of the RDoC. In that paper, disorders such as major depression, obsessive-compulsive disorder, and post-traumatic stress disorder are all represented by explanatory models emphasizing dysfunctions in brain circuits. For instance, Insel describes obsessive-compulsive disorder thusly:

> Neuroimaging studies of patients with OCD have discovered abnormal activity in an adjacent loop that includes the orbitofrontal cortex, which is involved in complex tasks such as decision making, the ventral caudate nucleus within the basal ganglia, and the thalamus, which relays and integrates sensory information. (Insel 2010, p. 48)

Furthermore, the explanatory figures he used to illustrate his models could be described as "brain centered," because they situate the source of mental disorders exclusively in brain circuits. By so doing, it seems that he is neglecting environmental factors that are also important causes of mental disorders. Since information about environmental effects exists,[8] why does Insel not bother to represent them in his figures? Is it because RDoC is epistemologically reductionist, as Gold suggests? We beg to differ.

---

[8] See, for instance, Wilkinson and Pickett (2009).

## If Not Reductionism, Then What?

As we saw in the previous section, RDoC advocates sometimes give ground to the accusation of (epistemological) reductionism. We will argue in this section that RDoC is in fact better conceived of as a non-reductionist, multi-level, and integrative enterprise in which the proper level of explanation cannot be decided in an *a priori* manner.

That advocates of RDoC do not see their project as reductionist is captured in the following quote:

> With a strong focus on biological processes, and emphasis on neural circuits at the outset, the RDoC effort could be construed as reductionist. However, a focus on lower level mechanisms does not necessitate that 'higher' level constructs be dismissed. Most researchers agree that causal influences are multidirectional across level (e.g., across genes, molecules, cellular systems, neural circuits, and behavior), leading some (Kendler 2005) to consider 'explanatory pluralism' or 'patchy reductionism' as an alternative to reductionism. (Sanislow et al. 2010, p. 633)[9]

Rather than being a form of straightforward reductionism, the project is conceived of as a form of 'explanatory pluralism' or 'patchy reductionism'. Given that these two concepts are terms of art, they require some explanation.

Let us start with *explanatory pluralism*. Reductionism is a form of monism. It claims that one level of explanation is more fundamental than other(s), and because of that, this level should be preferred. Alternatives to reductionism are holism, non-reductionism, and explanatory pluralism. *Holism* is the view that a complete explanation of a phenomenon should always take into account both higher level and more global factors (typically understood as non-reductive factors, such as social ones, like inequality of income or level of education) and lower level and more local factors (typically understood as reductive factors, like genes or molecules). An example of such an approach in psychiatry is Engel's 'biopsychosocial model' (1977). *Non-reductionism* (sometimes called "anti-reductionism" in the literature) is the view that a phenomenon should be explained only in terms high-level and global explanatory factors (while the previous model was accepting both reductive and non-reductive factors); the psychodynamic approach might be an example of such view. Finally, *explanatory pluralism* consists of the "claims that (i) the best form (and level) of explanation depends of the kind of question one is trying to answer by the explanation and (ii) that in order to answer all explanation

[9]One editor of this volume asked us about how serious we thought that these claims were. After all, he remarked, in psychiatry, as well as in society in general, reductionism is often perceived as a bad thing, so maybe what RDoC advocates are saying is just lip service. We agree that pressures of that kind against crude reductionism exist, but we like to think that the RDoC'S claims proceed instead from a heightened degree of "philosophical sophistication" in the understanding the scientific goings-on of psychiatric research (that idea have been suggested to us by Peter Zachar, personal communication). So in this section, we will explain the epistemological position that is explicitly adopted by the RDoC and its advocates. In section "Real Fears," we will indicate potential slip backs due to ideological convictions that are independent of their epistemological position.

seeking-questions in the best way possible we will need more than one form (and level) of explanation" (Van Bouwel et al. 2011, p. 36). Explanatory pluralism thus presented is a form of what is sometimes called "compatible pluralism" (Mitchell 2002, 2003), where different explanatory theories or models are thought to be necessary to explain different aspects of a phenomenon. What form (or level) of explanation is required is not an *a priori* question, fixed with finality, but rather it depends on the researcher's epistemic interests. Inside "compatible pluralism" one can identify a type of "interlevel compatible pluralism" (Faucher 2012), where theories at different levels cooperate in order to produce a more complete picture of a phenomenon. In such a case, one would have an "integrative" form of pluralism. Though this form of pluralism looks quite a lot like holism, it is not, because it does not specify in advance, in an *a priori* fashion, that all phenomena will need to be explained by reference to factors at every explanatory level. This is what Kendler called in a recent paper his "empirically based pluralism" (we will explain what precisely this expression means later in this section). Another form of interlevel pluralism which is defended by Van Bouwel et al. (2011) could be called the "merely compatible interlevel pluralism": in that form of pluralism, theories at different levels are required to answer different types of question—they don't necessarily need to be integrated in a coherent whole.

Kendler, in his (2005) paper, proposes that theories in psychiatry are likely to be of the "interlevel integrative pluralism" variety. In the pluralism he advocates for psychiatry, "active efforts are made to incorporate divergent levels of analysis. This approach assumes that, for most problems, single-level analysis will lead to only partial answers. However, rather than building large theoretical structures, integrative pluralism establishes small 'local' integrations across levels of analysis" (437). He goes on to cite, among others, work by Caspi that integrates genetics and environmental adversity in the latter's epidemiology of depression and antisocial behavior (for a review, see Caspi and Moffitt 2006).

In a more recent series of papers (2012a, b), Kendler further develops his explanatory pluralism and appears to move closer to Van Bouwel et al.'s position. For instance, in the explanation of alcoholism disorders (AD), Kendler invokes risk factors at four levels (Kendler et al. 2003, 2007b): biological (genetic variants that influence the metabolism of ethanol, or functional deficits in brain structures in charge of incentive sensitization), psychological (personality traits, perhaps externalizing tendencies), social (early childhood adversities), and cultural (alcohol price or religious beliefs concerning alcohol use). All of these factors are 'difference makers', i.e., manipulation of, or intervention with, one of these factors influences the probability of the presence of a disorder (without the difference-maker needing to be a necessary or sufficient condition). Kendler claims that with psychiatric disorders like AD, the difference-makers are sprinkled across levels (for a similar claim, see Woodward 2008). Yet different psychiatric disorders have different "causal signatures" in that for different psychiatric disorders, difference-makers at various levels explain a proportion of variance: for instance, schizophrenia "has more difference-makers in the biological arena, [major depression] in the psychological arena and [alcoholism disorder] in the higher-order domains. However, each of these three disorders has important difference-makers across all domains" (Kendler 2012b, p. 282). As stated earlier, this form of pluralism is 'empirically-based' in that pluralism is not adopted in an *a priori* fashion, but rather is driven by what is revealed by research: some disorders, like Huntington's disease or Alzheimer's disease, are explained mainly in terms of genes. Others such as schizophrenia, major depression, and AD are explained in terms of different difference-makers distributed differently across various levels.

Kendler also claims that the choice of one level of explanation for an etiological basis depends on the researcher's (and, sometimes, a whole community of researchers') epistemological interests. To make a rather long story short (see Kendler 2012a for the full story), according to Kendler, different types of causes have different types of epistemic properties (strength, generalizability, specificity, manipulability, proximity, generativity, causal role). Since researchers' goals vary, they will choose different levels of explanation by virtue of their epistemic properties: for instance

> [a] basic biobehavioral researcher will be interested in aetiology and would value explanations with high levels of causal confidence, strength and proximity. For a clinical researcher, specificity and manipulability will likely be important as she would want to know how treatment of condition X should differ from that of Y or Z. Those from a public health perspective who are focused on prevention will be interested in causal confidence, strength, generalizability and manipulability. They want to find true strong causes that apply across most conditions and which they can alter thereby reducing risk. Low specificity might even be a virtue as you might reduce risks for a broad area of disorders. For someone identifying a high-risk population, strength will be important but not causal confidence. As long as the risk factors reflect liability to illness, it will not be critical that the association is causal. (Kendler 2012a, p. 17)

Due to the variety of needs of different groups of researchers, Kendler concludes that "there is no *a priori* way to pick a single level of explanation on which to base an etiological nosology, we could [only] try to argue it out on pragmatic grounds" (2012a, p. 17). This is exactly the conclusion of De Vreese et al. (2010), p. 381; see also Van Bouwel et al. 2011). If the goal is prevention, often focus will be on non-reductive, higher-level explanations (such as poverty or immigration); if the goal is efficient treatment of someone that showed up in the office or a reliable diagnosis and prognosis of a condition, one might think that one has to turn to reductive factors (such as genes or bio-markers).[10]

Let us now turn to the concept of *patchy reductionism*, a concept developed by philosopher Kenneth Schaffner in a series of papers (see, among others, his 2006 and 2008). From his point of view, reductionism has been traditionally thought of as 'sweeping'. By this he means that according to models of reductionism such as that which Ernest Nagel champions (1961, Chap. 11), reduction has been thought of as involving a whole low-level theory which would "cannibalize" a whole higher-level theory, i.e., reduce it. As Schaffner remarks, this kind of reduction might

[10]This is not to say that ideological convictions cannot be responsible for the focus on these factors. We will come back on that in section "Real Fears."

occasionally happen in physics, but it is not characteristic of biology. In biology, reductions are 'partial', 'creeping', or 'patchy'. What does that mean? A thought experiment proposed by Kendler (2005), perfectly illustrates what this kind of reductionism amounts to (notice in passing the resemblance between what Kendler describes, and the RDoC structure):

> Imagine that there are 15 discrete levels, with the mind brain system between DNA on one hand and clinical manifestations of schizophrenia on the other. Researcher 1 is conducting linkage and association studies that attempt to directly relate levels 1 and 15 but would provide no insight into the intervening levels. Researcher 2 is trying to understand, at a basic molecular level, the actions of a putative altered gene transcript, thereby trying to move from level 1 to level 2 or 3. Meanwhile, researcher 3 is seeking to understand the neuropsychological deficits in schizophrenia, trying to clarify the link between levels 13 and 15. (438)

In this thought experiment, there is no encompassing reduction of one 'grand theory' to another; instead, there are partial, patchy reductions. These reductions have several properties that are not typically associated with classical reduction:

1. They are incomplete: more often than not, they will not explain all instances of a phenomenon. For instance, genetic factors will explain some cases of schizophrenia or major depression, but not all of them. Or a researcher may identify only part of a pathway thought to contribute to the explanation of a phenomenon (Schaffner calls such explanations *causal-sketch explanations*; Schaffner 2008, p. 75). In this kind of explanation, the pathway from gene to behaviour is said to be "gappy" (idem).
2. They skip levels: this is another way that reductions are incomplete. For instance, a linkage between a gene mutation and behaviour or a disorder can be established, but some key mechanisms, which gene mutation impacts in order to produce the disorder, are not necessarily known or acknowledged.
3. They are interlevel: the explanatory mechanisms that are produced to explain a condition are typically interlevel. Etiological models of a condition will not be purely molecular, they will be mixed, multi-level mechanisms including genes,— but also cells, brain systems, behaviour, and sometimes social factors.
4. They are bushy: at the end of the day, when one provides a mechanism that explains a phenomenon, there is no tidy or discreet one-to-one relationship between a gene and a behaviour, but a rather messy and bushy causal 'thicket' where many genes act together, in a particular (epigenetic) context, with causal loops that might include 'outside-the-skin' pathways. There is nothing that resembles a nice axiomatic theory.

Now that we better understand what explanatory pluralism and patchy reductionism are, let us return to the RDoC. First, let us begin with interlevel explanatory pluralism. RDoC has never claimed that lower level explanations should be preferred; the RDoC explicitly states that environment is a critical element in the research it fosters (though, one might wonder why this dimension is not represented on its matrix). Therefore, in theory, RDoC does not obliterate the role of the environment. Indeed, one of RDoC's preoccupations is to develop "a more mechanistic understanding of how such factors as life events and the social environment interact with development to produce a range of observed outcomes" (Cuthbert 2014, p. 30). Research, such as Michael Meaney's, on the effect of maternal care on individual differences in hypothalamic-pituitary-adrenal responses to stress (Hellstrom et al. 2012; see also Champagne 2010), or the description of the epigenetic mechanisms through which factors like maternal infection, chronic stress, urban environment, or migration at particular key moments of the development have an impact on the development of schizophrenia (Rutten and Mill 2009), are examples, quoted by RDoC advocates, of the kind of research that help us to understand how environmental factors have an effect on the brain. In these cases, researchers are trying to develop a better-integrated multi-level theory or model of a disorder or of the development of a disorder by including the environment factors in the explanation.

Why have brain circuits been selected as the research hub? It is not—as Gold posits—because projects like RDoC hold that higher-level theories or models are explanatorily impotent, the advocates of RDoC would claim. This decision is not motivated by an *a priori* belief concerning the fundamental nature of some levels of explanation.[11] It is motivated instead by researchers' epistemological interests (mainly, manipulability, generalizability, specificity, proximity, causal role: all interests deemed essential for treatment and drug's research). Given researchers' interests, the brain is the best level to focus on. For instance, because mental disorders are unlikely to be monogenic (like Huntington's disease), and are instead polygenic, valid natural kinds are not to be found at the gene level. Rather, it is a better idea to attempt to identify how "many genetic mutations, epigenetic changes and other cellular or morphological brain lesions can all converge on disturbing a given brain circuit and result in shared clinical manifestations [....] that lead to the same clinical diagnosis" (Akil et al. 2010, p. 1580). The brain is also the site where environment factors responsible for mental disorders reveal their effects. An improved understanding of the way these effects are manifest has to focus on the brain. Finally, given that advocates of the RDoC are interested in the diagnosis and treatment of mental disorders, and further, that they want to emulate the success of medicine, the brain appears to be the correct level to focus on. Firstly, medicine was successful in focusing on disorders affecting functions and organs, so it makes sense to look for a cerebral equivalent (psychic functions and brain circuits). Secondly, given that there is too much noise at the clinical level, it is logical to look for valid constructs at a lower level like that one (for a similar conclusion, see Van Praag 2008).

[11]As we stated earlier, we think that the decision to focus on one level can be warranted by epistemological considerations. We do not want to claim that it is always warranted by such considerations (see section "Real Fears").

## Problems with the RDoC

Once we better understand the purported nature of RDoC, it is easier to evaluate what its genuine issues are. In this section, we will first mention a few problems that we think do not affect the RDoC (section "False Fears"). Then we will conclude our paper with problems that, from our point of view, legitimately do affect the RDoC (section "Real Fears").

### *False Fears*

In this section, we identify two false fears concerning the RDoC.

The first fear concerns RDoC's epistemological reductionism. If the RDoC is reductionist, will it not disqualify research completed on higher-level factors (in psychology, for instance)? As was seen in the previous section, RDoC reductionism is of the "patchy," not the "sweeping" variety, and in a way, it is much more humble in its ambitions than the sweeping variety. Firstly, it does not loose sight of the fact that theories or models in psychiatry are typically interlevel, i.e., "they describe idealized causal interactions among entities at various levels of aggregation (molecules, cells, neural pathways, behaviors)" (Harris and Schaffner 1992, p. 146). Secondly, in this context one can talk of reduction "when a *more* biochemically oriented theory provides explanations for a *more* behaviorally based theory" (idem, 146). The RDoC is definitively aiming at building such theories or models and therefore is reductionist, yet as we have repeatedly stated, it does not claim that disorders can be fully explained by the theories of molecular genetics or biology. Other factors will typically have to enter into the explanation of disorders and there are no *a priori* decisions concerning their respective explanatory weight. The RDoC aims to provide a more complete etiological picture of disorders, and maintains that in order to so do, knowledge from genetics and neural sciences are required in the proverbial mix. Higher levels (psychology, environment, development, etc.) are not excluded from the explanatory picture, and are to be included in a more integrated picture with lower level explanatory factors.

The second fear relates to the fact that RDoC research might necessarily lead to over-medicalization, or that it might be driven by the pharmaceutical industry (or work at the service of such an industry). It is true that there is hope that by producing better knowledge of at-risk individuals or of the sub-threshold or prodromal states of mental disorders, it will allow earlier interventions, thus preventing the progression and worsening of some conditions. It is also true that RDoC seems to be at least partly motivated by the fact that pharmaceutical research has recently stalled and that current nosologies (like DSM) are obstacles to research for the discovery of pharmaceutical treatments of mental disorders (as mentioned in section "DSM vs. RDoC: The Problem and Its Solution"). From this point, it's an easy leap to the conclusion that RDoC will lead to over-medicalization and/or be driven by 'big pharma.'

Concerning over-medicalization, we think that the danger of blurring the lines between health and disease might be balanced with the benefit of treating individuals before their symptoms worsen. These benefits have apparently been corroborated for some conditions (see for instance, Obiols 2012, pp. 286–287). If there are such benefits to the introduction of sub-threshold and prodromal states for a particular disorder, it would be foolish to ignore them.[12] We recognize that there are costs associated with the introduction of such states, for instance, a large number of false positives which might translate into individuals being subjected to unnecessary treatment (that can eventually worsen their condition or prolong it), or who might be stigmatized as result of a mental condition diagnosis. Yet, as we have stated, costs need to be balanced against benefits and it is society as a whole (and not by a group of scientists, but by scientists belonging to different groups, government, patient groups, citizens, etc.) who must decide if these costs are equal or inferior to the benefits (we are thinking here along the lines of the kind of process which took place with hypertension; Kotchen 2011). In brief, it should not be assumed in an *a priori* fashion and for all conditions that early intervention will always translate into better prognosis, but for those conditions in which early intervention has been shown to translate into better prognosis, a decision should be taken (based on the gravity of the condition, the accuracy and reliability of the screening instruments, etc.) concerning the appropriateness to intervene or not.

Concerning influence from big pharma, Hickie and colleagues have noted that it is totally unclear whether improved knowledge of aetiological factors will necessarily lead to more drug-oriented therapy: "the evidence from the clinical staging of ischemic heart disease largely demonstrates the opposite. The greatest public health emphasis was placed on reducing smoking rates, promoting exercise and addressing other modifiable risks" (Hickie et al. 2013, p. 4). Similarly, Rutten and Mill (2009), whose research is mentioned favourably by RDoC advocates, insist on conceptualizing schizophrenia as an epigenetic disorder. As we mentioned earlier, such a model emphasizes the role of environmental factors in the emergence of the disorder.[13] As Rutten and Mill note, "these aspects are of key relevance for clinical psychiatry given the potential reversibility of epigenetic makers under the influence of nutrition, social factors, behavioral interventions, and drugs" (1045). So while there may be concerns that more reductionistic knowledge will lead to an increase in drug-related therapy, one should at least note that this link is not necessarily firm. Moreover, RDoC advocates (Morris et al. 2014) also approvingly quote recent research on the neurobiological effects of psychotherapies such as cognitive behavioural therapy (see for instance, Clark and Beck 2010). What this kind of research has revealed is that psychological therapy genuinely modifies

[12]To be sure, research has to show that an early intervention does have an impact on a condition, which might not be the case with all early interventions; see for instance, Wagner 2005.

[13]Such models refer to "the reversible regulation of various genomic functions, occurring independently of DNA sequence, mediated principally through changes in DNA methylation and chromatin structure." (Rutten and Mill 2009, p. 1045).

faulty brain circuits. If such is the case, there are no theoretical reasons why pharmaceutical interventions should be privileged over psychotherapy.

## *Real Fears*

As we have just seen, some fears concerning the RDoC project are unfounded. However, we assert that others are legitimate; we'll mention three in this section.

The first legitimate fear is the following: the DSM has been rejected on the basis that it is founded on categories that were not validated, but instead established by expert consensus and became reified. The fear is that the RDoC might be the DSM all over again. After all, like in the case of the DSM's categories, the RDoC's new categories (the constructs that figure in the matrix) are established through the consensus of a small group of experts and there is fear that its categories—units of analysis like "acute threat" or "negative valence systems"—will themselves become reified. The solution is to keep an open mind, to let different groups define different sets of dimensions[14] (some of which are not RDoC's official ones) and—perhaps, why not—to sponsor research done in the DSM framework as well. As Sartorius observed recently, "research using categories created on the basis of observations of behavior and the development of the disorder over time is as justified as other approaches" (2014, p. 51).

The second fear is the following: consider a man who suffers a depression after his wife leaves him for another, taking half of his meagre salary with her. His symptoms might be "directly" caused by an imbalance of certain chemical substances in the brain, but the (more remote) source of his problem is that his wife has left him, that he is penniless and that he feels powerless to change this fact. You can make him 'better' by giving him pills, but his problem won't disappear. He might instead need to differently frame his situation (think that he has a chance to be more happy now, that money is not that important, etc.). Given this, some might argue that at least some disorders are psychological in essence, and the RDoC's focus on the brain does not consider the relevant level of explanation in the above case. To be clear, we are not stating that the fact that his wife left him has not had an effect on this man's brain, and that, perhaps, cognitive therapy won't modify his brain chemistry (as we have seen earlier, cognitive therapy does have effects on the brain). Yet it seems bizarre to speak of his problems as categorically being the result of a faulty brain instead of saying that they are also, and maybe foremost, psychological in nature.

A related problem is that it is not at all clear that each and every—or even most!—of the disorders now listed in the DSM will be appropriately explained through the

[14]Cuthbert himself is aware of this problem and suggests that RDoC will succeed only if it takes an open-minded approach: "[p]erhaps the outcomes for RDoC might be accessed by the number of research programs that, freed from the strictures of current diagnostic guidelines, outstrip the RDoC matrix to move in entirely new directions that transcend the organization of the current system" (2014, p. 35).

kind of matrix proposed by RDoC. One way to illustrate this is to consider that many disorders might first and foremost result from "software" and not "hardware" problems. Some cases of anorexia or antisocial disorders might result from faults in the "mental programming" of subjects, which result from select social norms (like ideals of thinness or an emphasis on competitiveness) or social arrangements (such as the presence of gangs or the absence of stable jobs). If such is the case, we cannot expect much from the RDoC concerning these disorders. Thus, one fear is that the RDoC strategy will not be able to provide us with a complete picture of disorders.

A third, and again related, fear is that by focusing on brain systems, we might lose sight of psychological or social factors that are important in the explanation of disorders.[15] The problem might be seen as having both a theoretical and a pragmatic aspect. The theoretical aspect is that by identifying the brain as the hub of psychiatric research, we might end up, perhaps quite unintentionally, making social factors appear less important, or as less real than other factors in the explanation of disorders. In the end, these factors might just be neglected or ignored in favour of an individual-centered science.[16] Think of the following example in the domain of medicine: there is an association between tobacco smoking and lung cancer. One can focus on individual factors that lead to the development of cancer (genes, psychological profile, etc.); but one could also focus on population level factors (tobacco production, advertising, economic factors, etc.). One could argue that by focusing research on individual factors, it is more likely that we will end up with a solution defined by individual terms. Note that we are not claiming that this is necessarily the case, rather that this is one danger of such an approach. Similarly, in terms of mental disorders, we might, for instance, identify some individuals who have a disposition to developing addictive behaviours, and we might understand how their brains work, whether the problem might be that these at-risk individuals have easy access to drugs or gambling/games. The latter kind of social phenomena crucially contributes to the burden of diseases that Stein calls "exophenotypes" (2014, p. 52). For instance in Quebec, studies have shown that video poker machines (which are government-owned) are more numerous in certain disadvantaged neighbourhoods, thus exposing a vulnerable part of the population to known risks factors of addiction (Fortier 2013). Thus, it seems to us that the characterization of disorders in terms of faulty brain systems might leave important explanation-seeking questions aside: to what extent is addiction a social problem? Can changes in society bring about a reduction in addiction cases? To be clear, we do not want to take issue with research on individual causes of mental disorders. Instead, we seek to attract attention to the potential danger of defining mental disorders as "faulty circuits." Sometimes, so we argue, such a definition might act as a blinder on researchers who, convinced that all disorders have neurobiological causes, will overlook different (higher-level) types of

---

[15]Lewontin (1992) voiced such complaints about genome research.

[16]Some complain that a similar phenomenon occurred in the domain of epidemiology earlier in the century (see for instance De Vreese et al. 2010, p. 379).

explanation or different (non-pharmaceutical) forms of treatment. In such situations, researchers are not guided by valid epistemological reasons, but by some sort of ideological conviction.

We will describe the pragmatic aspect of the problem only briefly, though it is by no means unimportant. It is that, monetary resources being limited, the decision to focus on the brain and its molecular basis might, as a result, be privileged in lieu of funding for research on psychological or sociological factors, in the explanation of the disorders.[17]

## Conclusion

In this paper, we have described the context in which the RDoC has appeared and demonstrated that, if it is not a disavowal of the DSM-5's work, it marks the abandonment of a specific means of trying to establish a valid nosology—a sort of nosology paradigm shift. At present, the RDoC is at too early a stage for us to predict how exactly it will affect the DSM. As we have mention earlier, in the short term, the DSM's diagnostic categories can be used in a non-reified way, as the heuristic devices that they were meant to be. We have also demonstrated that the RDoC is not reductionist in a strong and naïve sense (the 'sweeping' variety of reductionist), but rather that it is reductionist in a weaker sense (the 'interlevel' and 'patchy' type of reductionism). We argued that this weaker form of reductionism does not possess the problems that stronger forms of reductionism do (i.e. it does not render higher level of explanation impotent). However, RDoC reductionism might generate some of its own disputes, of which researchers should be aware. As a result, we have attempted herein to delineate some of these potential issues.

## References

Akil, H., Brenner, S., Kandel, E., Kendler, K. S., King, M. C., Scolnick, E., & Zoghbi, H. Y. (2010). The future of psychiatric research: Genomes and neural circuits. *Science, 327*(5973), 1580–1581.

Andreasen, N. C. (1995). The validation of psychiatric diagnosis: New models and approaches. *The American Journal of Psychiatry, 152*(2), 161–162.

[17]A similar complaint has been made in a recent editorial of the journal *Nature* in which attention was attracted to the fact that "studies on psychological treatments [were] 'scandalously under-supported', despite their 'potential to make a substantive difference to patients'. It concluded that 'many funding agencies around the world are too keen solely to support mechanistic investigations with potential long-term payoffs, and too unwilling to appreciate that part of their portfolio should be oriented towards identifying immediately effective psychological interventions" (quoted by Fava 2014, p. 49).

Anttila, S., Huuhka, K., Huuhka, M., Rontu, R., Hurme, M., Leinonen, E., & Lehtimäki, T. (2007). Interaction between 5-HT1A and BDNF genotypes increases the risk of treatment-resistant depression. *Journal of Neural Transmission, 114*(8), 1065–1068.

Aragona, M. (2009). The role of comorbidity in the crisis of the current psychiatric classification system. *Philosophy, Psychiatry, and Psychology, 16*(1), 1–11.

Bechtel, W. (2009). Looking down, around, and up: Mechanistic explanation in psychology. *Philosophical Psychology, 22*(5), 543–564.

Belluck, P., & Carey, B. (2013). Psychiatry's guide is out of touch with science, experts say. *New York Times*. http://www.nytimes.com/2013/05/07/health/psychiatrys-new-guide-falls-short-expertssay.html. Retrieved 28 Feb 2014.

Bickle, J. (2003). *Philosophy and neuroscience: A ruthlessly reductive account*. Norwell: Kluwer Academic Press.

Bogdan, R., & Hariri, A. R. (2012). Neural embedding of stress reactivity. *Nature Neuroscience, 15*(12), 1605–1607.

Bunge, M. (1977). Levels and reduction. *American Journal of Physiology. Regulatory, Integrative and Comparative Physiology, 233*(3), R75–R82.

Campbell, H. (2013). NIMH delivers a kill shot to DSM-5. *Science* 2.0. http://www.science20.com/science_20/blog/nimh_delivers_kill_shot_dsm5-111138. Retrieved 3 May 2013.

Caspi, A., & Moffitt, T. E. (2006). Gene–environment interactions in psychiatry: Joining forces with neuroscience. *Nature Reviews Neuroscience, 7*(7), 583–590.

Champagne, F. A. (2010). Early adversity and developmental outcomes interaction between genetics, epigenetics, and social experiences across the life span. *Perspectives on Psychological Science, 5*(5), 564–574.

Clark, D. A., & Beck, A. T. (2010). Cognitive theory and therapy of anxiety and depression: Convergence with neurobiological findings. *Trends in Cognitive Sciences, 14*(9), 418–424.

Cuthbert, B. N. (2014). The RDoC framework: Facilitating transition from ICD/DSM to dimensional approaches that integrate neuroscience and psychopathology. *World Psychiatry, 13*(1), 28–35.

Cuthbert, B. N., & Insel, T. R. (2012). New diagnosis in diagnostic research domain criteria and global mental health. In E. Sorel (Ed.), *21 st century global mental health* (pp. 55–70). Jones & Bartlett Publishers.

Cuthbert, B. N., & Insel, T. R. (2013). Toward the future of psychiatric diagnosis: The seven pillars of RDoC. *BMC Medicine, 11*(1), 126–133.

De Vreese, L., Weber, E., & Van Bouwel, J. (2010). Explanatory pluralism in the medical sciences: Theory and practice. *Theoretical Medicine and Bioethics, 31*(5), 371–390.

Engel, G. L. (1977). The need for a new medical model: A challenge for biomedicine. *Science, 196*(4286), 129–136.

Faucher, L. (2012). Unity of science and pluralism: Cognitive neurosciences of racial prejudice as a case study. In O. Pombo, J. M. Torres, J. Symons, & S. Rahman (Eds.), *Special sciences and the unity of science* (pp. 177–204). Netherlands: Springer.

Fava, G. (2014). Road to nowhere. *World Psychiatry, 13*(1), 49–50.

Feighner, J. P., Robins, E., Guze, S. B., Woodruff, R. A., Winokur, G., & Munoz, R. (1972). Diagnostic criteria for use in psychiatric research. *Archives of General Psychiatry, 26*(1), 57–63.

Fortier, M. (2013). Loteries vidéo : Loto-Québec cible les pauvres. *Le Devoir* the 21 of November. http://www.ledevoir.com/politique/quebec/393288/loteries-video-loto-quebec-cible-les-pauvres. Retrieved 2 Mar 2014.

Gazzaniga, M. S. (2009). *The cognitive neurosciences*. Cambridge, MA: The MIT Press.

Gold, I. (2009). Reduction in psychiatry. *Canadian Journal of Psychiatry/Revue canadienne de psychiatrie, 54*(8), 506–512.

Griffiths, P., & Stotz, K. (2013). *Genetics and philosophy an introduction*. Cambridge: Cambridge University Press.

Harris, H. W., & Schaffner, K. F. (1992). Molecular genetics, reductionism, and disease concepts in psychiatry. *Journal of Medicine and Philosophy, 17*(2), 127–153.

Hellstrom, I. C., Dhir, S. K., Diorio, J. C., & Meaney, M. J. (2012). Maternal licking regulates hippocampal glucocorticoid receptor transcription through a thyroid hormone–serotonin–NGFI-a signalling cascade. *Philosophical Transactions of the Royal Society, B: Biological Sciences, 367*(1601), 2495–2510.

Hickie, I. B., Scott, J., Hermens, D. F., Scott, E. M., Naismith, S. L., Guastella, A. J., & McGorry, P. D. (2013). Clinical classification in mental health at the cross-roads: Which direction next? *BMC Medicine, 11*(1), 1–14.

Horgan, J. (2013).Psychiatry in crisis! Mental health director rejects psychiatric 'Bible' and replaces with... nothing|cross-check, Scientific American Blog Network. http://blogs.scientificamerican.com/cross-check/2013/05/04/psychiatry-in-crisis-mental-health-director-rejects-psychiatric-bible-and-replaces-with-nothing/. Accessed 28 Feb 2014.

Hyman, S. E. (2010). The diagnosis of mental disorders: The problem of reification. *Annual Review of Clinical Psychology, 6*, 155–179.

Hyman, S. E. (2011). Diagnosis of mental disorders in light of modern genetics. In D. A. Regier (Ed.), *The conceptual evolution of DSM-5* (pp. 3–18). Washington, DC: American Psychiatric Pub.

Hyman, S. E., & Fenton, W. S. (2003). What are the right targets for psychopharmacology. *Science, 299*(5605), 358–359.

Insel, T. R. (2010). Faulty circuits. *Scientific American, 302*(4), 44–51.

Insel T. R. (2013). Director's blog: Transforming diagnosis. *NIMH website*. http://www.nimh.nih.gov/about/director/2013/transforming-diagnosis.shtml. Retrieved 27 Oct 2013.

Insel, T. R. & Lieberman, J. A. (2013). «DSM-5 and RDoC: Shared Interests». *NIMH website*. http://www.nimh.nih.gov/news/science-news/2013/dsm-5-and-rdoc-shared-interests.shtml. Retrieved 27 Oct 2013.

Insel, T., Cuthbert, B., Garvey, M., Heinssen, R., Pine, D. S., Quinn, K., & Wang, P. (2010). Research Domain Criteria (RDoC): Toward a new classification framework for research on mental disorders. *American Journal of Psychiatry, 167*(7), 748–751.

Insel, T. R., Sahakian, B. J., Voon, V., Nye, J., Brown, V. J., Altevogt, B. M., & Williams, J. H. (2012). Drug research: A plan for mental illness. *Nature, 483*(7389), 269.

Kandel, E. R., Schwartz, J. R., & Jessell, T. (Eds.). (2000). *Principles of neural science* (4th ed.). New York: McGraw-Hill.

Kapur, S., Phillips, A. G., & Insel, T. R. (2012). Why has it taken so long for biological psychiatry to develop clinical tests and what to do about it? *Molecular Psychiatry, 17*(12), 1174–1179.

Kendell, R., & Jablensky, A. (2003). Distinguishing between the validity and utility of psychiatric diagnoses. *American Journal of Psychiatry, 160*(1), 4–12.

Kendler, K. S. (2005). Toward a philosophical structure for psychiatry. *American Journal of Psychiatry, 162*(3), 433–440.

Kendler, K. S. (2012a). Levels of explanation in psychiatric and substance use disorders: Implications for the development of an etiologically based nosology. *Molecular Psychiatry, 17*(1), 11–21.

Kendler, K. S. (2012b). The dappled nature of causes of psychiatric illness: Replacing the organic–functional/hardware–software dichotomy with empirically based pluralism. *Molecular Psychiatry, 17*(4), 377–388.

Kendler, K. S., Prescott, C. A., Myers, J., & Neale, M. C. (2003). The structure of genetic and environmental risk factors for common psychiatric and substance use disorders in men and women. *Archives of General Psychiatry, 60*(9), 929–937.

Kendler, K. S., Gardner, C. O., Gatz, M., & Pedersen, N. L. (2007a). The sources of co-morbidity between major depression and generalized anxiety disorder in a Swedish national twin sample. *Psychological Medicine, 37*(3), 453.

Kendler, K. S., Myers, J., & Prescott, C. A. (2007b). Specificity of genetic and environmental risk factors for symptoms of cannabis, cocaine, alcohol, caffeine, and nicotine dependence. *Archives of General Psychiatry, 64*(11), 1313–1320.

Kotchen, T. A. (2011). Historical trends and milestones in hypertension research a model of the process of translational research. *Hypertension, 58*(4), 522–538.

Koven, S. (2013). A Q&A with Dr. Thomas Insel, Director of the National Institute of Mental Health–The Boston Globe. *BostonGlobe.com*. http://www.bostonglobe.com/lifestyle/health-wellness/2013/12/16/with-thomas-insel-director-national-institute-mental-health/nZ9Qjw88QmJtXWNQft1CFN/story.html. Accessed 28 Feb 2014.

Krishnan, V., & Nestler, E. J. (2008). The molecular neurobiology of depression. *Nature, 455*(7215), 894–902.

LeDoux, J. (2012). Rethinking the emotional brain. *Neuron, 73*(4), 653–676.

Lee, Y. S., & Silva, A. J. (2011). Modeling hyperactivity: Of mice and men. *Nature Medicine, 17*(5), 541–542.

Lewontin (1992, May 28). The dream of the human genome. *New York Review of Books* (pp. 31–39).

Meehl, P. E. (1995). Bootstraps taxometrics: Solving the classification problem in psychopathology. *American Psychologist, 50*(4), 266–275.

Miller, G. (2010). Beyond DSM: Seeking a brain-based classification of mental illness. *Science, 327*(5972), 1437–1438.

Mitchell, S. D. (2002). Integrative pluralism. *Biology and Philosophy, 17*(1), 55–70.

Mitchell, S. D. (2003). *Biological complexity and integrative pluralism*. Cambridge: Cambridge University Press.

Morris, S. E., & Cuthbert, B. N. (2012). Research domain criteria: Cognitive systems, neural circuits, and dimensions of behavior. *Dialogues in Clinical Neuroscience, 14*(1), 29–35.

Morris, S. E., Rumsey, J. M., & Cuthbert, B. N. (2014). Rethinking mental disorders: The role of learning and brain plasticity. *Restorative Neurology and Neuroscience, 32*(1), 5–33.

Nagel, E. (1961). *The structure of science. Problems in the logic of explanation*. New York: Harcourt, Brace & World.

NIMH. (2008). The national institute of mental health strategic plan. *NIMH website*. http://www.nimh.nih.gov/about/strategic-planning-reports/index.shtml. Retrieved 27 Oct 2013.

NIMH. (2011a). NIMH Research Domain Criteria (RDoC). *NIMH website*. http://www.nimh.nih.gov/research-priorities/rdoc/nimh-research-domain-criteria-rdoc.shtml. Retrieved 5 Apr 2014.

NIMH. (2011b). RDoC constructs. *NIMH website*. http://www.nimh.nih.gov/research-priorities/rdoc/rdoc-constructs.shtml#acute_threat. Retrieved 27 Oct 2013.

Obiols, J. E. (2012). DSM 5: Precedents, present and prospects. *International Journal of Clinical Health & Psychology, 12*(2), 281–290.

Parnas, J. (2014). The RDoC program: Psychiatry without psyche? *World Psychiatry, 13*(1), 46–47.

Poland, J., Von Eckardt, B., & Spaulding, W. (1994). Problems with the DSM approach to classifying psychopathology. In G. Graham & G. L. Stephens (Eds.), *Philosophical psychopathology* (pp. 235–260). Cambridge, MA: MIT Press.

Ressler, K. J., & Mayberg, H. S. (2007). Targeting abnormal neural circuits in mood and anxiety disorders: From the laboratory to the clinic. *Nature Neuroscience, 10*(9), 1116–1124.

Robins, E., & Guze, S. B. (1970). Establishment of diagnostic validity in psychiatric illness: Its application to schizophrenia. *American Journal of Psychiatry, 126*(7), 983–987.

Rosenberg, A. (2011). *The atheist's guide to reality: Enjoying life without illusions*. New York: W.W. Norton.

Rutten, B. P., & Mill, J. (2009). Epigenetic mediation of environmental influences in major psychotic disorders. *Schizophrenia Bulletin, 35*(6), 1045–1056.

Sanislow, C. A., Pine, D. S., Quinn, K. J., Kozak, M. J., Garvey, M. A., Heinssen, R. K., & Cuthbert, B. N. (2010). Developing constructs for psychopathology research: Research domain criteria. *Journal of Abnormal Psychology, 119*(4), 631–639.

Sartorius, N. (2014). The only one or one of many? A comment on the RDoC project. *World Psychiatry, 13*(1), 50–51.

Schaffner, K. (2002). Clinical and etiological psychiatric diagnoses: Do causes count? In J. Sadler (Ed.), *Descriptions and prescriptions: Values, mental disorders, and the DSMs* (pp. 271–290). Baltimore: Johns Hopkins University Press.

Schaffner, K. (2006). Reduction: The Cheshire cat problem and a return to roots. *Synthese, 151*(3), 377–402.

Schaffner, K. (2008). Etiological models in psychiatry: Reductive and nonreductive. In K. Kendler & J. Parnas (Eds.), *Philosophical issues in psychiatry: Natural kinds, mental taxonomy and causation* (pp. 48–90). Baltimore: Johns Hopkins University Press.

Spitzer, R. L., Endicott, J., & Robins, E. (1975). Clinical criteria for psychiatric diagnosis and DSM-III. *The American Journal of Psychiatry, 132*(11), 1187–1192.

Stein, D. J. (2014). An integrative approach to psychiatric diagnosis and research. *World Psychiatry, 13*(1), 51–53.

Van Bouwel, J., Weber, E., & De Vreese, L. (2011). Indispensability arguments in favour of reductive explanations. *Journal for General Philosophy of Science, 42*(1), 33–46.

Van Praag, H. M. (2008). Kraepelin, biological psychiatry, and beyond. *European Archive of Psychiatry and Clinical Neuroscience, 258*(suppl. 2), 29–32.

Wagner, R. (2005). Problems with early and very early intervention in psychosis. *British Journal of Psychiatry, 187*, s104–s107.

Wardenaar, K. J., & de Jonge, P. (2013). Diagnostic heterogeneity in psychiatry: Towards an empirical solution. *BMC Medicine, 11*(1), 201.

Wilkinson, R. G., & Pickett, K. E. (2009). *The spirit level*. London: Penguin.

Woodward, J. (2008). Cause and explanation in psychiatry: An interventionist perspective. In K. Kendler & J. Parnas (Eds.), *Philosophical issues in psychiatry: Explanation, phenomenology, and nosology* (pp. 132–184). Baltimore: Johns Hopkins University Press.

# DSM-5 and the Reconceptualization of Obsessive-Compulsive Disorder

## An Anthropological Perspective from the Neuroscience Laboratory

**Baptiste Moutaud**

> *"How are the clinician and researcher to understand this enigmatic and often disabling condition? Is OCD a thought disorder? Is it a disturbance of repetitive behavior? Is it a problem with impulse control? Is it one of anxiety? How one answers these questions has major implications for how they conceptualize OCD, how they study the disorder, and how they evaluate and treat affected individuals."*
>
> (Abramowitz and Deacon 2005a, p. 119)

**Abstract** In the DSM-5, obsessive-compulsive disorder (OCD) has been removed from the category of "Anxiety Disorders" and classified into a new separate diagnostic category of "Obsessive-Compulsive and Related Disorders." According to the diagnostic criteria, anxiety, which was the central emotional component of the OCD definition in the DSM-IV and that articulated the relation between obsessions and compulsions, will no longer have to be systematically identified in OCD patients at all. The importance of this transformation is visible in the debates that took place about the possibility of introducing a new category in the DSM-5, "Obsessive-Compulsive Spectrum Disorders." Although it was ultimately rejected, this category was supposed to include OCD as well as Tourette syndrome, stereotypic movement disorders, or impulse control disorders. The anxious component that defined OCD as an anxiety disorder in the DSM-IV would have been used only to distinguish OCD from other diseases within the spectrum.

This chapter will clarify how this transformation is characterized conceptually by the shift in the criteria of OCD from affect to behavior, and institutionally by the entrance in the DSM of a diagnostic category that originates mostly from a research,

B. Moutaud (✉)
Centre de recherche, médecine, sciences, santé, santé mentale, société (CERMES3), INSERM U988, CNRS UMR 8211, École des hautes études en sciences sociales, Université Paris Descartes, 45, rue des Saints-Pères, cedex 06, 75270 Paris, France
e-mail: moutaud.baptiste@sfr.fr



S. Demazeux, P. Singy (eds.), *The DSM-5 in Perspective*, History, Philosophy and Theory of the Life Sciences 10, DOI 10.1007/978-94-017-9765-8_13

rather than a clinical, setting. My analysis will be based on an ethnographic research of a French neuroscience team that developed experimental therapeutics for OCD.

## Introduction

Among the controversies that have surrounded the development of the DSM-5 and the changes in classification that it validated, the fate of obsessive-compulsive disorder (OCD) went relatively unnoticed. While in the DSM-IV-TR OCD it was included in the category of "Anxiety Disorders," in the DSM-5 it was moved to the "Obsessive-Compulsive and Related Disorders" (OCRD). This may not be the most spectacular shift, nor was it the one most discussed in the media. Yet it raises questions because, as the work group in charge of revising its diagnostic criteria reminds us, the previous classification of the disorder had been working just fine since 1994, both in clinical and research settings (Leckman et al. 2010, p. 510). In addition, OCD is, clinically, a relatively stable and well-delimited class: it has not been the subject of any major controversy regarding its definition or treatment, nor has it been a subject of dispute among the various mental-health professionals, patients' organizations, or users' movements. This makes the reasons for this shift and the epistemological and anthropological stakes involved all the more intriguing.

Despite the lack of transparency in the content of the discussions that led to the development of the new DSM, this chapter will focus on the rationale that has presided over the emergence of OCRD. I will argue that what is at stake is the continuity and connections between mental and motor functions, as well as the biological substratum of these functions. This substratum is today a major line of research in neuroscience, in which OCD appears as a key entry point. For the past century, OCD has undergone—and is still undergoing—all the twists and turns of the history of the ideas, treatments, and practices that have stirred psychiatry. It has been given every possible definition: formerly a classic Freudian neurosis (Castel 2008), it then became an anxiety disorder, responded to psychopharmacology, and in the 1960s was taken up by cognitivists and behaviorists (Healy 1997, 2002) while becoming one of the latest indications for psychosurgery.[1] Since the turn of the 1990s, it has been, along with schizophrenia, one of the psychiatric disorders for which biologization has most progressed and it continues to feed, preferentially, therapeutic innovations (Moutaud 2009). This unique position of OCD in psychopathology is in my opinion due to the interest of researchers and clinicians who hope to understand better complex phenomena such as doubt, intentionality, or anxiety, and above all, who want to explore the origin of human behavior and the relations between motion and emotion, between thought and action. This is why, although the removal of OCD from anxiety disorders is significant in the general DSM-5 project (and its relative failure), I will deal more broadly with the question

[1] For a detailed conceptual history of OCD, see Castel (2012).

of the relation between psychiatry and neuroscience, and with the reconfiguration of psychiatric morbid entities and knowledge under the neuroscience regime of knowledge and practices. It is in the contemporary reconfigurations of OCD that we can observe the profound transformations imposed by neuroscience not solely in the ways of knowing human beings but also how to modify their thoughts and behaviors.

First and foremost, I will look at the history of the OCRD category in order to show that it is only a truncated emanation of the initial project to establish a broader category of "Obsessive-Compulsive Spectrum Disorders." OCRD is the result of a consensus reached during the development of the DSM-5 and it highlights the failure of the new edition to establish a dimensional classification based on neuroscience data (Demazeux 2008). I will show that OCRD is a research category, i.e., a category derived from research and designed to feed other research on the mechanisms of OCD and the clinical and biological interaction of its behavioral and emotional components.

Second, I will take a detour through the experimentation performed by a neuroscience team to justify a novel neurosurgical treatment of OCD. My aim is to shed light on the anthropological stakes of the nosological transformation of OCD, which is revealing of the recent transformations of psychopathology produced by neuroscience. I will show how the new classification partially justifies the neurologization of the disorder by tipping the identification criteria of the core symptoms of OCD from affect to behavior, from mental to motor. In this movement, OCD will no longer be considered an anxiety disorder but will become an urge-control disorder.

## From Morality to Absurdity: OCD in the DSM-IV

In DSM-IV and DSM-IV-TR, OCD was classified under the category of anxiety disorders. It combined obsessions, intrusive anxiety-producing thoughts (e.g., religious, aggressive or incestuous thoughts, or thoughts of fear of contamination), and compulsions (repetitive behaviors that people feel compelled to engage in order to reduce the anxiety caused by obsessions like washing one's hands several times in a row or repeating a series of numbers or phrases before performing an action). For this condition to be classified as pathological, the obsessions and compulsions had to have a negative impact on the person's life and/or that of their entourage.

The origin of this construction of OCD can be traced back to the deep conceptual transformations in the disorder starting in the 1960s (Moutaud 2009, pp. 229–247; Castel 2008, 2012). Under the concomitant prompting of biological theories of psychopharmacology and, above all, of the cognitive and behavioral sciences, the dismemberment of the old Freudian obsessive neurosis was accomplished. More specifically, the cognitive and behavioral sciences emphasized two basic points.

First, obsessions are normal thoughts that anyone can experience on a regular basis (Rachman and da Silva 1978). The fear of driving over someone, imagining pushing someone onto the subway tracks, having incestuous thoughts or homosexual desires, fearing contamination from microbes, etc., are thoughts that can come to anyone. A "normal" person, however, pushes them aside and finds them absurd, inappropriate, immoral, contrary to his or her values, or irrational. What is pathological is therefore the invasiveness of these thoughts due to a failure in their cognitive interpretation. A person suffering from obsessive thoughts attributes too much attention to them and cannot get rid of them.

Second, compulsions are also "normal" motor behaviors, be they innate or acquired. We all check, wash, collect, and count. This would explain why the same compulsive symptoms are found all over the world. What is characteristic of OCD is the patient's inability to control the pathological repetition of such behaviors to the point that they no longer make any sense.

By postulating the normality, or even the "banality," of obsessional thoughts and compulsive behaviors (Bouvard 2006, pp. 15, 39), cognitive-behavioral theories invalidated the moral and personal significance of the symptoms, which became the consequences of cognitive and behavioral processes running out of control, the mechanical effects of conditioning and habituation and/or of a neurochemical dysfunction. No longer was it the patient who was producing them; they came from a learned or innate behavioral and cognitive repertory that had to be restructured (this is the basis of cognitive behavioral therapy).

Although responsibility for the acts as well as guilt about their repercussions or shame of the thoughts were still ubiquitous in the patients' discourse and were partly the cause of their suffering (Moutaud 2011a), this was alleviated by these theories, for which the overestimation of morality or responsibility is a cognitive bias (Harrison et al. 2012). The suffering shifted away from the moral, or guilt-producing and responsibility-inducing, content of the symptoms (this is what made them "obsessions" and "compulsions"), to the awareness of the absurdity of the thoughts and behaviors, and of their impact on the patient's life (this is what made OCD "pathological"). In other words, OCD lost its semantic content. This is a shift in the content and value of obsessive and compulsive symptoms, which I qualify as a shift from morality to absurdity (Moutaud 2011a).

I will show that in the DSM-5 we are witnessing an attempt to transform OCD once again, this time from affect to behavior, from emotion to motor control. This new conceptualization is the recent, "neuroscience" version of the earlier cognitive-behavioral concept. It implies a new hierarchy between the disorder's emotional and behavioral attributes.

## "Keep It Narrow": A Truncated and Consensual Category

The DSM-5 did not change much of the diagnostic criteria for OCD. The fundamental transformation is the removal of OCD from the category of "Anxiety Disorders" and its inclusion under a new category, OCRD, which developed around it. The rationale for this change was that a number of people suffering from OCD feel no anxiety. The new OCRD category includes disorders that share a clinical proximity with OCD insofar as they associate compulsive behaviors with a mental state, or secondary obsessive-compulsive symptoms with another disorder, or "substance-/medication-induced" obsessive-compulsive symptoms.[2]

The OCRD category is a compromise between (1) keeping OCD as an anxiety disorder and (2) creating a broader category, "Obsessive-Compulsive Spectrum Disorders" (OC spectrum). If we look at the thread of discussions that took place since the mid-2000s about the classification of OCD in the DSM-5, we see that a main point of discussion was the possible creation of an OC spectrum. This is attested to by the name of the work group in charge of this category ("DSM-V Anxiety, Obsessive-Compulsive Spectrum, Post-Traumatic, and Dissociative Disorders Work Group"),[3] by the title of the 2006 conference organized prior to discussions among the experts in the field ("Obsessive-Compulsive Spectrum Disorders Conference"),[4] and by various publications on the topic (e.g., Hollander et al. 2011; Phillips et al. 2010; Regier 2007). It seems, however, that a consensus on the notion of an OC spectrum was never reached. For instance, to the question of where to classify OCD, the surveyed professionals replied that OCD should be taken out of the anxiety disorders because "obsessions and compulsions, rather than anxiety, are the fundamental features of the disorder," but that it should nevertheless be kept in a "narrower" category than this OC spectrum (Mataix-Cols et al. 2007, p. 1313). In the end, the experts commissioned by the work group suggested a middle-of-the-road solution: to keep OCD under anxiety disorders but also to include it in a broad category of "Anxiety and Obsessive-Compulsive-Spectrum Disorders" (Stein et al. 2010). But this suggestion was rejected. This vacillation demonstrates that the OCRD category is a truncated outcome (a "narrower" version) of the very notion of OC spectrum. In order to clarify what OCD is in DSM-5, I will describe the OC spectrum and its basis.

## A Spectrum to Consider the Continuity from Motor to Mental

Eric Hollander, who today remains its main advocate, has developed the association of OCD with a spectrum of related disorders since the 1990s.[5] "Obsessive-Compulsive Spectrum Disorders" would include OCD as well as Tourette Syndrome

[2]The disorders which associate a mental state with compulsive behaviors are Body Dysmorphic Disorder, Trichotillomania (hair-pulling disorder), and two new categories: Hoarding Disorder and Excoriation (skin-picking) Disorder.

[3]http://www.dsm5.org/progressreports/pages/0904reportofthedsm-vanxiety,obsessive-compulsive-spectrum,posttraumatic,anddissociativedisordersworkgroup.aspx. Accessed 13 June 2013.

[4]http://www.dsm5.org/research/pages/obsessivecompulsivespectrumdisordersconference%28june20-22,2006%29.aspx. Accessed 13 June 2013.

[5]It seems that Michael Jenike was the first author to introduce the idea of a "spectrum" concerning OCD (or rather concerning OCD "related disorders": 1989, 1990). Hollander then appropriated the concept (Hollander and Stein 1993) and produced countless publications on the topic. The journal *CNS Spectrum*, of which he is editor-in-chief, provided him with a particularly timely forum during the development of the DSM-5. Researchers such as R. K. Pitman or J. Leckman

(TS) and all disorders featuring impulsive, compulsive, stereotyped symptoms, as well as all disorders related to behavior control. It is based on a variety of clinical and etiological organic criteria: frequent comorbidity between OCD and the spectrum disorders; similarities and overlap of clinical features; similarities in the age when they begin and in their clinical evolution, in genetic factors or family histories, and in the disrupted neurobiological circuits; and responses to the same chemical treatments. These criteria are not exclusive, but inclusive, which has the effect of creating fuzzy boundaries between the disorders of the spectrum. If we follow Hollander et al. (2005), according to these criteria three groups would cut across OCD: (1) "impulse control disorders," which include compulsive and impulsive elements as well as repetitive behaviors accompanied by subjective elements when they occur, such as pleasure, disgust, guilt, etc. (sexual compulsions, trichotillomania, compulsive gambling, kleptomania, self-injurious behaviors, etc.); (2) "Body Image, Body Sensitization, and Body Weight Concern Disorders," characterized by obsessive thoughts and concerns about the body and compulsive behavior (eating disorders, hypochondriasis, body dysmorphic disorder, etc.); and (3) neurological disorders that have obsessive and compulsive features or cause stereotyped behaviors, such as autism or Sydenham's chorea, but mostly TS, which is the other central disease of this spectrum. The list varies depending on the predominant criterion or criteria and can integrate addictions, neurodegenerative diseases (Parkinson's, Huntington's, Alzheimer's), borderline-personality disorder, autism, schizophrenia, etc.

This dimensional approach, which relies on clinical description, prolongs the dismemberment of OCD begun by the cognitive-behavioral theories. It adds, to the normal/pathological continuum of obsessions and compulsions, an impulsive-compulsive continuum along which disorders are distributed, depending on the authors, according to dimensions that blur out category-based partitioning, as for instance: symptoms predominantly expressed as "coercive" versus "disinhibitionary," or as "risk-aversive" versus "impulsive," or even as "sensorimotor" versus "cognitive-affective." One of the consequences of this approach would be to devise OCD as a heterogeneous category, leading to its fragmentation into symptomatic subcategories spread all along the OC spectrum with distinct responses to treatments and possibly even different etiological causes. The topic of the unitary nature of OCD was fiercely discussed in the run-up to the DSM-5 and was partially validated with the removal of hoarding disorder and its entry as a specific diagnosis in the OCRD category (Mataix-Cols et al. 2005).

Several researchers have criticized the clinical and scientific validity of this OC spectrum and its coherence. Led by Jonathan Abramovitz, these psychiatrists have first and foremost emphasized the weakness, variability, and superficiality of the

---

also worked on this approximation through their research on the links between OCD and Tourette Syndrome (Leckman et al. 1994/1995; Pitman et al. 1987) (see Moutaud 2009, p. 295).

arguments provided and insisted on the complete pointlessness of this spectrum for treating and understanding these disorders and their links (Starcevic and Janca 2011). The group would have no identified common pathophysiological and genetic basis. Each side then mobilized the research that substantiated its arguments (for an overview, see the exchange of comments and responses between the two sides in Abramowitz and Houts 2005). It then seems more fruitful to clarify the theoretical stakes involved in this opposition.

According to Abramowitz (Abramowitz and Houts 2002; Abramowitz and Deacon 2005a, b), the notion of OC spectrum would be based on a weak perception of OCD focused on a dimensional approach that overestimates symptomatic proximity and co-occurrence (a notion otherwise central to the dimensional approach). Instead, he advocates a clinical approach that emphasizes the structure of the disorder and the function of its various components. He underscores the centrality of anxiety, which is the pivot articulating obsessions and compulsions: the patient performs the compulsions in order to calm the anxiety caused by the obsessions. This articulation between mental state and behavior would be common to all the different anxiety disorders (e.g., social phobia), the ultimate argument being that the goal of cognitive behavioral therapy is precisely to restructure this connection and that all the disorders in this category respond favorably to this type of therapy.

The OC spectrum category instead sets up a new hierarchy in the clinical components of OCD. Within the OC spectrum, the anxious component, and more broadly any emotional or subjective component that underpins the behavior, serves to distinguish OCD from other disorders, while in DSM-IV it was used for the purpose of identification and inclusion in the category of "Anxiety Disorders." The emotional state, anxiety, is no longer the key feature of OCD. The patients do not have to systematically express this feeling. This is a major change in the conceptualization of this disorder, tipping the identification criteria of the core symptoms of OCD from affect to motor control, from its emotional component to the visible, behavioral component.[6] OCD becomes an "urge"-control disorder (as DSM-5 now defines obsessions they are like the premonitory sensations that precede tics in TS), a disorder of autonomy of action colored with secondary emotional or subjective components.

Here we see, then, how the question of emotional causality, the action-to-emotion relationship, becomes the core of the debate about the OC spectrum and its epistemological structure. As its partisans explain, the notion of "Obsessive-Compulsive Spectrum Disorders" and its derivatives appears as a "research category": it is a malleable category, in its boundaries and contents, created mainly to study the

[6]Interestingly, the category was temporarily renamed "Obsessive-Compulsive and *Movement-Related* Disorders" (my emphasis) in its description on the DSM-5 Web site. The page was deleted following publication of the DSM-5.

relationships between mental and motor functions and their production mechanisms, not for diagnosis and care.[7] I will examine these points by exploring the translation of this OC spectrum and the idea of OCD in clinical practice and research.

## Can Monkeys Be Anxious? How to Put OCD into a Brain

Between 2004 and 2008 I conducted an ethnographic study of a French team of neuroscientists and clinicians who have been developing the application of an experimental neurosurgical technology, Deep Brain Stimulation (DBS)[8] for treatment of OCD and TS. I have described elsewhere the experimental and intellectual path that led this team to this field of research (Moutaud 2008, 2011b). In this chapter I will briefly consider how the French team scientifically legitimizes the practice of putting electrodes deep inside the brains of OCD patients. Drawing on the example of one of their fundamental research findings, I will argue that the rationale supporting this controversial therapeutic technique is underpinned by a process of reconceptualizing the disorder that in turn support the operationalization, in research and clinical settings, of the notion of OC spectrum. I will then demonstrate through this ethnographic vignette how the reconceptualization of OCD opens up, in practice, a field of neuropsychiatric research for neuroscientists and allow them to explore the links between human emotion and action and their biological bases.

The team (made up of neurologists, psychiatrists, neurosurgeons, neuroanatomists, neuropsychologists, etc.) was already using this neuromodulation technology with Parkinson's disease (PD) since 1996. Having found clinical behavioral and emotional side effects among the stimulated patients (such as hypomania, pathological gambling, depression, or hypersexuality) as well as unexpected improvement in comorbid OCD symptoms for two patients diagnosed with Parkinson's disease, the team speculated at the turn of the millennium on the

[7]This is what Pallanti and Hollander seem to acknowledge when they use, fairly obscurely, the term "scientific metaphor" to qualify the prospects opened up by this spectrum (and its weaknesses) (2008). In addition, one of the direct consequences of the entry of the OCRD category in the DSM was the launch in 2012 of a journal devoted to it called *Journal of Obsessive-Compulsive and Related Disorders*, of which Abramowitz is the founding editor-in-chief. The journal observes, in its first editorial, that the first effect of the new classification will be to "expand the scope of what is considered an 'obsessive-compulsive disorder' and fuel increased funding and research on OCD and these putatively related conditions" (Abramowitz 2012, p. iii).

[8]This invasive neurosurgical technology involves the insertion of two small electrodes inside the patient's brain linked to a pulse generator placed in the chest, allowing physicians to modulate its activity through ongoing electrical stimulation. Invented in 1986 in France for the symptomatic treatment of Parkinson's disease and essential tremor, DBS was rapidly considered at the end of the 1990s as a potential experimental treatment for severe forms of various neurological and psychiatric conditions (such as dystonia, OCD, depression and TS). DBS quickly became the subject of major social, scientific and ethical concerns, raising hope and fears about its increasingly widespread use.

possibility of modulating emotions and behaviors and treating psychiatric disorders with this innovative technology. Their premise, which was intended both to explain the empirical effects observed in patients with PD and to justify its application for OCD and TS, was that the brain structures that they stimulated, the basal ganglia (BG), are involved in controlling emotions and behavior, and are therefore also likely to be involved in the pathophysiology of OCD and TS.[9]

To demonstrate this, the team developed an animal model of compulsive and impulsive symptoms. The idea was to reproduce a symptomatology in monkeys, most similar to that of TS and OCD, and then to demonstrate the effectiveness of DBS by eliminating the clinical signs thanks to stimulation.

The model was designed as follows: by injecting a substance into the BG, the researchers aimed at artificially causing stereotyped behaviors and movements in the monkeys, such as "stereotypies" (the monkeys appeared to bite and lick their fingers), a "hyperactive" state accompanied by an apparently compulsive behavior (the monkeys were agitated, licked and touched the elements in front of them, and repeated grooming acts), or hemiballismus (abnormal movement of their arms). In a second step, they stimulated the same brain structure but a few millimeters away and observed the disappearance of these behaviors. Setting aside debates about the scientific validity of animal models of behavior, this experiment allowed the team to constitute what were close to animal models of TS and OCD.[10] They then demonstrate the effectiveness of DBS in making these symptoms disappear, which then came up as a prospect for experimental therapeutics in humans.

But the results also raised a question: why were the researchers getting different types of effects depending on stimulation sites that were only a few millimeters apart within the same structure? This question had occurred already with regard to humans when the researchers had reproduced hypomanic symptoms in two stimulated PD patients: depending on the stimulation site within the same BG structure, they had perceived a variety of clinical effects combinations. For example, upon the stimulation of one site a patient's motor symptoms were improved while at the same time he experienced behavioral and cognitive side effects. Interestingly, when the same site was stimulated, a different patient's cognitive abilities were not altered even though she was also experiencing a hypomanic state.

---

[9]This theory had long been supported by various experimental and clinical data, particularly neuroimaging data or that indicating the high comorbidity of TS and Parkinson's disease with OCD (Moutaud 2008).

[10]For an overview of the scientific debate about animal models in neuroscience, see: Rose and Abi-Rached 2013, chapter 3. From an anthropological perspective, as Schlich and colleagues put it ( 2009, pp. 330–331), animal experiments in medicine are a "manifestation of pure form" of the naturalistic ontology described by Descola (2005): if non-human animals and humans differ in interiority, they have a shared physicality which make animals pertinent as a human body substitute in order to experimentally manipulate its biological functions (Bynum 1990; Löwy 2000). But if the denial of human-like interiority to animals would give them the lower ethical status that allows for these experiments, it also becomes the core of the epistemological controversy concerning their use as human behavioral models in psychiatry.

The team had just demonstrated experimentally that the BG process, within a very small volume, not only motor information (which is what they were known for), but also emotional and cognitive information. More precisely, the functional territories seemed not to be segregated (with each structure anatomically divided into a motor area, a limbic area, and an associative area). Instead, different functional networks appear to converge, explaining why, depending on which area is stimulated, researchers were getting a range and combination of side effects that were more elaborate or complex: purely motor phenomena (abnormal movement such as hemiballismus among monkeys), but also cognitive, behavioral, and emotional symptoms (such as the hypomania caused by DBS in Parkinsonian patients or the monkeys' stereotypies).

The team then developed a functional model of the BG that supported a theory of the genesis and regulation of human behaviors. Roughly, they argued that the BG are a set of structures known to interact with the cortex via a system of functional loops transmitting information and initiating various brain functions through successive processes of activation or inhibition of nerve activity. The assumption was that when certain BG structures are hyperactive, they slow down the cortical programs, resulting in slowness of action or choice, or even blockage. When they are, on the contrary hypoactive, this results in impulsive behavior in individuals.

The model offered a new reading of the pathophysiological mechanisms of all the disorders involving motor, impulsive, compulsive, and stereotypical components, and it also explained all the therapeutic and side effects obtained by applying DBS. Thus, the same hyperactivity or hypoactivity mechanisms are involved in explaining neurological motor signs such as akinesia (hyperactivity inhibited by DBS) or hemiballismus and stereotypies (hypoactivity caused by DBS), as they are in behavioral or emotional symptoms such as hypomania (consequences of hypoactivity leading to impulsivity) and in OCD symptoms. In the latter case, the symptomatology would correspond to a hyperactive state of the BG causing inability to select an appropriate action. This indecision would result in endless repetition of the same behavior. Thanks to the use of DBS, OCD is therefore no longer understood as an isolated phenomenon brought about by fortuitous clinical observations made on Parkinsonian patients. It fits into a broader research program within which it becomes a behavior-control disorder caused by a dysfunction in the regulation loops within the BG.

The first immediate effect of this conception was to move OCD symptomatology right to the center of a nosographic spectrum of impulse-control disorders and compulsions mixing motor, emotional, and cognitive symptoms. These include hypomania and gambling addictions as well as stereotypies, TS-related tics, or eating disorders that, in the same movement, all become potential indications for DBS. In other words, DBS, due to its ability to regulate BG functioning, becomes a treatment for the disorders of this spectrum, with the clinical specificity of the disorders and the patients reappearing in the forms of care (Moutaud 2012).

The DBS technology has therefore permitted clinicians to operationalize the concept of "Obsessive-Compulsive Spectrum Disorders" by offering it some pathophysiological and therapeutic depth, and by placing it in the body and in the experience of individuals. This is also a reason to distinguish this model clearly from the concept defended by Hollander. Although both concepts are concurrent, share many components, and are based on overlapping data, that of the team is rooted in the local history of the application of DBS. It is DBS that is both the origin and the finality of the team's concept of spectrum.

It is nonetheless important to keep in mind the common trend, which has been to defend the idea of a spectrum of damaged-action pathologies that focuses on the inability to control actions or thoughts. DBS is there to restore this inability. In its work, the team freed itself from category-based partitioning and focused on exploring combinations—the complex entanglement—of the emotional, cognitive, and motor dimensions of a continuum of repetitive, compulsive, and stereotyped behaviors. For the team, the goal was to investigate the hierarchy of complexity from simple movement (e.g., motor tic) to more elaborate behavior (e.g., compulsion), depending on the nature of the associated subjective experience. We can thus grasp the importance of the stakes that fascinated the team, which debated them at length while developing the models of monkeys bursting with stereotypies and compulsions: Were these monkeys anxious when they compulsively licked and chewed on their hands and tail? More precisely, were the hyperactivity and stereotypies triggered in them part of a mental condition? The issue of emotional causation, inaccessible here, became central because it made possible the determination of the nature of these motor, impulsive, or compulsive phenomena, and the distinction between "simple" abnormal movements and more elaborate neuropsychiatric disorders, or between hemiballismus and stereotypies, tics, or compulsions.

This summarizes the whole of the clinical stakes involved in the issues structuring the team's research during my fieldwork, but also those inherent in the "Obsessive-Compulsive Spectrum Disorders," which are to impose a nosographic repartitioning that would pave the way to the exploration of the continuity between motor behaviors and mental states, between the internal subjective, emotional, cognitive, or psychic dimension and its motor consequences, and between action and the intentionality underpinning it.

Intellectual ambition is completed with a disciplinary project: to lay the foundations for a proper "neuropsychiatric" research program. The OC spectrum would provide a new space of convergence and of symmetrical questioning of hybrid, in-between entities, for neurology and psychiatry (Galison 1997, pp. 783–784; Löwy 1992). DBS would give consistency to this project, it would make it more robust: not only by opening shared spaces and times of scientific exploration (such as animal models or patients who are experiencing behavioral side effects), making it possible to intervene in bodies and to manipulate these different functional dimensions experimentally (Hacking 1983), but also in the care and treatment of patients (Moutaud 2011b, 2012).

## Conclusion

The OCD classification concentrated the main debates that led to the epistemological development of the DSM-5: the debate on a category-based versus a dimension-based approach to the classification of disorders; and the one on the entry of families of disorders no longer exclusively developed on the basis of clinical descriptions and epidemiological data (the foundations of the two previous DSM editions), but also on the basis of data gathered from research in neuroscience. The motivation to introduce an "Obsessive-Compulsive Spectrum Disorder," which was ultimately rejected and became the truncated OCRD category, sums up this aborted project. This failed shift is the consensual version of a change that was either too controversial or not stable enough in its scientific foundations.

I should note that taking OCD out of the class of anxiety disorders entailed a new order in its clinical components where behaviors or motor control are more important than affects and emotions and (partially) justifies its neurologization. This move relegates the subjective, emotional component—i.e. that which individualizes behavior and makes it properly psychiatric—to the background (or may even make it optional). So this is also a new hierarchy of knowledge and form of intervention that more deeply questions the relation between psychiatry and neuroscience and the disciplinary demarcation between psychiatry and neurology, especially in the renewed contemporary ideal of instituting a field of neuropsychiatric activity (Moutaud 2012). What is at stake is the identity of psychiatry as a type of medicine and science, the status of its knowledge and its objects (the diseases and the patients), and its modalities of intervention.

## References

Abramowitz, J. S. (2012). Editorial. *Journal of Obsessive-Compulsive and Related Disorders, 1*(1), iii.

Abramowitz, J. S., & Deacon, B. J. (2005a). Obsessive-compulsive disorder: Essential phenomenology and overlap with other anxiety disorders. In J. S. Abramowitz & A. C. Houts (Eds.), *Concepts and controversies in obsessive–compulsive disorder* (pp. 119–136). New York: Springer Science + Business Media.

Abramowitz, J. S., & Deacon, B. J. (2005b). The OCD spectrum: A closer look at the arguments and the data. In J. S. Abramowitz & A. C. Houts (Eds.), *Concepts and controversies in obsessive–compulsive disorder* (pp. 141–149). New York: Springer Science + Business Media. Reply to Hollander et al.

Abramowitz, J. S., & Houts, A. C. (2002). What is OCD and what is not: Problems with the OCD spectrum concept. *Scientific Review of Mental Health Practice, 1*, 139–156.

Abramowitz, J. S., & Houts, A. C. (2005). *Concepts and controversies in obsessive–compulsive disorder*. New York: Springer Science + Business Media.

Bouvard, M. (2006). *Les troubles obsessionnels compulsifs: Principes, thérapies, applications.* Paris: Masson.

Bynum, W. (1990). "C'est un malade": Animal models and concepts of human diseases. *Journal of the History of Medicine and Allied Sciences, 45*(3), 397–413.

Castel, P.-H. (2008). De la névrose obsessionnelle aux TOC: Remarques sur le passage du paradigme psychanalytique au paradigme cognitivo-comportementaliste. In C. Françoise (Ed.), *Les psychothérapies aujourd'hui* (pp. 186–202). Paris: Armand Colin.

Castel, P.-H. (2012). *La Fin des coupables: Obsessions et contrainte intérieure de la psychanalyse aux neurosciences*. Paris: Ithaque Éditions.

Demazeux, S. (2008). Les catégories psychiatriques sont-elles dépassées? *Psychiatrie, Sciences Humaines Neurosciences, 6*(1), 17–25. doi:10.1007/s11836-008-0049-z.

Descola, P. (2005). *Par delà nature et culture*. Paris: Gallimard.

Galison, P. (1997). *Image and logic*. Chicago: University of Chicago Press.

Hacking, I. (1983). *Representing and intervening: Introductory topics in the philosophy of natural science*. Cambridge: Cambridge University Press.

Harrison, B. J., Pujol, J., Soriano-Mas, C., Hernández-Ribas, R., López-Solà, M., Ortiz, H., Alonso, P., Deus, J., Menchon, J. M., Real, E., Segalàs, C., Contreras-Rodríguez, O., Blanco-Hinojo, L., & Cardoner, N. (2012). Neural correlates of moral sensitivity in obsessive-compulsive disorder. *Archives of General Psychiatry, 69*(7), 741–749. doi:10.1001/archgenpsychiatry.2011.2165.

Healy, D. (1997). *The anti-depressant era*. Cambridge: Harvard University Press.

Healy, D. (2002). *The creation of psychopharmacology*. Cambridge: Harvard University Press.

Hollander, E., & Stein, D. J. (1993). The spectrum of obsessive-compulsive related disorders. In E. Hollander (Ed.), *Obsessive-compulsive related disorders* (pp. 241–271). Washington: American Psychiatric Press.

Hollander, E., Friedberg, J. P., Wasserman, S., Yeh, C.-C., & Iyengar, R. (2005). The case for the OCD spectrum. In J. S. Abramowitz & A. C. Houts (Eds.), *Concepts and controversies in obsessive–compulsive disorder* (pp. 95–118). New York: Springer Science + Business Media.

Hollander, E., Zohar, J., Sirovatka, P. J., & Regier, D. A. (2011). *Obsessive-compulsive spectrum disorders: Refining the research agenda for DSM-V*. Arlington: American Psychiatric Association.

Jenike, M. A. (1989). Obsessive-compulsive and related disorders: A hidden epidemic. *New England Journal of Medicine, 321*(24), 539–541.

Jenike, M. A. (1990). Illnesses related to obsessive-compulsive disorders. In M. A. Jenike, L. Baer, & W. E. Minichiello (Eds.), *Obsessive-compulsive disorders: Theory and management* (pp. 45–75). Chicago: Year Book Medical Publishing.

Leckman, J. F., Grice, D. E., Barr, L. C., de Vries, A. L., Martin, C., Cohen, D. J., McDougle, C. J., Goodman, W. K., & Rasmussen, S. A. (1994/1995). Tic-related vs. non-tic-related obsessive-compulsive disorder. *Anxiety, 1*(5), 208–215.

Leckman, J. F., Denys, D., Simpson, H. B., Mataix-Cols, D., Hollander, E., Saxena, S., Miguel, E. C., Rauch, S. L., Goodman, W. K., Phillips, K. A., & Stein, D. J. (2010). Obsessive-compulsive disorder: A review of the diagnostic criteria and possible subtypes and dimensional specifiers for DSM-V. *Depression and Anxiety, 27*(6), 507–527. doi:10.1002/da.20669.

Löwy, I. (1992). The strength of loose concepts. Boundary concepts, federative experimental strategies and disciplinary growth: The case of immunology. *History of Science, 30*(90), 371–396.

Löwy, I. (2000). The experimental body. In R. Cooter & J. Pickstone (Eds.), *Medicine in the twentieth century* (pp. 435–449). London: Harwood Academic Publishers.

Mataix-Cols, D., Rosario-Campos, M. C., & Leckman, J. F. (2005). A multidimensional model of obsessive-compulsive disorder. *American Journal of Psychiatry, 162*, 228–238. doi:10.1176/appi.ajp.162.2.228.

Mataix-Cols, D., Pertusa, A., & Leckman, J. F. (2007). Issues for DSM-V: How should obsessive-compulsive and related disorders be classified? *American Journal of Psychiatry, 164*, 1313–1314. doi:10.1176/appi.ajp.2007.07040568.

Moutaud, B. (2008). "C'est un problème neurologique ou psychiatrique?" Anthropologie de la neurostimulation intracrânienne appliquée à des troubles psychiatriques. In M. Jean-Nöel (Ed.), *Les maladies Mentales* (pp. 175–214). Paris: Presses Universitaires de France.

Moutaud, B. (2009). *"C'est un problème neurologique ou psychiatrique ?" Ethnologie de la stimulation cérébrale profonde appliquée aux troubles psychiatriques*. PhD dissertation, Université Paris-Descartes.

Moutaud, B. (2011a). Are we receptive to naturalistic explanatory models of our disease experience? The applications of deep brain stimulation to obsessive-compulsive disorders and Parkinson's disease. In I. van Keulen & M. Pickersgill (Eds.), *Sociological reflections on the neurosciences* (pp. 179–202). Bingley: Emerald.

Moutaud, B. (2011b). From the experimental body to the experimental patient. Medical experimentation in neuroscience. In F. Sylvie & H. Claudie (Eds.), *Of bodies and symptoms. Anthropological perspectives on their social and medical treatment* (pp. 239–259). Tarragona: Publicacions URV.

Moutaud, Baptiste. (2012). Une "convergence forcée" ? Ethnographie d'une collaboration entre neurologie et psychiatrie dans un centre de neurosciences en France. *Anthropologie & Santé, 4*, 927. doi:10.4000/anthropologiesante.

Pallanti, S., & Eric, H. (2008). Obsessive-compulsive disorder spectrum as a scientific "metaphor." *CNS Spectrums, 13*(9 Suppl. 14), 6–15.

Phillips, K. A., Stein, D. J., Rauch, S. L., Hollander, E., Fallon, B. A., Barsky, A., Fineberg, N., Mataix-Cols, D., Ferrão, Y. A., Saxena, S., Wilhelm, S., Kelly, M. M., Clark, L. A., Pinto, A., Bienvenu, O. J., Farrow, J., & Leckman, J. (2010). Should an obsessive-compulsive spectrum grouping of disorders be included in DSMV? *Depression and Anxiety, 27*(6), 528–555. doi:10.1002/da.20705.

Pitman, R. K., Green, R. C., Jenike, M. A., & Mesulam, M. M. (1987). Clinical comparison of Tourette's disorder and obsessive-compulsive disorder. *The American Journal of Psychiatry, 144*(9), 1166–1171.

Rachman, S. J., & Da Silva, P. (1978). Abnormal and normal obsessions. *Behaviour Research and Therapy, 16*, 233–248.

Regier, D. A. (2007). Obsessive-compulsive behavior spectrum: Refining the research agenda for the DSM-V. *CNS Spectrum, 12*, 343–344.

Rose, N., & Abi-Rached, J. M. (2013). *Neuro: The new brain sciences and the management of the mind*. Princeton: Princeton University Press.

Starcevic, V., & Janca, A. (2011). Obsessive-compulsive spectrum disorders: Still in search of the concept-affirming boundaries. *Current Opinion in Psychiatry, 24*(1), 55–60. doi:10.1097/YCO.0b013e32833f3b58.

Stein, D. J., Fineberg, N. A., Bienvenu, O. J., Denys, D., Lochner, C., Nestadt, G., Leckman, J. F., Rauch, S. L., & Phillips, K. A. (2010). Should OCD be classified as an anxiety disorder in DSM-V? *Depression and Anxiety, 27*(6), 495–506. doi:10.1002/da.20699.

Thomas, S., Mykhalovskiy, E., & Rock, M. (2009). Animals in surgery – surgery in animals: Nature and culture in animal–human relations and modern surgery. *History and Philosophy of the Life Sciences, 31*, 321–354.

MIX
Papier aus verantwortungsvollen Quellen
Paper from responsible sources
FSC® C105338

If you have any concerns about our products,
you can contact us on
**ProductSafety@springernature.com**

In case Publisher is established outside the EU,
the EU authorized representative is:
**Springer Nature Customer Service Center GmbH**
**Europaplatz 3, 69115 Heidelberg, Germany**

Printed by Libri Plureos GmbH
in Hamburg, Germany